"Misticall Wordes and Names Infinite":
An Edition and Study of Humfrey Lock's Treatise on Alchemy

Medieval and Renaissance
Texts and Studies
Volume 367

"Misticall Wordes and Names Infinite":
An Edition and Study of Humfrey Lock's Treatise on Alchemy

Peter J. Grund

ACMRS
(Arizona Center for Medieval and Renaissance Studies)
Tempe, Arizona
2011

Published by ACMRS (Arizona Center for Medieval and Renaissance Studies)
Tempe, Arizona
© 2011 Arizona Board of Regents for Arizona State University.
All Rights Reserved.

Library of Congress Cataloging-in-Publication Data

Lock, Humfrey, fl. 1560-1570.
  Misticall wordes and names infinite : an edition and study of Humfrey Lock's Treatise on alchemy / Peter J. Grund.
       p. cm. -- (Medieval and Renaissance texts and studies ; v. 367)
  Revision of Grund's thesis (doctoral)--Uppsala universitet, 2004.
  Includes bibliographical references.
  ISBN 978-0-86698-415-7 (acid-free paper)
  1. Alchemy--Early works to 1800. 2. Lock, Humfrey, fl. 1560-1570--Manuscripts. 3. Alchemy--Manuscripts. I. Grund, Peter. II. Title.
  QD25.L63 2011
  540.1'12--dc22
                                                                    2011002930

Cover Credit:
Trinity College, Cambridge, MS. R. 14. 45 (fol. 67r).
Reproduced with the permission of the Master and Fellows of Trinity College, Cambridge.

∞
This book is made to last. It is set in Adobe Caslon Pro,
smyth-sewn and printed on acid-free paper to library specifications.
Printed in the United States of America

TABLE OF CONTENTS

Acknowledgments   *vii*

Abbreviations   *ix*

Introduction   *xi*

Sociohistorical Context   1

Sources   29

Language and Alchemy: The Case of the Treatise   71

Manuscripts   95

Collation   117

Editorial Principles   125

Dedication   129

Edition   137

Appendix 1: Letter from Humfrey Lock to Robert Dudley, 19 May 1572   239

Appendix 2: Collation Patterns   243

Explanatory Notes   247

Glossary   313

References   339

Acknowledgments

Like many alchemical procedures, the writing of this book has involved several stages. Throughout these different stages, I have enjoyed the help and advice of many people. Merja Kytö expertly supervised the dissertation on which this book is based, and she has unfailingly encouraged my subsequent work. Linda E. Voigts, who acted as the faculty examiner for my dissertation, has liberally shared with me her vast knowledge of scientific writing in early English; without her groundbreaking research, many of the mysteries of Lock's alchemical treatise would have remained unsolved. Other people have painstakingly read various chapters and drafts and provided invaluable suggestions, or discussed with me the sometimes intractable problems of alchemical texts: Ralph Hanna, Monica Hedlund, Lauren Kassell, George R. Keiser, Päivi Pahta, Erik Smitterberg, and the participants in the research seminars in English linguistics at the Department of English, Uppsala University. My colleagues at the Department of English, University of Kansas, especially Amy J. Devitt and James W. Hartman, have provided wholehearted support and encouragement. You all have my sincerest gratitude. I also wish to thank the anonymous reviewers, whose suggestions helped greatly in revising the original dissertation into the present book. I am grateful to Robert Bjork and the board of MRTS for accepting my book for publication, and the staff of MRTS, especially Roy Rukkila, who has patiently answered numerous queries.

I am indebted to the staff of various libraries, who have been of great help to me during my research trips: the Bodleian Library, the British Library, the Cambridge University Library, the Library of Corpus Christi College (Oxford), the Glasgow University Library, the Royal Library of Copenhagen, the Library of Trinity College (Cambridge), and the Wellcome Library for the History and Understanding of Medicine (London). I am especially grateful to Bodleian Library, University of Oxford, for allowing me to reproduce images from MS Ashmole 1490, and to the Master and Fellows of Trinity College Cambridge, for allowing me to reproduce an image from MS. R. 14. 45 (fol. 67r).

I gratefully acknowledge financial support from the Faculty of Languages, Uppsala University; from the Swedish Foundation for International Cooperation in Research and Higher Education (STINT), which made it possible for me to spend the academic year 2001–2002 at Corpus Christi College, Oxford; from Ograduerade forskares fond, Department of English, Uppsala University, which enabled me to attend conferences; and from the Herbert and JoAnn Klemmer Fund (Department of English, University of Kansas), which helped to defray costs in connection with reproduction of images.

Throughout the various permutations of this book, one thing has remained constant: my wife Molly Zahn's unflagging moral and intellectual support. For her meticulous care in reading, her incisive comments and gentle prodding toward "less pedestrian prose," and for simply being there when I needed to talk, she has my gratitude and love.

I dedicate this book to my parents, Bengt and Gull-Britt, who have unreservedly accepted my choices and supported my interests.

November 2009
Peter Grund

Abbreviations

DML	*Dictionary of Medieval Latin from British Sources*
EETS ES	Early English Text Society, Extra Series
EETS OS	Early English Text Society, Original Series
EN	Explanatory Note(s)
eVK	*Scientific and Medical Writings in Old and Middle English*, ed. Linda E. Voigts and Patricia D. Kurtz
MED	*Middle English Dictionary*
ODNB	*Oxford Dictionary of National Biography*
OED	*Oxford English Dictionary*
OLD	*Oxford Latin Dictionary*

Alchemical sigils

☉	Gold, the sun
☽	Silver, the moon
♀	Copper, Venus
♂	Iron, Mars
♃	Tin, Jupiter
♄	Lead, Saturn
☿	Mercury, Mercury
🜍	Sulfur
🜁	Air
🜂	Fire
🜃	Earth
🜄	Water
℞	Take, Recipe/Recipite (Latin)
℥	Ounce
li	Pound, Libra (Latin)

Introduction

In May 1568, the English builder and military engineer Humfrey Lock wrote to Queen Elizabeth's principal secretary William Cecil, expressing regrets that he had ever decided to go to Russia to work for Tsar Ivan Vassilivitch (Ivan the Terrible). Embroiled in battles between factions of English merchants in Russia, Lock reminds Cecil that, if he had been given financial support, he "cold haue done proffytable thing*es* in England," and exclaims: "Nowe I ame in Rusland; God send me ones ought of yt ageyne."[1] Four years later he was still there, and his situation had turned even worse. As a last and desperate attempt at enlisting Cecil's help to return to England, Lock compiled a long alchemical treatise in English as a gift to Cecil. The text describes procedures for the production of the philosophers' stone, a miraculous elixir that would turn base metals into silver and gold. Lock prefaced his text with a verse dedication to Cecil, which illustrates both the virtues of alchemy and Lock's dire straits.

This book is an edition of Lock's untitled alchemical text (henceforth: the Treatise), and a study of its sociohistorical context, manuscripts, sources, and language. Although the dearth of editions of English vernacular texts on alchemy from the fifteenth and sixteenth centuries would alone seem to warrant an edition, there are other, more specific features that recommend the Treatise. The peculiar circumstances of its origin (hinted at above and explained in more detail below) point to the importance of alchemy in late sixteenth-century Elizabethan England.[2] Once in circulation, the Treatise attracted a great deal of attention among alchemists: although not an innovative text, it survives in seven more or

---

[1] 20 May 1568: National Archives, London, State Papers 70/98, fols. 63v–64r.

[2] The definition of *alchemy* is not an easy one. Although it has historically been regarded as the pursuit of the transmutation of base metals into silver and gold, recent research has pointed out that *alchemy* was used (near-) synonymously with *chemistry* (or *chymistry*) in the sixteenth and seventeenth centuries, covering a whole range of chemical, alchemical, and pharmacological approaches. When I use *alchemy*, it should thus not be taken exclusively to mean the quest for transmutation, although the focus of the Treatise is very much on this aspect of the art. For some readers of the Treatise, such as Simon Forman, the medical applications of the philosophers' stone may have had much more attraction. See p. 22 below, and also William R. Newman and Lawrence M. Principe, "Alchemy vs. Chemistry: The Etymological Origins of a Historiographic Mistake," *Early Science and Medicine* 3 (1998): 32–65.

less complete manuscript copies, in five extracts (in three manuscripts), and in an abridged version (in two manuscripts) under the title of "Beniamin Lock his Picklock to Riplye his Castle" (see the table below). Several of the manuscripts were copied, owned, or read by such sixteenth- and seventeenth-century notables as Simon Forman, Richard Napier, Arthur Dee, Thomas Harriot, Thomas Browne, and Elias Ashmole. Compiled from a surprisingly large number of source texts, the Treatise also illustrates the recycling and adaptation of earlier alchemical texts to form a new composition. As it draws on sources that were written in English as well as Latin, it underscores the growing importance of the vernacular as a language of science in the early modern period.

The edition presents Lock's Treatise as it is found in MS. Ashmole 1490, which was copied out by the astrological physician and alchemist Simon Forman in 1590. For reasons that will be discussed below, I have made no attempt to reconstruct Lock's original or the last common ancestor of the now surviving manuscripts. Instead, the text provided in this book gives the Treatise as it appeared at one point in time, as copied by a known copyist. As such, it will serve as material not only for further explorations of Lock's text but also for investigations of Forman's interests in alchemy and his prodigious copying activities.

Table 1.1. Manuscripts of the Treatise

Text versions	Folios/Pages	Sigla
Treatise		
Oxford, Bodleian Library MS. Ashmole 1490	294r–325r	A
Glasgow, Glasgow University Library MS. Ferguson 216	pp. 7–80	F
Copenhagen, Royal Library of Copenhagen MS. Gamle Konglige Samling (Gl. Kgl. Saml.) 1784	4r–43v	G
London, British Library MS. Sloane 288	1r–32v	S1
London, British Library MS. Sloane 299	23r–42v	S2
London, British Library MS. Sloane 3180	1r–11v	S3
London, British Library MS. Sloane 3684	1r–36r	S4
Extracts of the Treatise		
Oxford, Bodleian Library MS. Ashmole 1407	a) pp. 17–21 (Part 2) b) pp. 39–50 (Part 2)	E1
Oxford, Bodleian Library MS. Ashmole 1424	a) 31v–34r b) 38r–41v	E2
Oxford, Bodleian Library MS. Ashmole 1494	160r–160v	E3
The Picklock		
Oxford, Bodleian Library MS. Ashmole 1507	158r–176r	Ba
London, Wellcome Institute MS. 436	1r–30v	Bw

Sociohistorical Context

Author and Date of Composition

The Treatise is attributed to one Humfrey Lock in four manuscripts of the text.[1] Very little has been written on Humfrey Lock, and the few sources that do mention him provide contradictory information. For example, Joseph Ritson classifies Lock as a fifteenth-century poet,[2] Robert Schuler places Lock in the sixteenth century,[3] and Lyndy Abraham lists Lock as a seventeenth-century writer.[4] In order to get a clearer picture of Lock and his circumstances it is necessary to consider a wide variety of primary sources, including the dedication in verse that precedes the Treatise, marginalia, and epistolary evidence. These sources present Lock as a man of many talents: he was a builder, engineer, inventor, and alchemist. He had dealings with some of the most prominent personages of the late sixteenth century: William Cecil Lord Burghley, Robert Dudley, and the Russian Tsar Ivan Vassilivitch (Ivan the Terrible). On some aspects of Lock's life, such as the year and place of birth or even his early life or final days, the sources

---

[1] The name is found in different forms and spellings in the manuscripts: Humfrey Lock or H. L. (A), Humphrey Lock/e/ or H. L. (F), Humphry Locke (S2), H. Lock (G). I will use the A spelling *Humfrey Lock*, since this is the version of the Treatise presented in the edition. However, see also n. 14.

[2] Joseph Ritson, *Bibliographia Poetica: A Catalogue of Engleish* [sic] *Poets of the Twelfth, Thirteenth, Fourteenth, Fifteenth, and Sixteenth, Centurys* [sic] (London, 1802). The basis of Ritson's claim is unclear. His entry on Lock states that "LOCK, HUMPHREY wrote 'Verses on alchymy, to sir William Cycil; 1490;' among Ashmoles MSS, at Oxford (Num. 18)"; Ritson, *Bibliographia*, 65. This manuscript is today A, which is dated 19 June 1590. Perhaps Ritson misread the date as "1490."

[3] Robert M. Schuler, ed., *Alchemical Poetry 1575–1700, from Previously Unpublished Manuscripts* (New York: Garland, 1995), xxix. Schuler seems to base his dating on the dedication to Cecil. He states that the dedication was written "sometime before 1590," and he refers to an earlier study of his, which contains an entry on Lock: Robert M. Schuler, *English Magical and Scientific Poems to 1700: An Annotated Bibliography* (New York: Garland, 1979), 55. The only version of the Treatise and its dedication referred to in that work is S2, which he dates to the late sixteenth century.

[4] Lyndy Abraham, *A Dictionary of Alchemical Imagery* (Cambridge: Cambridge University Press, 1998), 245. Abraham seems to have consulted only the copy of the Treatise in F, a seventeenth-century copy; *A Dictionary*, 230.

are mostly silent.[5] The part of Lock's life that can be reconstructed, however, highlights power struggles among English merchants in Russia and the importance of alchemy in sixteenth-century England.

When Lock compiled his Treatise, he was stranded in Russia, as he himself reveals in a marginal note in the Treatise (henceforth "the Russia note"):

> ffor when I compiled it, I ment to haue sent it into Ingland as a pre*sent* & mediator to help me home out of Russia, wherfore I made it the more darke that I might the sonner be sente for home for to doe it myselfe [316v: 27][6]

The dire situation in Russia that Lock hints at in the note was the result of a series of events. He first enters public records in 1559, when he was appointed the surveyor and chief carpenter of the construction of Upnor Castle, on the river Medway (Kent).[7] This was a significant piece of construction, which was intended to protect navy ships when they were at anchorage close to Rochester Bridge. The designer of the castle was the prominent architect and military engineer Sir Richard Lee (1501/2–1575), and the paymaster and purveyor of the construction was Richard Watts (*c*. 1529–1579). Lock appears to have been "the professional expert on site."[8] Perhaps his role was what Eric Ash has recently referred to as an "expert mediator," a person who "served as the intellectual, social, and managerial bridge between the central administrators who were his patrons on the one hand, and the various and far-flung objects of their control on the other."[9] Throughout the construction, Lock must have been in intimate contact

---

[5] Humfrey Lock is not included in the pedigree of the Lock family found in the 1792 issue of the *Gentleman's Magazine*, which was later augmented by Clay and Locke; *Gentleman's Magazine* 62 (1792): 798–801; John W. Clay, ed., *Familiae Minorum Gentium*, 4 vols., Publications of the Harleian Society 37–40 (London: The Harleian Society, 1894–1896), 4: 1306–9; John G. Locke, *Book of the Lockes: A Genealogical and Historical Record of the Descendants of William Locke of Woburn with an Appendix Containing a History of the Lockes in England* (Boston: James Munroe, 1853). The Lock family pedigree in the *Gentleman's Magazine* and in the later sources is far from exhaustive. For example, Sir William Lock (d. 1550) is said to have had eight sons and daughters, but not all are mentioned by name. Humfrey's absence from the pedigree is thus not necessarily an indication that he was not a member of the influential Lock family.

[6] Unless otherwise stated, references in examples are to the edited text (folio and line number).

[7] A. D. Saunders, "The Building of Upnor Castle, 1559–1601," in *Ancient Monuments and their Interpretation*, ed. M. R. Apted, R. Gilyard-Beer, and A. D. Saunders (London: Phillimore, 1977), 269–71; H. M. Colvin, John Summerson, Martin Biddle, J. R. Hale, and Marcus Merriman, *The History of the King's Works*, vol. 4, *1485–1660 (Part II)* (London: Her Majesty's Stationery Office, 1982), 478.

[8] Saunders, "The Building," 270.

[9] Eric H. Ash, *Power, Knowledge, and Expertise in Elizabethan England* (Baltimore: Johns Hopkins University Press, 2004), 8–9.

with Sir William Cecil (later Lord Burghley, 1520–1598). In 1562, Lock sent a letter to Cecil, outlining a new and improved way of procuring stone for the building from the nearby Rochester Castle.[10] He entreats Cecil to persuade the Lord Admiral, presumably Edward Fiennes de Clinton, first Earl of Lincoln, to pursue Lock's plan.[11] It is clear that Lock has broached the topic with Cecil before in person:

> [. . .] wheras I movid your Hono*ur* at my last being at London concerning the taking downe of the rest of the stone at Rochester Castell to end the work*es* at Vpnore, at the whiche tyme your Hono*ur* promysyd to move my Lorde Admyrall, ffor trustinge your Hono*ur* haith it still in remembraunce and for thies causes folowy*ng*, I haue byne so bold as to troble your Hono*ur* wi*th* it ageyne [. . .] [June 1562; State Papers 12/23, fol. 102r]

Lock's obviously close connections with Elizabeth's court, especially with Cecil, were to become increasingly important to him almost ten years later when he was facing accusations of treason.

Lock never had the opportunity to see the Upnor project finished. By 8 November 1567, his assignment had been taken over by Richard Watts.[12] However, this was hardly a question of demotion; rather, Lock had taken on a novel task. In 1566, Tsar Ivan Vassilivitch requested permission from Queen Elizabeth for English craftsmen to enter his service. The request was granted, and in 1567 Lock went to Russia together with his assistant John Fenton and a number of other craftsmen.[13] It is unclear whether Lock signed up willingly for the assignment in Russia, or whether he was appointed by the Queen. Presumably, Elizabeth would have preferred to have subjects that were very loyal to her employed by a foreign sovereign, and it is thus tempting to see the appointment as an honor bestowed on Lock. But even if Lock went to Russia at the Queen's command, there may also have been other reasons compelling him to go. The dedication of the Treatise hints at financial problems:

> and showe himselfe to his country
> a member worthy place,

---

[10] National Archives, London, State Papers 12/23, fol. 102r–v. The signature in this letter matches the signature in his later letters from Russia (see below).

[11] *ODNB* s.n. Clinton, Edward Fiennes de.

[12] Colvin et al., *The History*, 479.

[13] Joseph Hamel, *England and Russia; Comprising the Voyages of John Tradescant the Elder, Sir Hugh Willoughby, Richard Chancellor, Nelson, and Others, to the White Sea* (London: Richard Bentley, 1854), 177; T. S. Willan, *The Early History of the Russia Company 1553–1603* (Manchester: Manchester University Press, 1956), 91–92. An English translation of the Tsar's offer to the craftsmen is given in George Tolstoy, ed., *The First Forty Years of Intercourse between England and Russia. 1553–1593* (St. Petersburg, 1875), 37.

which froward fortune in tyme paste
out of that land did chase,

by wante, the cruell enimie
of every perfecte minde
that vnto arte and sciences
moste aptly ar inclinde.
[*Dedication*, ll. 18–25]

Once in Russia, Lock quickly got into trouble. Four letters from Lock[14] in Russia have survived, dated 20 May 1568, 1 July 1568, 23 June 1569, and 19 May 1572.[15] These letters chronicle Lock's fall from grace: they provide rich details on the power struggle of English merchants in Russia, and how Lock was caught up in this struggle. The 1568 letters, which are both addressed to Cecil, are basically catalogues of complaints about the behavior of English merchants in Russia.[16] Lock emphasizes that the English merchants are "estemyd as moste gredye car-

---

[14] The letters are all written in one and the same hand, most likely Lock's own, except for the 1569 letter, which is written in two hands, one of the hands being the hand found in the other letters. The signature is the same in all the letters (including the 1562 letter from Upnor cited earlier), given as *Humffrid* or possibly *Humffridus* (depending on whether a florished 'd' is interpreted as a 'd' and an abbreviation or simply as a 'd'). For *Humffrid* or *Humffridus*, which may be used as a more formal version of the name, see W. S. B. Buck, *Examples of Handwriting 1550–1650* (London: Society of Genealogists, 1996), 43.

[15] 20 May, 1568: National Archives, London, State Papers 70/98, fols. 62r–64r. In the beginning, this letter mentions that a letter was sent on 4 March (1568). Similarly, the July letter states that two letters have been sent previously. Hamel, *England*, 177, also notes the March letter, but it is not listed by Crosby, and I have not been able to locate it: Allan J. Crosby, ed., *Calendar of State Papers, Foreign Series, of the Reign of Elizabeth, 1566–8* (London: Longman, 1871). It is uncertain whether Hamel has actually consulted the letter or whether he relies on the mention of this letter in the May and July 1568 correspondence. 1 July 1568: National Archives, London, State Papers 70/100, fol. 4r–v. 23 June 1569: National Archives, London, State Papers 70/147/2, fols. 355r–56r. (I am indebted to Terry Walker for drawing my attention to this letter.) 19 May 1572: National Archives, London, State Papers 70/123, fols. 148r–49r. See also Crosby, *Calendar*, 463, 492–93; Allan J. Crosby, ed., *Calendar of State Papers, Foreign Series, of the Reign of Elizabeth, 1572–74* (London: Longman, 1876), 111–12; Hamel, *England*, 177; Armand J. Gerson, "The Organization and Early History of the Muscovy Company," in *Studies in the History of English Commerce in the Tudor Period* (New York: D. Appleton, 1912), 1–122; Mildred Wretts-Smith, "The English in Russia during the Second Half of the Sixteenth Century," *Transactions of the Royal Historical Society*, 4th ser. 3 (1920): 72–102; Willan, *The Early History*, 92, 97–98.

[16] The May 1568 letter is also signed by Lock's assistant, John Fenton, about whom I have failed to find any additional information.

myraunt*es* and devowerers of co*m*man weales whose covytouse desyre all the ryches in the world cannot sattysffye."[17] He complains about his "evill intreating" by these merchants, and expresses the wish to leave the Tsar's service and return to England. At the same time, he admits being deeply in debt to the merchants, a fact that probably fueled his caustic comments about them. Reluctant to live like a beggar in England, Lock even considers residing in a different country:

> ffor we are ffor the moste p*a*rte so indettyd to the marchaunt*es* that almost we dare not come home otherwyse than able to pay them. And yf yo*u*r Honou*r* shall heare that I shall happen to stay in any other countrye, I hu*m*blye beseache yo*u*r Honou*r* so to consydre of me that necessytie hath compellyd me so to doo; ffor I had rather lyve in a straunge land thane begge in myne owne countrye [July 1568; State Papers 70/100, fol. 4v]

Lock frequently refers to "the companye," making clear that the merchants that he castigates belong to the Muscovy Company. Chartered as early as 1555 by Queen Mary and granted a monopoly on the Anglo-Russian trade, the Muscovy Company was a powerful trading organization, with influential investors and backers, including, among others, Cecil.[18] As we shall see later, the company merchants fiercely protected their monopoly.

Apart from his grievances against the merchants, Lock is not very forthcoming about his life in Russia. He describes himself as an artificer, but provides no details about his commission. However, in a passage where he tries to establish his moral superiority, he provides a rare glimpse of his background when he exclaims:

> I myselffe yf I wold caste the ffeare of God behynd me and desyre riches more than Godes glorye I cold do for the Emp*er*rou*r* suche thing*es* and make hym suche engynes for his warres that he might therby subdewe any prynce that wold stand ageynst hym. [May 1568; State Papers 70/98, fol. 63v]

Although the statement undoubtedly contains a great deal of self-aggrandizement, it indicates that Lock was no mere builder. The same impression is given in a later passage (in which he mildly rebukes both Cecil and Robert Dudley (1532–1588), the earl of Leicester):

> I cold haue done proffytable thing*es* in England abought the making of salt but I dyd well p*er*ceave yf I shold make it manyfest another man shold repe the ffrewt*es* of my labours. For whan I wold not be vsyd lyke a boye at comaundement and was causles complaynyd vppon, my Lorde of Lecestre and yo*u*r Honou*r* bothe gave credyte vnto the complayner. I had devysyd a

---

[17] National Archives, London, State Papers 70/98, fol. 62v.
[18] Willan, *The Early History*.

way wherby I wold haue made more salt in one weke than was made in two wit*h* the burning of lesse wood in twoo wekes ˄{than} they burnyd in one. And su*m*me thing*es* else I cold haue done of moche more proffette yf mayntena*u*nce had be gyven [May 1568; State Papers 70/98, fols. 63v–64r]

Both passages are very elusive, the second hinting at an episode about which I have not been able to uncover any additional information. But they do provide some circumstantial evidence. Lock appears to be a man of many talents: builder, engineer (in the sixteenth-century definition of 'one who designs and constructs military works for attack or defence'),[19] and inventor. His status as a military engineer may already be indicated by his involvement in the building of Upnor Castle, a defensive, yet military fortification. But whether he actually had warfare experience and was a mathematical practitioner like more famous engineers such as Thomas Bedwell (*c.* 1547–1595) is unknown.[20]

Besides testifying to his wide-ranging interests, the reference to salt making in the second note reveals Lock's engagement in a topic of great national import in late sixteenth-century England. Primarily to reduce England's dependence on France for salt, Elizabeth appears to have tried to boost the domestic industry in the 1560s. Through Cecil, she issued a salt licence that basically gave the holder a monopoly on making white salt in England.[21] After being in the hands of several foreign experts, once the licence promised to be lucrative, it came into the hands of wealthy courtiers, including Cecil and Dudley. However, the lucrativeness of the venture hinged upon discovering new and cheap methods to extract salt. In particular, several salt makers in the Netherlands, whom the licence holders were in contact with, were seeking to reduce the amount of artificial fuel used at the extraction stage. When these experiments failed, the courtiers seem to have lost interest in the venture.[22] Exactly how Lock fits into this picture is difficult to assess, but he was obviously aware of the problems involved in salt making. Perhaps the note about his novel method, which had not previously been revealed to Cecil, was intended as a pointed hint at Lock's allegedly cutting-edge inventions and talents: Lock was clearly a resource that could be profitably employed by Cecil.

---

[19] *OED* s.v. *engineer* 2a.

[20] Stephen Johnston, "Mathematical Practitioners and Instruments in Elizabethan England," *Annals of Science* 48 (1991): 320–22. For the growing interest in and importance of mathematics in early modern England, see Ash, *Power*. Benson Bobrick calls Lock "an outstanding military engineer," which he seems to base solely on Lock's own statement in the first extract: *Fearful Majesty: The Life and Reign of Ivan the Terrible* (New York: G. P. Putnam's Sons, 1987), 215, 259.

[21] Edward Hughes, *Studies in Administration and Finance 1558–1825, with Special Reference to the History of Salt Taxation in England* (Manchester: Manchester University Press, 1934), 31–38; idem, "The English Monopoly of Salt in the Years 1563–71," *English Historical Review* 40 (1925): 334–50.

[22] Hughes, *Studies*, 34–35; "The English Monopoly," 339–40, 343.

After these hints in the letter of May 1568, we learn very little about Lock's previous scientific pursuits and background. Instead, the 1569 and 1572 letters, which are addressed to Cecil and Dudley respectively, are more illuminating as regards his reasons for compiling his alchemical Treatise.[23] In the 1569 letter, Lock again complains vehemently about the treatment he has received in Russia, but this time his complaints are aimed at a very specific target: Queen Elizabeth's ambassador to Russia, Thomas Randolph. Randolph was sent to Russia in 1568 with the special commission of winning back the Tsar's favor for the Muscovy Company, which had fallen into disrepute.[24] Lock's letter reveals that the Muscovy Company's monopoly had been challenged by a number of independent merchants or "interlopers," who tried to pursue trade with Russia without Queen Elizabeth's sanction.[25] Lock was caught up in this struggle between the two factions of merchants, and he does not hide where his sympathies lie. He mentions that he has been criticized for not avoiding "Mr Glou*ers* company," presumably Thomas Glover, who was accused of trading independently with Russia, and that Randolph "hathe reportyd agaynst me to thEmpero*u*r to be a moste malycyous, spytefull, & kankaryd of nature."[26] Lock further hints that a number of alleged interlopers have been accused by the English ambassador of being traitors to their country, and Lock expresses the fear that he too will be accused of treason, which will lead to his exile: "Pleasith it yo*u*r Honour, I feare my Lorde by vntrewe report*es* will as moche as in hym lyeth procure my banyshement owt of England."[27] Indeed, the whole point of Lock's 1569 letter to Cecil appears to be an attempt to anticipate and refute the accusations that may be brought against him by Randolph.

From the 1572 letter, it is clear that Lock's fears are coming true: he reports having received information from his friends in England that Dudley, to whom the letter is addressed, has "conceavyd great displeasure ageynste" him. Although the exact reason for Dudley's anger is unclear,[28] Lock presumes that he has been accused of unpatriotically criticizing English merchants in Russia and defends himself by saying:

> I and suche as I ame haue not brought the Emp*erroures* displeasure vppon the m*a*rchauntes but there owne garboyle, there stryvinge, there hatinge one another, there entreprysynge thing*es* vnlawffull, there discredytinge one another, ye and there gredye sekinge to robbe one another, in suche

---

[23] A transcription of the 1572 letter is given in Appendix 1.
[24] Willan, *The Early History*, 95–111.
[25] Willan, *The Early History*, 95–96.
[26] National Archives, London, State Papers 70/147/2, fol. 355v; Willan, *The Early History*, 86–87.
[27] National Archives, London, State Papers 70/147/2, fol. 355v.
[28] National Archives, London, State Papers 70/123, fol. 148r.

> a conffusyd ordre that it was {not} lyke to contenewe, and all seamyd to the Russes to be do*n*ne by the Quenes Maiesties auctorytie, whiche was, as I thought, rather dishono*u*r to her Maiestie and slaunder to her coun-trye than otherwyse, so that very zeale constrayned me to ffynd falt with the abuses and somwhat to speke therof not to Russes and straungers but to Englyshemen; wherffore I was callyd a traytou*r* and an ennymye to my countrye bycause I cold not prayse there evyll doing*es*, flattre them, and say they dyd well. [May 1572; State Papers 70/123, fols. 148v–49r]

Lock is not certain who his accusers are, but he supposes them to be "Mr Randall by his reportes and the m*a*rch*a*unte Banester by his l*ett*res."[29] "Mr Randall" is presumably Randolph, the queen's ambassador, and "the m*a*rch*a*unte Banester" Thomas Bannister, a merchant who accompanied Randolph to Russia.[30] However, whether either of them ever reported on Lock remains unclear.

Lock stresses that he wishes to return to England in order to answer his accusers and asks Dudley (and Cecil) to obtain letters from Queen Elizabeth requesting his dismissal from the Tsar's service. This is a slightly peculiar, but at the same time telling request. On 14 May 1572, the new English ambassador to Russia, Anthony Jenkinson, had obtained the Tsar's permission for English craftsmen to leave Russia if they so wished.[31] Lock is clearly aware of this recent development since he hints at the "ffre passing of Englyshemen of all sortes into Russia and fourth ageyne at there pleasure."[32] But he doubts that he will be allowed to leave "yf the Emp*e*rrou*r* contenewe his buyldinge."[33] This formulation is significant for two reasons: it shows that Lock was involved in a building project at the time, but, more importantly, it implies that Lock's role in the project must have been fairly prominent, if the Tsar was reluctant to let him go. As early as 1569, Lock informed Cecil that "thEmperou*r*s Ma*ies*tie of late hathe comyttyd to my doinge a peace of worke somewhat chargeable," but did not provide any details.[34] Unlike his other letters, however, the 1569 letter was not sent from Moscow, but from Vologda (approximately 300 miles northeast of Moscow). The change of venue may provide a clue to the possible building project that Lock had undertaken. In 1567, the Tsar had commissioned the building of a fortress or castle in Vologda.[35] Although it is unlikely that Lock was put in charge of the whole enterprise, it is quite possible that he was set to supervise some aspect of it. This would not have been an unjustified move by the Tsar: Lock had a great deal

---

[29] National Archives, London, State Papers 70/123, fol. 148v.
[30] Willan, *The Early History*, 145.
[31] Willan, *The Early History*, 123.
[32] National Archives, London, State Papers 70/123, fol. 148v.
[33] National Archives, London, State Papers 70/123, fol. 148v.
[34] National Archives, London, State Papers 70/147/2, fol. 355v.
[35] Hamel, *England*, 176.

of previous experience with building a castle because of his involvement in the construction of Upnor Castle in 1559 to 1567, as we saw earlier.

If Lock's 1572 letter met with a favorable reply from Dudley (or Cecil), there appear to be no extant records of it. In fact, like previous attempts, the 1572 letter may have failed Lock, and he seems to have resorted to a last (and desperate) solution: presenting Cecil with an alchemical treatise written by himself that promises to provide the key to the philosophers' stone, a marvelous substance that would transform base metals into the more noble silver and gold. In "the Russia note" cited above, Lock even plainly admits that he made the text obscure so that Cecil (and Elizabeth) would permit him to return and carry out the procedures himself.

There are several indications that the Treatise was not finished until 1572 or shortly after. The dedication of the Treatise addresses Cecil as *Lord Burghley* and *High Treasurer of England*, titles that Cecil did not receive until 25 February 1571 and 15 July 1572, respectively.[36] Although this would seem to point to a post-July 1572 date, there are, unfortunately, complications. The Treatise was most likely revised before it was circulated. Again, the formulation of "the Russia note" is significant: "for when I compiled it, I *ment* to haue sent it into Ingland" (my emphasis). The use of "ment" signals two things: first, although Lock intended to send the Treatise, he never did; second, the note must be a later addition. If Lock made other changes at this secondary stage, in addition to this note, he would most probably have updated Cecil's titles, to reflect his new status and not risk offending him. On the other hand, the very existence of "the Russia note" suggests that Lock did not expect his Treatise to circulate widely: he admits that he made the text intentionally obscure so that he would be allowed to return to England. Surely, an admission of fraudulent behavior, even for justifiable reasons (in Lock's eyes), would never have been intended to be too widely disseminated.[37] That Lock would have bothered to revise Cecil's titles thus seems like a fairly remote possibility, though one that cannot be wholly dismissed. But even if we treat Cecil's nomenclature with some suspicion, it seems likely that the Treatise was finished sometime in or shortly after 1572, since the Treatise appears to have been compiled in response to the accusations brought against Lock. At least, it was in circulation before 1590 when Simon Forman copied out the A manuscript. Lock may have been ready to send it as early as 1572, but something must have happened that anticipated this move. Perhaps his pleas were finally heard; perhaps he obtained the Tsar's permission to leave Russia. He may even

---

[36] *ODNB*, s.n. Cecil, William.

[37] Both the S3 and S4 manuscripts of the Treatise append the following comment (or perhaps indignant exclamation) to "the Russia note": "Nota impostorem" (S4, fol. 24v), 'Note an impostor.' Although Lock's name is not attached to either of these copies, they do demonstrate how subsequent readers may have interpreted "the Russia note."

have been banished together with a group of interlopers in 1572–1573.³⁸ On this and other aspects of Lock's subsequent life, extant records are silent.

Although Lock never seems to have presented Cecil with a copy of the Treatise, his choice to compile an alchemical text and send it as a gift to Cecil is highly significant, and it highlights the extraordinary status of and widespread interest in alchemy not only in Elizabeth's England but also in Ivan's Russia. If Lock was a practicing alchemist before his departure for Russia, I have failed to find any records of it. In the dedication of the Treatise, Lock even cryptically declares:

> Yet repent I not; my longe absence
> in vain yet hath not byn:
> such knowledg here to me hath come
> as at home I haue not seen,
> [*Dedication*, ll. 26–29]

This stanza may of course point to new building techniques or other crafts that were not practiced in England, but a more intriguing possibility is that it alludes to alchemy. There would have been plenty of opportunities for Lock to acquire or develop his knowledge and practice of alchemy in Russia. Although there was apparently no real interest in alchemy in Russia until the latter half of the sixteenth century, alchemy flourished right at the time of Lock's residence there. Tsar Ivan Vassilivitch himself employed a number of alchemists at his court, most, if not all, of them foreigners.³⁹ Some were even Englishmen.⁴⁰ Figurovski goes as far as to state that it was to a large extent owing to the presence of Englishmen that

---

³⁸ Hamel, *England*, 221–22; Tolstoy, *The First Forty*, 153. It has not been within the scope of this study to consult possible archival records in Russia, which may throw further light on Lock's employment there.

³⁹ N. A. Figurovski, "The Alchemist and Physician Arthur Dee (Artemii Ivanovich Dii)," *Ambix* 13 (1965): 35–51, here 36; William F. Ryan, "Alchemy and the Virtues of Stones in Muscovy," in *Alchemy and Chemistry in the 16th and 17th Centuries*, ed. Piyo Rattansi and Antonio Clericuzio (Dordrecht: Kluwer, 1994), 149–59, here 154–56; idem, "Magic and Divination: Old Russian Sources," in *The Occult in Russian and Soviet Culture*, ed. Bernice Glatzer Rosenthal (Ithaca: Cornell University Press, 1997), 35–58, here 47. There is a Russian translation (in several copies) of the pseudo-Aristotelian text *Secreta secretorum* probably dating back to the late fifteenth century. It shows that the notion of alchemy was not completely new in Russia in the late sixteenth century. However, it was not until then that alchemy, and especially the medical aspects of alchemy, attracted attention in Russia, in particular at the court of the Tsar: Ryan, "Alchemy," 151; idem, *The Bathhouse at Midnight: An Historical Survey of Magic and Divination in Russia* (Phoenix Mill, UK: Sutton, 1999), 359.

⁴⁰ Ryan, "Alchemy," 155.

alchemy was spread in Russia.[41] The duties of the court-employed alchemists seem primarily to have involved preparing different medicinal potions for the members of the court, but most of them were undoubtedly involved in transmutation experiments as well.

Most of Ivan's alchemists remain nameless. However, there is one very important exception: the Cambridge-educated Eliseus Bomelius (*d.* 1579). Bomelius enjoyed the Tsar's favor as a physician/alchemist from the second half of 1570, when he arrived in Russia from England, to around 1574. According to William F. Ryan, Bomelius "was reputed to have been the official alchemist and poisoner of those who fell under Ivan's disfavour."[42] Since both Lock and Bomelius were employed by the Tsar, and since Lock seems to have frequented Ivan's court with regular intervals (as his letters suggest), it is not unlikely that they met, and, if Lock had alchemical aspirations, he would presumably have taken the opportunity to query the famous alchemist. (Whether Bomelius actually deigned to answer is of course a different issue.) Another significant detail that may connect Lock with Bomelius or with other English alchemists at the Russian court is the fact that some of the sources of Lock's Treatise appear to have been written in English (as will be discussed in greater detail below). Unless Lock brought these texts himself from England, he would probably have had easy access to them through English alchemists in Russia.

Although Lock would certainly have had the opportunity to be initiated into the secrets of alchemy in Russia, the opportunities for alchemical pursuits would have been even more ample in England. Alchemical practice had been officially banned since as early as 1403–1404, but little attention appears to have been paid to this law.[43] According to Deborah Harkness, as many as "[s]eventy-four alchemists are known to have practiced in the city [of London] during the reign of Elizabeth, and because this number is based on written remains it may well underrepresent the number of actual practitioners."[44] Such a large number of alchemists is perhaps at least partly explained by the "entrenched interest within court circles" in the potential of alchemy. Courtiers such as Sir Edward Dyer, Sir Walter Ralegh, and Robert Dudley, among many others, dabbled in alchemy, or

---

[41] Figurovski, "The Alchemist," 36.

[42] Ryan, "Alchemy," 155. See also Hamel, *England*, 202–6; *ODNB* s.n. Bomelius, Eliseus.

[43] D. Geoghegan, "A Licence of Henry VI to Practise Alchemy," *Ambix* 6 (1957): 10–17, here 10.

[44] Deborah E. Harkness, "'Strange' Ideas and 'English' Knowledge: Natural Science Exchange in Elizabethan London," in *Merchants and Marvels: Commerce, Science, and Art in Early Modern Europe*, ed. Pamela H. Smith and Paula Findlen (New York: Routledge, 2002), 137–60, here 151.

invested in alchemical enterprises.⁴⁵ In fact, alchemy seems to have been one of the most popular, if not *the* most popular, avocation in court circles.

Two of the staunchest supporters of alchemical experimentation were Cecil and Queen Elizabeth. They appear to have actively encouraged and financially sponsored alchemical experimentation. Again, Harkness remarks that "[f]or decades Cecil vetted and supervised natural science projects like an early prototype of today's National Science Foundation, judging the merits of each proposal [. . .]."⁴⁶ These "science projects" included alchemical experiments. In 1565, two years before Lock's departure for Russia, Cornelius de Lannoy (or Cornelius Alnetanus) was set up at Somerset House by Queen Elizabeth, who seems to have accepted de Lannoy's offer to produce 50,000 marks of gold for her on an annual basis.⁴⁷ More than a decade later, in 1588, John Dee and Edward Kelley, who were then in Prague, were sent for after Edward Dyer had informed Elizabeth and Cecil about Dee's and Kelley's success in transmuting base metals into gold.⁴⁸ Several alchemical treatises, in addition to Lock's, were also dedicated to Cecil or Elizabeth, some by alchemists who sought maintenance for their practice. In 1565–1566, Thomas Charnock pursued Elizabeth's support for his alchemical studies with the gift of a long alchemical treatise.⁴⁹ In the 1570s, Francis Thynne dedicated a number of alchemical writings to Cecil; and Edward Cradock, an Oxford professor, addressed two alchemical treatises to Elizabeth sometime in the latter half of the sixteenth century.⁵⁰ Having such high hopes of a favorable reception of the Treatise, Lock must have been well aware of Cecil's and Elizabeth's attitude toward alchemy.

---

⁴⁵ Charles Webster, "Alchemical and Paracelsian Medicine," in *Health, Medicine and Mortality in the Sixteenth Century*, ed. idem (Cambridge: Cambridge University Press, 1979), 301–34, here 306–7.

⁴⁶ Harkness, "'Strange' Ideas," 142.

⁴⁷ Robert Steele, "Alchemy," in *Shakespeare's England*, 2 vols. (Oxford: Clarendon Press, 1917), 1: 462–74, here 472–73; Harkness, "'Strange' Ideas," 151.

⁴⁸ Conyers Read, *Lord Burghley and Queen Elizabeth* (London: Jonathan Cape, 1960), 474–75.

⁴⁹ Allan Pritchard, "Thomas Charnock's Book Dedicated to Queen Elizabeth," *Ambix* 26 (1979): 56–73, here 56, 61.

⁵⁰ David Carlson, "The Writings and Manuscript Collections of the Elizabethan Alchemist, Antiquary, and Herald Francis Thynne," *Huntington Library Quarterly* 52 (1989): 203–72, here 208; Mordechai Feingold, "The Occult Tradition in the English Universities of the Renaissance: A Reassessment," in *Occult and Scientific Mentalities in the Renaissance*, ed. Brian Vickers (Cambridge: Cambridge University Press, 1984), 73–94, here 86.

Circulation, Manuscript Environment, and Title

Although it is uncertain whether Lock ever returned to England, his Treatise certainly did, and it attracted a great deal of attention among alchemists. As mentioned above, seven more or less complete copies have survived, and extracts are found in three manuscripts. A collation of these extant manuscripts makes clear that several other copies must have been in circulation in the late sixteenth and seventeenth centuries. The Treatise may thus have circulated much more widely than Lock ever envisaged. As I suggested earlier, the incriminating note about purposeful obscurity would presumably not have been intended to be public knowledge (at least not too public).[51] Furthermore, Lock probably did not foresee a later writer taking pains to make an abridged version of his text. Circulating under the title "Beniamin Lock his Picklock to Riplye his Castle" (henceforth *The Picklock*), this redaction was copied and read by such luminaries as Arthur Dee, Thomas Harriot, Thomas Browne, and Elias Ashmole.

The Treatise in all its permutations is thus an illustrative example of the widespread circulation and vibrant copying of alchemical manuscripts in early modern England. Research into manuscript circulation in this period has revealed that scribal copying and scribal publication did not cease with the invention of the printing press and advances in printing technique.[52] Instead, poetry, newsletters, and political pamphlets, among others, continued to appear primarily in manuscript well into the seventeenth century. To these text categories must also be added alchemical texts, as has recently been noted by David McKitterick and George R. Keiser.[53] Very few alchemical texts were printed before the seventeenth century in England, and they become frequent only in the second half of that century.[54] But they were readily available earlier in manuscript, as the vast

---

[51] See p. 9 above.

[52] Harold Love, *Scribal Publication in Seventeenth-Century England* (Oxford: Clarendon Press, 1993); Arthur F. Marotti, *Manuscript, Print, and the English Renaissance Lyric* (Ithaca: Cornell University Press, 1995); H. R. Woudhuysen, *Sir Philip Sidney and the Circulation of Manuscripts 1558–1640* (Oxford: Clarendon Press, 1996); Peter Beal, *In Praise of Scribes: Manuscripts and their Makers in Seventeenth-Century England* (Oxford: Clarendon Press, 1998); David McKitterick, *Print, Manuscript, and the Search for Order, 1450–1830* (Cambridge: Cambridge University Press, 2003).

[53] McKitterick, *Print*, 27; George R. Keiser, "Preserving the Heritage: Middle English Verse Treatises in Early Modern Manuscripts," in *Mystical Metal of Gold: Essays on Alchemy and Renaissance Culture*, ed. Stanton J. Linden, AMS Studies in the Renaissance 42 (New York: AMS Press, 2006).

[54] Rudolf Hirsch, "The Invention of Printing and the Diffusion of Alchemical and Chemical Knowledge," in *The Printed Word: Its Impact and Diffusion*, ed. idem (London: Variorum Reprints, 1978), 115–41. Among the most important vernacular exceptions is e.g. George Ripley's *Compound of Alchymy*, printed in 1591; see Stanton J. Linden, ed., *George Ripley's* Compound of Alchymy *(1591)* (Aldershot: Ashgate, 2001).

number of handwritten texts attests (preserved in e.g. the Ashmole collection at the Bodleian Library and the Sloane collection at the British Library). The exact dynamics of this scribal culture remain to be explored in detail. Texts were undoubtedly passed around within networks of alchemical adepts, who made their own copies or had their personal secretaries or amanuenses copy out the texts.[55] A clear example of this is *The Picklock*, to which I will return later. However, whether copying was also undertaken by professional scribes on commission, whether alchemical texts could be bought at copying shops, or whether alchemical writers actually "published" their writings in manuscript form is not clear.[56] A key project, as Keiser points out, would also be to discover what part John Dee's famous library at Mortlake played as a possible copying center and scholarly haven for interested alchemists.[57]

Although most of the copyists of the Treatise are anonymous and although most early owners are unidentified, the manuscripts of the Treatise have much to offer in illustrating how an alchemical text circulated and how it was read. In this section, I will look at the manuscript context in which the Treatise was couched, suggesting that the Treatise was part of a dynamic collection of alchemical writings that was added to and subtracted from in accordance with the tastes of the copyists and owners. In the subsequent sections, I will return to the sociohistorical context of the Treatise, and investigate two cases where the environment in which the Treatise was copied can be reconstructed in some detail.

During its lifetime, the Treatise circulated in different forms, together with different texts. Most importantly, Lock's verse dedication has been preserved only in three copies, A, F, and S2, while it has been omitted or lost in four: G,

---

[55] For secretaries and amanuenses copying texts, see Woudhuysen, *Sir Philip Sidney*, 66–87.

[56] For professional copying of other kinds of manuscript texts, see, for example, Peter Beal's discussion of the "feathery scribe": *In Praise of Scribes*, 58–108. The Treatise in F, S1, and S2, in particular, is written by neat professional-looking hands, but the circumstances of these copies are unfortunately unknown. The connection between S1 and S2 is particularly interesting, since the paper of the two volumes shares the same watermark; see p. 110. This may be an indication that they represent commissioned work. However, more research is needed on the additional texts in the two volumes to see if there is any supporting evidence of this. Cf. G, which is clearly an alchemical practitioner's copy, written in a rapid and irregular (primarily) secretary hand.

[57] Keiser, "Preserving the Heritage." An interesting discussion along these lines is a recent investigation by Penny Bayer, who suggests that the person that compiled Lady Margaret Clifford's (1560–1616) alchemical recipe collection had access to and copied texts at Dee's library; "Lady Margaret Clifford's Alchemical Receipt Book and the John Dee Circle," *Ambix* 52 (2005): 271–84. See also Julian Roberts and Andrew G. Watson, *John Dee's Library Catalogue* (London: Bibliographical Society, 1990); William H. Sherman, *John Dee: The Politics of Reading and Writing in the English Renaissance* (Amherst: University of Massachusetts Press, 1995), 29–52.

S1, S3, and S4.[58] The copies that include the dedication are of particular significance because in all of these manuscripts the Treatise is accompanied by a group of short alchemical texts in prose and verse. Some of these short texts also appear in S1 (though not directly in connection with the Treatise), and in the two copies of *The Picklock*. Table 2.1 lists all the items that are found in two or more manuscripts.[59]

TABLE 2.1. MANUSCRIPT ENVIRONMENT

Manuscript item	A	F	S2	S1	Bw[C]
"I haue also hearvnto added certayne abreuiacions in miter . . ."	7	1	1	–	–
"Of fermente	When thy ferment thow wilt prepare . . ."	8	2	2	–	1
"Of the stone & fermente &c"	9	3	3	–	1[B]
"Of aurum potabile &c"	10	4	4	–	2
"Ripplies castell"	11	5	5	–	3
"The way to calcine perfectly . . ."	12	6	6	11	4
"The standinge of the glas . . ."	1	10	8	–	5
"Heer followeth the sawes of philosophy . . ."	2	11	9	–	6
"To calcine lune &c"	3	12	10	8	–
"The calcynacion of {the} salte"	4	16	11	9	–
"Hereafter followeth a bal for alkamistes to play withall &c"	–	19	12	1	–
"To make flos eris"	–	20	13	2	–
"To rubify flos aeris"[A]	–	21	14	3	–
"The sublimacion of sall armoniac for the redd tincture followinge &c"	–	22	15	4	–
"Heer followeth the 5intaecencie of antimony &c"	5	23	16	5	–
"A redd tincture by the projection of oyles aforesaid"[A]	–	24	17	6	–

---

[58] Since the beginning of the Treatise (Chapter 1 and most of Chapter 2) is missing in S1, it is impossible to determine whether S1 may at one point have contained the dedication.

[59] The headings and the order of the items have been taken from S2, since this manuscript contains the most items that are also found in at least one other manuscript. The number in the column of each manuscript represents the place in the manuscript where the particular item is found after the Treatise.

Manuscript item	A	F	S2	S1	Bw^C
"To make sall armoniac"	6	25	18	7	–
"After the putriffac{c}*i*on and congelacion{s} of the ferment"	–	–	19	10	–

   ^A Since the titles of these items have been cropped away in the upper part of the manuscript in S2, the readings are taken from F.
   ^B Bw presents "Of ferment" and "Of the stone & fermente" as one item.
   ^C Since Bw is the exemplar of Ba and the two have the exact same content, I cite only Bw.

The additional items consist of a brief introduction (1 in S2), poems (2–8 in S2), a short tract on the sayings of different alchemists (9 in S2), and alchemical recipes (10–19 in S2). The order and number of these additional texts vary in each manuscript, and the manuscripts have varying affinities to each other. F and S2 show clear similarities in the order of items, but F contains quite a few, presumably interpolated, items that are not found in the other manuscripts.[60] While the order in Bw (*The Picklock*) is similar to that of F and S2, all the recipes are absent in this copy. Manuscript A presents a significantly different order from that of F and S2 (cf. S1),[61] and it also contains substantially fewer recipes (5) than the other two manuscripts. S1, on the other hand, includes all the recipes, but it excludes all the poems except "The way to calcine p*er*fectly."[62]

Although the manuscripts do not correlate precisely, the sheer number of items that several manuscripts share clearly demonstrates that the Treatise and the concomitant items were copied as a collection. What is more, this collection may even have been put together by Lock himself. Several features support such a hypothesis. As mentioned above, the additional items almost exclusively appear in copies that also feature the dedication. It is reasonable to assume that copies that do not include the dedication represent a later adaptation, since the

---

   [60] F also includes a recipe that is followed by "Probatum est H. L." 'Proved/Tested H. L.' (p. 107), probably referring to Humfrey Lock. This recipe is not found in any other copy. In his description of the poem "Of fermente" in S2, Schuler claims that it is s poem of 244 stanzas, occupying fols. 42v–50v; Schuler, *English Magical and Scientific Poems*, 55. However, these folios are in fact occupied by a number of poems, from "Of fermente" to "The way to calcine p*er*fectly" in fols. 42v–43v. Fols. 43v–50r contain an abbreviated version of George Ripley's *Compound of Alchymy*, followed by "The standinge of the glas." For a discussion of the copy of Ripley's *Compound* in S2, see p. 59.
   [61] The collation of A shows that this order is not a result of a misplacement of quires (see the description of manuscript A).
   [62] S1 does not present the recipes and poem immediately after the Treatise: while the Treatise is found in fols. 1r–32v, the poems and recipes do not occur until fols. 199r–201v, though in the same hand as the Treatise (see also the description of the S1 manuscript).

dedication was such an integral part of the Treatise. But even more important is a note that precedes the poems and recipes in F and S2, and occurs slightly later in A (item 7). This note makes clear that a number of texts were appended to the Treatise at some point of the transmission:

> I haue also hearevnto added sertayne abreviaciones in myter w*hi*ch I had drawne owt long before I beganne this booke bycause I woulde not let slippe owt of my rememberance these things that are therein contayned fforasmoche as thaye be bothe nessessary and very nedfull to be integratelye pervsed of all such as shall heareaffter begynne and practice to worke in this most worthy scyence of the ph*ilozofe*rs stone & of ferment [A, fol. 329r]

The formulation of the note is admittedly ambiguous. The 'I' may refer to a later copyist, or the whole note may have been added with the intention that subsequent readers should believe the poems and recipes to be Lock's. Such an attribution would have attempted to capitalize on Lock's presumed prestige as an alchemist and grant authority to the additional material. However, the most straightforward interpretation seems to be that Lock himself added the note and at least some of the poems and/or recipes.[63]

Even if some or all of the items were not originally appended by Lock, at least some copyists and owners believed that they had been written by him. Simon Forman, who copied out A, and the anonymous copyist of F explicitly attribute the poem "The standinge of the glas" to Lock.[64] In S2, the index, which was probably made by the seventeenth-century owner Gabriel Gostwyk, labels the Treatise and the concomitant texts "Humfry Locks Collections of Alchime," showing that the texts were seen as an entity.[65]

---

[63] Although perhaps it is only incidental, the note refers to the Treatise as a book, which is also Lock's term for his Treatise in the dedication (l. 73).

[64] This poem is included in Ashmole's *Theatrum Chemicum Britannicum* as anonymous: Elias Ashmole, *Theatrum Chemicum Britannicum, Containing Severall Poeticall Pieces of our Famous English Philosophers* (London, 1652; repr. Kila, MT: Kessinger Publishing, 1997), 421–22. Referring to Ba (one of the manuscripts of *The Picklock*), C. H. Josten attributes this and some of the other poems to the famous English alchemist George Ripley, but it is unclear on what grounds this ascription is made: *Elias Ashmole (1617–1692): His Autobiographical and Historical Notes, his Correspondence, and Other Contemporary Sources Relating to his Life and Work*, 5 vols. (Oxford: Clarendon Press, 1966), 2: 758. Basing her argument on the same manuscript, Lyndy Abraham points out that it is unlikely that these poems are by Ripley, since he is addressed in at least one of them: *Harriot's Gift to Arthur Dee: Literary Images from an Alchemical Manuscript*, The Durham Thomas Harriot Seminar, Occasional Paper 10 (Durham: Durham University Library, 1993), 3.

[65] Possibly guided by this note, Schuler also attributes some of the poems and recipes in S2 to Lock: Schuler, *English Magical and Scientific Poems*, 55.

If we accept that Lock was the originator of some of the additional texts, it is still unclear precisely what items are his and what items represent later supplements, and when Lock would have appended these short texts in verse and prose. Somewhat surprisingly, the note cited earlier only refers to "abreviaciones in myter" (abbreviations in meter), despite over half of the items being prose compositions. If the statement is taken literally, it would imply that the items in prose were not added by the same person or at the same time as the poems. Supposing that this is indeed the case, the different stages of additions and/or omissions cannot be charted with certainty. Support for the prose texts being later additions may come from the contents of the extant manuscripts. The varying number of poems and recipes is clear evidence of the waxing and waning of the corpus, illustrating that texts could be added or omitted at the whim of the copyists.[66] Manuscript F, which appears to incorporate a number of interpolated items, is a clear case in point, as noted above.

Unfortunately, the content of the poems and recipes does not show clear affinities to the content of the Treatise, which would help to tie them more closely to the Treatise and even to Lock. The correspondences tend to be very general, although the links do seem to be stronger between the Treatise and the poems than between the Treatise and the recipes. Such links may not be significant, however, if the additional items were intended to complement the Treatise (as the note indicates). We must even acknowledge that the note may have been included simply to introduce the poems, whereas the recipes, although added by the same person, were not mentioned specifically.

The texts that Lock is responsible for may have been supplied at several stages. He may have added them at the same time as he finished his compilation of the Treatise. However, there is also another possibility. Important to this discussion is "the Russia note." As I have argued above, the formulation of the note implies that it is a later addition made by Lock, presumably some time after he first compiled his Treatise, but before it was circulated.[67] Although there is direct evidence only for this one addition, it still opens up the possibility that Lock appended other texts at this second stage. Unfortunately, we cannot determine exactly when this second stage occurred, but it must have been between 1572 and June 1590 when Simon Forman dated manuscript A.

The status of the collection circulating together with the Treatise also has some bearing on the title of Lock's text. So far, I have referred to the text as the "Treatise," preferring this "neutral" designation in the belief that the Treatise first appeared untitled. Titles do appear in some manuscripts, but at least two of them appear to be descriptions of the content rather than a proper title. In S1, for example, the title "Philosophicall Manusc˄{r}ips Large ffolº" has been added in

---

[66] Early modern manuscript collections of poems seem to have been subjected to very much the same processes of addition and omission: Love, *Scribal Publication*, 181.
[67] See p. 9.

a later hand, and is probably a designation of the whole manuscript book rather than the title of the Treatise. This assumption is corroborated by the fact that the beginning of the Treatise, where the title would appear, is missing in S1. Three of the manuscripts, S2, S3, and S4, feature the unassuming title "Ex Semita Alberti Magni et Medulla Ripleyij" ('From Albertus Magnus's *Semita* and Ripley's *Medulla*'). This title is presumably unoriginal, reflecting a copyist's identification of two of the most important sources of the Treatise: a version of (Pseudo-)Albertus Magnus's *Semita recta*, and George Ripley's *Medulla*.[68] In S2, the title even represents a later addition in a seventeenth- or eighteenth-century hand: S2 seems originally to have appeared without a title.

The title of the G manuscript stands out. It is more elaborate than the other titles, and it also explicitly attributes the text to Lock: "The hole Compound of Allchimya or Elixzeres of the philosophres wrigtten by H Lock."[69] This title is also significant in that it appears to be referred to elsewhere. Glasgow University Library MS. Ferguson 91, which includes a version of the dialogue *Scoller and Master*, one of the sources of the second chapter of the Treatise, contains the following note: "This is taken out of the second chapter out of the bouk intitulated The whole compound of Alchymie or Elixirs of Philosophers" (fol. 19r).[70] This note is puzzling. The text contained in MS. Ferguson 91 is definitely not a version of the dialogue found in the second chapter of the Treatise, but a version of its source, *Scoller and Master*. The reference may be accounted for if the copyist had at his disposal the G manuscript or a relative of this copy with the same title. Noticing their similarities, he subsequently confused the independent tract *Scoller and Master* with the second chapter of the Treatise. Some additional notes in MS. Ferguson 91 may support this assumption, since the notes are very close to passages in Chapter 3 in the Treatise, and may thus have been extracted from it.[71] The reference in MS. Ferguson 91 is not direct evidence that G's title was widely spread, but it does suggest that the title may have appeared in other copies than G.

G's title may thus seem to be a prime candidate for the original title of the Treatise. However, despite its elaborateness and its appearance elsewhere, it is

---

[68] Probably owing to this title, Roberts includes S2, S3, and S4 under texts ascribed to Albertus Magnus. See Gareth Roberts, *The Mirror of Alchemy: Alchemical Ideas and Images in Manuscripts and Books from Antiquity to the Seventeenth Century* (London: The British Library, 1994), 120.

[69] The final 'a' in "Allchimya" is uncertain: it may be an 'e' or simply a flourish. My interpretation of G's title differs slightly from Irma Taavitsainen's. She transcribes it as "The hole compound of allchimy or elixzeres of the philosophres written by HL": *The Index of Middle English Prose, Handlist X: Scandinavian Collections* (Cambridge: D. S. Brewer, 1994), 12.

[70] See p. 36.

[71] Cf. MS. Ferguson 91 (fol. 21v) and 304v: 5–7, 304v: 28–30, 304v: 35–37.

most likely unoriginal. The collation of the manuscripts reveals that the G copyist (or the copyist of one of its ancestors) was one of the most inventive writers among the Treatise copyists. G contains numerous unique readings, some perhaps even supplied from consulting additional sources. One such instance suggests that a copyist had access to George Ripley's famous late-medieval verse tract *Compound of Alchymy*.[72] Although G's title is longer, its affinity to Ripley's title is unmistakable, and it may have been the inspiration for G's inventiveness.

Stronger than this speculation is the evidence found in other copies of the Treatise. Three manuscripts clearly suggest that the Treatise was originally untitled. In A and F, and even in S2 (if we disregard the unoriginal title in a later hand), the Treatise simply begins with a chapter heading, *Capitulum 1* (or something similar). Importantly, these are also the copies that contain the dedication as well as the additional poems and recipes. Since at least the dedication is intricately tied to the Treatise, this is a sign that Lock did not title his text, perhaps because he saw it as part of a larger collection. The Treatise would have been provided with a title when it was copied independently without the dedication and the other items, which is presumably what has happened in G, S3, and S4.[73]

Simon Forman and the Treatise

The multiple titles or lack of title, the existence or absence of the dedication, and the additions and subtractions from the collection of poems and recipes highlight the dynamic nature of the Treatise and its manuscript environment. The collation of the manuscripts also shows that the text of the Treatise was transformed over time, as may be expected in a text that was manually copied. Unfortunately, most of the copyists and readers of the Treatise remain unknown, and it is hence impossible to place the Treatise in specific sociohistorical contexts for the most part. There is one very important exception: manuscript A. This manuscript was copied out in 1590 by Dr Simon Forman (1552–1611), the astrologer, physician, and alchemist.[74] This copy was subsequently owned and read by Dr Richard Napier (1559–1634), Forman's friend and disciple, and Elias Ashmole (1617–1692), the collector of manuscripts and student of the sciences. Simon Forman has received a great deal of attention in recent years both in articles and in book-length

---

[72] See p. 58.

[73] Some support comes from the dates of at least two of the manuscripts. Both S3 and S4 are seventeenth-century copies, probably from the second half of the century. The date of G is unclear. See the descriptions of the manuscripts.

[74] Forman is responsible for copying about ninety-two percent of the Treatise. Three additional copyists produced the remaining eight percent. See further below.

studies.⁷⁵ To some extent, these studies redeem Forman's tarnished reputation as a quack, magician, and womanizer by pointing to the important legacy of Forman's writings, casebooks, and diary. This material illustrates the practice of an Elizabethan physician, who, like so many others in the period, had a profound interest in astrology, alchemy, and similar areas. Attested by the large number of manuscripts primarily in the Ashmole collection at the Bodleian library, Forman's intense copying and writing has left us numerous texts neatly copied, often dated and annotated. These manuscripts give us direct access to information on how different texts were received and understood by a sixteenth-century practitioner of medicine, astrology, and alchemy. Since manuscript A thus provides an excellent opportunity to tie the Treatise to a known copyist, I will contextualize the A manuscript, by giving a short outline of Forman's life, his copying and study of alchemical texts, and the place of the Treatise in his studies.⁷⁶ As I will argue later, the known context of the A manuscript is a powerful reason for selecting this manuscript as the basis for the edition of the Treatise.

Born in Quidhampton in Wiltshire on 31 December 1552, Simon Forman was the son of the farmers William and Mary Forman. Despite the humble circumstances of the Forman family, Simon was sent to school, first in the nearby town of Wilton and later on to grammar school in Salisbury. He did not finish his studies, however, mainly as a result of his father's death in 1564. After a period of apprenticeship with a Salisbury shopkeeper, Forman worked as a schoolmaster at his old primary school before becoming a poor scholar in Oxford in 1573–1574. To support himself, he offered his services to John Thornborough, a Wiltshire gentleman. Forman left Oxford for unknown reasons fairly soon after his arrival, possibly because his servant duties left him little time for studies.

During the latter half of the 1570s and the 1580s, Forman had a variety of occupations, mostly as a schoolmaster. He was imprisoned twice, once in 1579–1580 and once in 1587, and was entangled in several legal cases. From his diary, it is evident that Forman had begun his autodidact practice of medicine at least

---

⁷⁵ E.g. Judith Cook, *Dr Simon Forman: A Most Notorious Physician* (London: Chatto & Windus, 2001); Barbara H. Traister, *The Notorious Astrological Physician of London: Works and Days of Simon Forman* (Chicago: University of Chicago Press, 2001); Lauren Kassell, "How to Read Simon Forman's Casebooks: Medicine, Astrology, and Gender in Elizabethan London," *Social History of Medicine* 12 (1999): 3–18; eadem, "'The Food of Angels': Simon Forman's Alchemical Medicine," in *Secrets of Nature: Astrology and Alchemy in Early Modern Europe*, ed. William R. Newman and Anthony Grafton (Cambridge, MA: MIT Press, 2001), 345–84; eadem, *Medicine and Magic in Elizabethan London: Simon Forman: Astrologer, Alchemist, and Physician* (Oxford: Clarendon Press, 2005); Schuler, *Alchemical Poetry*, 49–70.

⁷⁶ This description is based on Traister, *The Notorious Astrological Physician*; Kassell, *Medicine*; and to a lesser extent Cook, *Dr Simon Forman*. I have used Cook with caution since it is mostly impossible to verify her claims, as no references to sources are given.

by the 1580s, and many of his problems during this period seem to stem from this practice. After living in various places, he settled permanently in London in 1591/92, establishing himself as a physician. Until he was granted a licence from the University of Cambridge in 1603 (and even after), Forman was continuously harassed by the College of Physicians, which summoned him to appear before a group of censors who challenged his knowledge of medicine. As a result of his conflicts with the College, he suffered further imprisonment on several occasions in the second half of the 1590s. Although practicing unlawfully for most of his life, Forman seems to have attracted a large clientele, as shown by his meticulous casebooks, and he prospered financially. He continued his successful practice up to his death in 1611. After his death, he was implicated in the fatal poisoning of Sir Thomas Overbury, an implication which has since blackened Forman's reputation even to the present day. Though Forman's alleged involvement in the murder was reviewed at the trials, there seems to have been no direct evidence of his participation.

Throughout his life, medicine and astrology were Forman's main interests. However, alchemy was also close to his heart, which is clear from the numerous alchemical texts that he copied. At least by 1594, as an entry in his diary shows, he was attempting to produce the philosophers' stone, and there are several other subsequent notes about alchemical experimentation. Lauren Kassell identifies two stages in Forman's copying of alchemical texts. The first was from approximately 1585 to 1594, when he copied numerous texts into twelve notebooks. In the second phase, he compiled three manuscript books of alchemical notes, the first book dating from around 1597.[77] Copied in 1590, the Treatise belongs to the first phase of his copying. There are no notes in A suggesting that the text was used as a basis for practical experimentation. Although the Treatise is heavily annotated in the margins, most, if not all, of the notes are such as to suggest that they came about in order to highlight passages rather than as a result of testing the procedures. Nevertheless, considering that Forman mostly copied texts for his own use, it is very likely that the Treatise either formed the basis for experimentation, perhaps after 1594, or informed his general knowledge of alchemy, possibly at an early stage of his alchemical endeavors. Kassell argues that Forman's alchemical interests were closely connected with his medical pursuits.[78] Since the Treatise provides references to medical applications of the philosophers' stone, it is possible that this aspect in particular attracted Forman's attention.

In 1590, when he was copying the Treatise, Forman lived in various places, including London and Salisbury. Kassell claims that Forman was staying with a Mr Dalles when he copied the Treatise.[79] In the same year, he also had dealings with Robert Parker or Parkes, the person mentioned in Forman's only printed

---

[77] Kassell, *Medicine*, 174.
[78] Kassell, *Medicine*, 189.
[79] Kassell, *Medicine*, 35–36.

work, *The Groundes of the Longitude,* and a Mr Cumber(s), who was apparently interested in Forman's ventures into magic.[80] There are no clues as to how Forman obtained a copy of the Treatise. Barbara Traister suggests that Forman preferred to borrow and copy books himself rather than buy them since there are no records of book purchases in his otherwise meticulous records.[81] It is very likely then that Forman borrowed a manuscript of the Treatise from someone, perhaps one of the previously named people, and made a copy for himself.

Forman's copy of the Treatise is peculiar in that, although responsible for by far the largest chunk, Forman did not copy out the whole text himself; instead, he had the help of three other copyists. The four writers (including Forman) seem to have worked almost as a tag-team in some parts, even switching in mid-sentence.[82] I have not been able to identify the three additional hands, which may belong to associates, amanuenses, or, perhaps most likely, assistants: unfortunately, the hands do not seem to appear elsewhere in Forman's voluminous manuscript production.[83] Indeed, hands other than Forman's rarely appear in other extant manuscripts, which clearly suggests that Forman preferred to do his own copying and writing.[84] I can only speculate as to why the Treatise should be such an unusual case in the Forman corpus. Since the Treatise was copied relatively early in Forman's career, perhaps it represents an experiment, employing assistants or paid copyists to write out a text. Unhappy with the result or finding greater satisfaction in copying out the text himself, Forman never returned to this model again (unless such examples remain unidentified). Another reason may have been a reluctance to give others access to his books and insights into his copying activities. This reluctance may stem from bad experiences with public knowledge of his ownership of certain books, which included volumes on such suspicious topics as alchemy, astrology, and magic. His books were inspected and confiscated several times, once in 1587 after he was caught in church with a book that, Kassell suggests, dealt with magic.[85] He would thus have been keenly aware that if people knew that he possessed "subversive" books, it could land him in trouble.[86]

---

[80] Kassell, *Medicine,* 36; Traister, *The Notorious Astrological Physician,* 159.

[81] Traister, *The Notorious Astrological Physician,* 122.

[82] The contributions of these copyists are charted in detail on pp. 102–3. For tag-team copying in early modern manuscripts, see Beal, *In Praise of Scribes,* 79.

[83] For some of Forman's assistants, see Kassell, *Medicine,* 147. I have consulted Dr Kassell, who is an expert on Forman's manuscripts, about these copyists. She informs me that she has not seen these hands in other Forman manuscripts: personal communication 2002.

[84] For some exceptions, see Kassell, *Medicine,* 147 n. 69.

[85] Kassell, *Medicine,* 28–29.

[86] Cf. Sherman, *John Dee,* 51–52.

The Picklock

If Forman's copy showcases how the Treatise was received at one point in time, *The Picklock* illustrates another facet of the circulation of alchemical texts: the redaction and transformation of a text into a new composition. The redactor that produced *The Picklock* substantially abbreviated the Treatise obviously with a clear goal (as we shall see), underscoring the fluid nature of many alchemical texts that circulated in manuscripts. Like Lock's original compilation, *The Picklock* redactor's revision also appears to have been very successful, since it attracted the attention of some very prominent alchemical adepts.

It is not exactly clear who the "author" of *The Picklock* was or when it was produced. A date in one of the copies of *The Picklock* makes clear that it must have been finished before 6 June 1602, but I have not been able to narrow down the time frame further. The question about the author is still more complicated. Both the manuscripts of the text, Ba and Bw, proclaim one Benjamin Lock to be the author of *The Picklock*. Although this may seem to be a peculiar coincidence, perhaps indicating a mistake for Humfrey Lock, the case may in fact be less straightforward. Based on two entries in the astrologer and mathematician John Dee's diary that mention a Benjamin Lock, Julian Roberts and Andrew Watson suggest that a Benjamin Lock was a disciple of Dee's from 13 September 1580 until at least 31 August 1582. They furthermore argue that this Benjamin Lock was the author of *The Picklock*.[87] Unfortunately, there seems to be no direct evidence for this claim. On the other hand, it cannot be wholly discarded, and I have found no other possible candidates. If the author was indeed Dee's student, he may have had easy access to Lock's Treatise through Dee's famous library at Mortlake, which contained numerous works on alchemy.[88] There is, however, no record that Dee did possess a copy of the Treatise.

There is of course a more intriguing possibility. Although I have found no evidence of it, it cannot be ruled out that Humfrey and Benjamin were somehow related, and that family relation may thus have put Lock's Treatise in Benjamin's hands. The missing, albeit speculative, link may be Michael Lock (or Lok; *c*. 1532–1620/22), the influential London merchant: the Benjamin Lock mentioned in Dee's diary was in all likelihood Michael's son.[89] This is particularly

---

[87] Roberts and Watson, *John Dee's Library*, 44; S. A. J. Moorat, *Catalogue of Western Manuscripts on Medicine and Science in the Wellcome Historical Medical Library*, vol. 1, *Mss. Written before 1650 A.D.* (London: Wellcome Historical Medical Library, Wellcome Institute of the History of Medicine, 1962), 290; Abraham, *Harriot's Gift*, 1. For the entries in Dee's diary, see James O. Halliwell, ed., *The Private Diary of Dr. John Dee and the Catalogue of his Library of Manuscripts* (London: John Bowyer Nichols and Son, 1842), 8, 17.

[88] Webster, "Alchemical and Paracelsian Medicine," 306; Roberts and Watson, *John Dee's Library*.

[89] *ODNB* s.n. Lok, Michael; Clay, *Familiae*, 1309.

striking because Michael was the London agent for the Muscovy Company during Humfrey's sojourn in Russia.[90] If Humfrey and Michael were indeed related, perhaps Humfrey presented his wealthy and influential relative with a copy, in the hope that Michael could persuade accusers with Company ties to retract their accusations against him. Another possibility is that one of the Company merchants acquired a copy in Russia, brought it back to England, and sold it or gave it to Michael.

Irrespective of who Benjamin was and how he initially found himself with a copy of the Treatise, the author of *The Picklock* thoroughly reworked the Treatise into a new text. His adaptation of the Treatise is first and foremost characterized by omission.[91] Table 2.2 illustrates the major omissions (fifty words or more in sequence) in *The Picklock*.[92]

TABLE 2.2. OMISSIONS IN Bw

Treatise chapters	Bw	Bw omission (folio and line number from the edited text)
1–2	–	
3	±	301v: 1–39, 302v: 36–46, 303r: 3–303v: 12, 303v: 23–28, 303v: 40–48, 306r: 3–8
4	±	307r: 36–41
5	–	
6–7	+	
8	±	308v: 23–34, 309r: 8–17
9	±	309v: 12–16
10	–	
11–15	+	
16	±	313r: 3–7, 313r: 12–18
17–20	+	
21	±	316r: 36–44[d], 317r: 12–25
22–24	+	
25	±	319r: 1–319v: 8

---

[90] *ODNB* s.n. Lok, Michael; Ash, *Power*, 125.

[91] There is very little difference between Ba and Bw (the two copies of *The Picklock*), only some forty disagreements, all of them of minor types of variation. In all cases, Bw is closer to the manuscripts of the Treatise. Considering the similarities, I will focus on Bw since it is the examplar for Ba. See further below and p. 123.

[92] A minus signals that the chapter is completely missing in Bw, while a plus-minus sign shows that the chapter is present but that *The Picklock* omits large parts. The extent of the omissions is specified in the third column by references to the edited text.

Treatise chapters	Bw	Bw omission (folio and line number from the edited text)
26	±	320v: 9–17, 321r: 31–37, 322r: 41–322v: 13
27	±	323r: 2–34
28–29	+	
30	±	324r: 34–324v: 14[B]
31–32	−	
33	+	

[A] Parts of this passage seem to be present in a restructured format.
[B] All lines except four of this chapter are absent in Bw.

Six of the Treatise's chapters (1, 2, 5, 10, 31, and 32) are completely missing. But this is where *The Picklock* clearly reveals its origin. The exclusions notwithstanding, the chapters that do exist in Bw are labeled as if the missing chapters were still there. Although Chapters 1, 2, 5, and 10 are missing, the chapters that should be 17 to 21 in Bw are labeled 21 to 25, and chapters 25–27 are labeled 29–31.

The redactor seemed particularly concerned to remove sections that deal with introductory material, or that explain fundamental assumptions of alchemical theory or practice. Clear examples of this strategy are the omissions of Chapters 1 and 2 of the Treatise, as well as a lengthy discussion of the properties of mercury from Chapter 3 (303r: 3–303v: 12). Metatextual passages that include comments from the author are also absent, such as in Chapter 21 (317r: 12–25); and historicizing sections dealing with alchemy as the gift of God, as in Chapter 25 (319r: 1–319v: 8), have been excised. The redactor did not add much material that is not already present in the Treatise, and he does not seem to have revised sections substantially (apart from omitting stretches of text). This strategy thus suggests that the redactor was especially interested in the advanced alchemical theories and practices that the text describes, and that he did not make particularly striking changes in order for the text to comply with his own experience of alchemical practice.

This redaction was obviously received very favorably by several prominent alchemists and scholars in the seventeenth century. Bw was copied out by Dr Arthur Dee (1579–1651), John Dee's eldest son.[93] In the copy, he notes that "[t]his booke I receaued from Mr Heriot at Sion Howse, who for many yeares instructed the Earle of Northumberland in the mathematicks when he liued in the Tower

---

[93] Moorat, *Catalogue*, 290–91.

| A° 1602 June 6."[94] The exemplar of *The Picklock* that Dee transcribed obviously belonged to the famous mathematician and astronomer Thomas Harriot (*c.* 1560–1621). From 1598, he was a pensioner of "the Earle of Northumberland" Henry Percy (1564–1632), the "wizard earl," who was imprisoned in the Tower in 1605 because of his alleged involvement in the Gunpowder Plot.[95] Perhaps Harriot gained access to *The Picklock* through Percy, who was an avid collector of books and as a result had a sizeable library, which included works on alchemy.[96]

The next stop for the Bw copy of *The Picklock* was Sir Thomas Browne of Norwich (1605–1682), who received the manuscript from Arthur Dee himself.[97] Browne was deeply interested in alchemy, amassing a collection of alchemical volumes.[98] In the late 1650s, Browne, in turn, lent the Bw manuscript to Elias Ashmole, who copied *The Picklock* into what is now Ba. Ashmole himself, whose alchemical interests are well known, records this transaction in Ba (fol. 181r): "9 Apri*l* 1659. Thus far I transcribed out of a MS lent me by Dr Browne of Norwich, written w*i*th Dr Arthur Dee's hand, & neately bound up in Rushian lether [. . .]." The small number of manuscripts of the *The Picklock* suggests that it was not widely copied. However, the circulation of a few copies within a network of alchemists and prominent intellectuals of the seventeenth century illustrates that a small number of extant copies should not necessarily be equated with a limited readership. The circulation of *The Picklock* is clearly indicative of how alchemical adepts exchanged manuscripts for copying or sometimes only for reading.

---

[94] The date has been interpreted in two ways: either the date refers to the time when Harriot provided Dee with the book, or the date indicates when the original manuscript was made. Irrespective of what the date refers to, the note must have been added later than 1602, since Henry Percy ("the Earle of Northumberland") was not imprisoned in the Tower until 1605. For the purposes of this discussion, settling this issue of what the date refers to is not of utmost importance: the dating is primarily of interest in that it makes clear that the reworking of the Treatise that resulted in *The Picklock* must have been done before 6 June 1602. See Roberts and Watson, *John Dee's Library*, 63; John H. Appleby, "Arthur Dee and Johannes Banfi Hunyades: Further Information on their Alchemical and Professional Activities," *Ambix* 24 (1977): 96–109, here 104; *ODNB* s.n. Percy, Henry, ninth earl of Northumberland.

[95] Abraham, *Harriot's Gift*, 1; Moorat, *Catalogue*, 290–91.

[96] G. R. Batho, "The Library of the 'Wizard' Earl: Henry Percy Ninth Earl of Northumberland (1564–1632)," *The Library*, 5th ser. 15 (1960): 246–61, here 247, 257; John W. Shirley, "The Scientific Experiments of Sir Walter Ralegh, the Wizard Earl, and the Three Magi in the Tower 1603–1617," *Ambix* 4 (1949): 52–66, here 57, 64, 66.

[97] Abraham, *Harriot's Gift*, 1; Lyndy Abraham, ed., *Arthur Dee:* Fasciculus Chemicus, *Translated by Elias Ashmole* (New York: Garland, 1997), xlvi–xlvii.

[98] Kevin Faulkner, "Scintillae Marginila [sic]: Sparkling Margins — Alchemical and Hermetic Thought in the Literary Works of Sir Thomas Browne." Available at http://levity.com/alchemy/sir_thomas_browne.html (As accessed in 2006.)

The reason why *The Picklock* attracted the attention of several prominent scholars may be twofold. As the title suggests, Benjamin Lock (whoever he was) or a subsequent copyist perceived the text to be the key to "Riplye his Castle." The text alluded to here is of course George Ripley's (1415?–1490?) famous *Compound of Alchymy*, also known as *The Twelve Gates*, allegedly finished in 1471.[99] This highly popular verse tract, which describes twelve procedures in the alchemical opus as twelve gates, brims with opaque language and obscure allusions. The promise of a key to its procedures would thus probably have been welcomed with open arms by interested readers. Peculiarly enough, however, there is no evidence to suggest that the Treatise was originally intended as a manual to or exposition of Ripley's *Compound*: if it was perceived as such, it must have been a later construct based on content or other features. One factor contributing to such an interpretation may have been the material that has been inherited from other texts allegedly written by Ripley: the Treatise and as a result *The Picklock* incorporate large chunks of text from *Concordantia Raymundi et Guidonis* and *Medulla alkemiae*, both ascribed to Ripley.[100] Indeed, perhaps what fueled the redactor's omissions when he pruned the Treatise to produce *The Picklock* was a sense that some parts helped the understanding of Ripley's *Compound*.

However, more important for the circulation of *The Picklock* may have been Arthur Dee's endorsement of the text (which also appears in Ashmole's copy, Ba). In an epistle to the reader, Dee underscores the important contributions that works like *The Picklock* can make. He advises the reader "not to contemne this or the lyke treatises in that they are written in homely English," because "I found som pointes in them notably conducing and directing an artist in sundry dangerous and secreat passages of the worke. Therefore whosoeuer meeteth with this booke and the lyke, [. . .] let hym not spare his labor to reade yt with a serious obseruation, and he may gather frute from yt beyonde exspectation."[101] At the same time, he instructs the reader to be well versed in "Raimunde, Riply, Turba, and Dastin" before grappling with works such as *The Picklock* and other works in English. *The Picklock* was clearly not seen as a beginner's manual, but rather as an advanced treatise that could be interpreted only in the light of previous studies of other alchemical authorities.

---

[99] See Abraham, *Harriot's Gift*, 2. One of the poems following *The Picklock* (and the Treatise) is also entitled "Replies castele" (A, fol. 330r–v). This poem gives a synopsis of the twelve gates in Ripley's *Compound*.

[100] See pp. 45 and 46–50. Chapter 21 of the Treatise also contains a passage (317r: 14–23) extolling the supreme importance of Ripley's *Compound*. However, this section has been omitted in *The Picklock*.

[101] Bw, fols. 1r–v.

Sources

One of the most remarkable features of the Treatise is that it draws upon and reuses a large number of earlier texts on alchemy. Using sources to construct a new text is of course not an uncommon strategy in alchemical literature. But what is conspicuous is that the Treatise does not simply allude or refer to earlier texts; rather, it incorporates long passages and, in at least two cases, it presents an originally independent composition in its entirety as part of a chapter without referring to the text's origin.[1] The table below provides an overview of the main sources of the Treatise.[2]

Table 3.1. Main source texts

Chapter	Source texts
1	*Mirror of Lights* (Pseudo-Albertus Magnus)
2	*Scoller and Master* (Anonymous)
	Perfectum magisterium (Pseudo-Arnold of Villanova)
	Unidentified (1)

---

[1] In my discussion, I am concerned only with sources of this kind, i.e. sources that the Treatise draws on directly by including large sections of text from the source. I do not include in my definition of *source* allusions, verbal echoes, or similar correspondences between texts. Similarities of this kind can be found between almost any texts on alchemy and do not necessarily indicate dependence. Many such similarities are simply part of the heritage of alchemical writing in general.

[2] Throughout my discussion, I will refer to medieval scholars and clergymen such as Albertus Magnus and Arnold of Villanova as the authors of alchemical writings. But it is vitally important to recognize that these names are pseudonyms. Pseudepigraphy was widespread in medieval and early modern alchemy, and scholars have conclusively shown that most, if not all, attributions to prominent personages are spurious; see e.g. Pearl Kibre, *Studies in Medieval Science: Alchemy, Astrology, Mathematics and Medicine* (London: Hambledon Press, 1984), 187–202 [no. III]; Michela Pereira, *The Alchemical Corpus Attributed to Raymond Lull*, Warburg Institute Surveys and Texts 18 (London: The Warburg Institute, 1989).

Chapter	Source texts
3	*Dicta* (Anonymous)
	Medulla alkemiae (George Ripley)
	Notabilia Guydonis Montaynor (Guido Montaynor?)
	Concordantia Raymundi et Guidonis (George Ripley)
4–9	*Scoller and Master* (Anonymous)
10–14	Unidentified (2)
15–20	*Medulla alkemiae* (George Ripley)
21–25	Unidentified (3)
26–27	*Medulla alkemiae* (George Ripley)
28–33	*Thesaurus philosophorum* (Anonymous)

This compilatory strategy reveals some very interesting information about the genesis of the Treatise. Very little of the text seems to have been the author's own invention. Although I have not been able to identify direct sources for all chapters or sections of the Treatise, the sources identified clearly suggest that Lock was more of a synthesizer or compiler than an original author.[3] It is also evident that Lock had a large number of texts at his disposal. This does not necessarily mean that he had access to anything resembling a library: late medieval and early modern manuscript books of alchemical texts commonly collect together large quantities of text (the manuscripts of the Treatise being good examples), and the selection of texts that the Treatise draws upon is not in any way peculiar.[4] Lock may thus simply have possessed a few volumes, which he himself brought to Russia or which he borrowed from alchemists there.

Although the selection of source texts is not particularly surprising, what is striking is that several of the sources were probably written in English (as we will see later). This is significant in that it points to the growing importance of the vernacular: not only did Lock choose to write his text in English (rather than Latin), but he also employed earlier material in English. This further underscores a trend that has been illustrated in several earlier studies: early modern alchemists extensively copied and adapted Middle English texts on alchemy in prose as well as verse.[5]

---

[3] Lock himself claims that he *compiled* the Treatise; see dedication, l. 7, and "the Russia note" (316v: 27). However, since *compile* had both the meaning 'to collect together material from different sources,' and 'to compose an original text' in early modern English, it is unclear exactly how Lock perceived his text; see *OED* s.v. compile.

[4] See p. 31.

[5] Peter Grund, "'ffor to make Azure as Albert biddes': Medieval English Alchemical Writings in the Pseudo-Albertan Tradition," *Ambix* 53 (2006): 21–42; John Reidy, ed., *Thomas Norton's* Ordinal of Alchemy, EETS OS 272 (Oxford: Oxford University

The existence of clearly identifiable sources also allows comparisons between the Treatise and the source texts. Such comparisons reveal (to some extent) how Lock engaged with his sources: whether he followed them closely or reworked them according to his own views or needs. Although such comparisons are not always straightforward, it is clear that the Treatise adheres closely to its sources in some parts of the text, whereas substantial differences appear in other parts. As we will see later, attempts appear only occasionally to have been made to make the text coherent, since contradictions and overlaps occur quite frequently. The limited engagement with the sources underscores the impression that Lock was more of a compiler than an original alchemical thinker.

Before the relationship between individual source texts and Lock's Treatise is explored in more detail, it is crucial to outline some of the methodological challenges and pitfalls inherent in the identification of sources in alchemical texts. Attention to these problem areas will help clarify the scope and limitations of my discussion.

Perhaps the most fundamental and obvious problem in the identification is that the Treatise never refers to its sources explicitly.[6] Needless to say, this lack of references makes identification particularly difficult. Although relying heavily on bibliographical works such as those of Dorothea Singer, Lynn Thorndike and Pearl Kibre, Willy Braekman, and George R. Keiser,[7] I have also supplemented previous research by consulting a variety of library catalogues and by carrying out searches on the powerful electronic tool compiled by Linda Voigts and Patricia Kurtz (*eVK*) and on the bibliographical list available at *The Alchemy Website*.[8] Despite these tools, there are still problems. It is evident from a study

---

Press, 1975); Linden, *George Ripley's* Compound; Keiser, "Preserving the Heritage." See also p. 33.

[6] References to alchemical authorities do occur in the Treatise, but as far as I have been able to ascertain these references are always to secondary sources, i.e. texts that the sources of the Treatise cite.

[7] Dorothea W. Singer, *Catalogue of Latin and Vernacular Alchemical Manuscripts in Great Britain and Ireland, Dating from before the 16th Century*, 3 vols. (Brussels: Maurice Lamertin, 1928, 1930, 1931); Lynn Thorndike and Pearl Kibre, *A Catalogue of Incipits of Mediaeval Scientific Writings in Latin: Revised and Augmented Edition* (London: Mediaeval Academy of America, 1963); Willy L. Braekman, ed., *Studies on Alchemy, Diet, Medecine [sic] and Prognostication in Middle English*, Scripta 22 (Brussels: Omirel, 1986); George R. Keiser, *A Manual of the Writings in Middle English 1050–1500*, vol. 10, *Works of Science and Information* (New Haven: Connecticut Academy of Arts and Sciences, 1998).

[8] Linda E. Voigts and Patricia D. Kurtz, *Scientific and Medical Writings in Old and Middle English: An Electronic Reference*, CD-ROM (Ann Arbor, MI: University of Michigan Press, 2000). For this indispensable research tool for studies of early scientific texts in English, see also Peter Grund, review, *ICAME Journal* 26 (2002): 160–64. The bibliographical component of *The Alchemy Website* (www.levity.com/alchemy/home.html) contains information on more than 4,000 manuscripts on alchemy in a variety of languages.

of the Treatise that Lock only occasionally adopted a source from beginning to end; instead, the source passage often stems from the middle of a text. This poses serious difficulties since most bibliographical tools provide the incipit, and sometimes the explicit, of a text, while passages in the middle of a text are not usually recoverable. Exceptions to this rule are Thorndike and Kibre and the *eVK*, which both record incipits of texts that appear as independent items in a manuscript, even though they may be extracts from other texts. But even these tools have limitations: they are, of course, only helpful in this respect if the particular text passage that was used as a source for the Treatise exists independently in a manuscript, which is far from always the case. Consultation of a wide range of alchemical source material has thus been necessary.

A related, and equally crucial, concern has to do with identifying the particular version of the source that was used in the compilation. Like other categories of medieval and early modern texts that circulated in manuscripts, alchemical writings commonly exist in a number of versions that may be more or less closely related to each other.[9] Editors have not paid enough attention to this fact: very few alchemical tracts written in Latin or English exist in modern editions that carefully consider the manuscript tradition of the texts.[10] Admittedly, we are fortunate in having several early printed, multi-author collections of alchemical writings, as well as collections of individual authors: these provide a starting point for identification.[11] But regrettably these editions present one canonical version of the text,

---

The descriptions of the manuscripts and of their contents commonly derive from the available printed or handwritten catalogues of the manuscripts. The information provided in the site is a good starting point for further investigations, and the electronic format facilitates searching and browsing. However, for some collections the information is scanty, and there are many errors inherited from the catalogues of the manuscripts. This tool should therefore be used with some caution.

[9] See e.g. Grund, "'ffor to make Azure'."

[10] For the problem of editions and manuscript variation, see Peter Grund, review of Stanton J. Linden, ed., *The Alchemy Reader: From Hermes Trismegistus to Isaac Newton* (Cambridge: Cambridge University Press, 2003), *Studia Neophilologica* 75 (2004): 211–12; idem, "Manuscripts as Sources for Linguistic Research: A Methodological Case Study Based on the *Mirror of Lights*," *Journal of English Linguistics* 34 (2006): 105–25. For editions of English writings, see e.g. idem, "In Search of Gold: Towards a Text Edition of an Alchemical Treatise," in *Middle English from Tongue to Text: Selected Papers from The Third International Conference on Middle English: Language and Text*, ed. Peter J. Lucas and Angela M. Lucas (Frankfurt am Main: Peter Lang, 2002), 265–79, here 265–66.

[11] The best-known and most useful alchemical compendia are probably Jean-Jacques Manget, ed., *Bibliotheca chemica curiosa*, 2 vols. (Geneva, 1702); Lazarus Zetzner, ed., *Theatrum chemicum, praecipuos selectorum auctorum tractatus de chemiae et lapidis philosophici antiquitate, veritate, iure, praestantia & operationibus, continens*, 4 vols. (Strasbourg, 1659). Single-author collections, such as *Georgii Riplaei Canonici Angli opera omnia chemica* (Kassel, 1649), are also very useful.

and hence do not give a complete picture of the variation among different versions that circulated in manuscript form. In my study, I have consulted both early printed editions and manuscripts. This procedure has proved crucial since the source employed in the Treatise must in most cases have been very much closer to a manuscript version than to that of a printed edition. Even more important, the Treatise often exhibits features that are found in different witnesses of the source text. This suggests that the direct source of the Treatise in these cases was a text that contained features that are now extant only in different copies. Needless to say, it has not been possible to consult all the identified manuscripts of a particular text; both time and availability were limiting factors. Most of the manuscripts consulted are from the Bodleian Library, Oxford, especially the Ashmole Collection, but I have also to a lesser extent consulted manuscripts from other locations such as the British Library, London, and Trinity College, Cambridge.

Another factor that needs to be considered is the language of the source text. During the Middle Ages, Latin was the language of choice for most alchemical writers in England, but English gained more and more currency in the fifteenth century.[12] This trend continued in the following century, when translators adapted numerous Latin alchemical writings into English. Comparisons between the Treatise and its sources reveal that an English translation or adaptation of a Latin text must have served as the source for some sections. In other cases, it remains unclear whether Lock appropriated an already existing vernacular version, or whether he himself made the translation. Whenever possible, I have therefore inspected both Latin and English manuscripts.

The degree of reworking of the source text can also make identification difficult. In many instances, the Treatise appears to incorporate material more or less verbatim, with few changes to the original wording of the source. Such a conservative treatment naturally facilitates identification. Even in the cases where sources have been reworked with additions of cross-references and explanations, they are still relatively easy to identify. When the Treatise features chapters and passages that only have parallels in other texts, the problem is more intractable: Lock (or a subsequent copyist) may have reworked the source substantially, hence blurring the relationship. Alternatively, the parallels between the two texts are just that, parallels, similar but not the same, both texts deriving from the same source. Such passages unfortunately elude straightforward classification.[13]

---

[12] See p. 74.

[13] Unless it proved necessary for the understanding of the Treatise, I have not attempted to identify secondary sources (i.e. texts that are referred to or cited in the actual source of the Treatise).

Mirror of Lights

The first chapter of the Treatise consists of a revised and restructured version of the *Mirror of Lights*. This pedagogical tract, which is commonly attributed to the renowned medieval scholar and theologian Albertus Magnus (*c.* 1200–1280), presents a handbook of alchemical practice in three parts.[14] A reworking of an even earlier and more famous pseudo-Albertan text, the *Semita recta*, the *Mirror of Lights* proceeds from a down-to-earth description of alchemical theory to a rigorously structured introduction of basic alchemical concepts and finally to a sequence of straightforward recipes. The *Mirror of Lights* appears to have been extremely popular, as it survives partly or wholly in at least thirteen Middle English and early modern English manuscripts, and fragments of a Latin version exist in three manuscripts. For my comparisons with the Treatise, I have used seven of the principal manuscripts.[15]

The Treatise draws upon the second part of the *Mirror of Lights*, which outlines three, or in some manuscripts four, "conditions" (or categories) of alchemical practice. However, in the same way that the *Mirror of Lights* reworked its source (the *Semita recta*), the Treatise substantially revises the structure and sometimes the content of the *Mirror of Lights* (illustrated schematically in Figure 3.1).

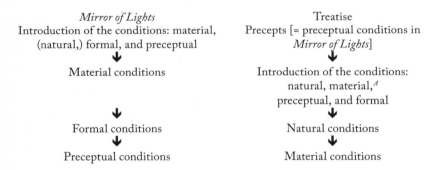

FIGURE 3.1. THE STRUCTURE OF THE *MIRROR OF LIGHTS* AND THE TREATISE
A The material conditions are not included in the enumeration in some manuscripts of the Treatise. See EN 295r: 11–21.

---

[14] For discussions of Albertus Magnus and alchemy, see e.g. Pearl Kibre, "Albertus Magnus on Alchemy," in *Albertus Magnus and the Sciences: Commemorative Essays 1980*, ed. James A. Weisheipl (Toronto: Pontifical Institute of Mediaeval Studies, 1980), 187–202; Dorothy Wyckoff, *Albertus Magnus: Book of Minerals* (Oxford: Clarendon Press, 1967).

[15] Cambridge, Trinity College MS. O. 2. 33, Part 2, pp. 1–37 (16th c.); Cambridge, Trinity College MS. R. 14. 37, fols. 115r–147r (15th c.); Cambridge, Trinity College MS. R. 14. 45, fols. 67r–77v (15th c.); Cambridge University Library MS. Kk. 6. 30, fols. 1r–10v (15th c.); London, British Library MS. Harley 3542, fols. 1r–14r (15th c.); London, British Library MS. Sloane 513, fols. 155r–168r (15th c.); British Library MS. Sloane 3580A, fols. 193v–208v (16th c.). For a study of the *Mirror of Lights*, see Grund, "'ffor to make Azure'," 32–38.

The *Mirror of Lights* begins by introducing the different conditions, and then proceeds to discuss their nature. The Treatise, by contrast, first establishes that it will begin with "*precept*es, admonitions, and co*mm*anedemen*tes*" (294r: 30–31), and subsequently lists seven precepts that the prospective alchemist should follow. This list corresponds to the seven preceptual conditions, which are not found until the very end of the *Mirror of Lights*. Peculiarly enough, the Treatise never makes clear that the precepts are indeed the preceptual conditions that it later mentions in the introduction of the conditions. In fact, in that introduction, the Treatise makes a point of saying that it will treat only the natural and material conditions.

Following the precepts in the Treatise is the introduction of the conditions, which mentions four categories: natural, material, preceptual, and formal. At the same time (as noted above), the Treatise underscores that only two types of condition are relevant, the natural and material, and that only these will be discussed. In including natural conditions, the Treatise shows a clear affinity to three manuscripts of the *Mirror of Lights*: MS. R. 14. 45, MS. Harley 3542, and MS. Sloane 3580A. These are the only manuscripts that mention natural conditions. But this is as far as the affinity goes; for where the three manuscripts of the *Mirror of Lights* surprisingly do not discuss the natural conditions at all, the Treatise finds a use for natural conditions in its taxonomy. It does this by redefining the material conditions. In the *Mirror of Lights*, the material conditions cover substances that are crucial in alchemical practice (e.g. mercury, sulfur, sal ammoniac). In the Treatise, these substances are instead assigned to the natural conditions; the material conditions become a superordinate category that comprises all the substances classified under the natural conditions as well as additional substances and equipment.[16]

After describing the natural conditions, the Treatise goes on to discuss the material conditions. In general, the discussion adheres closely to that of the *Mirror of Lights*, though some differences occur. For example, in the treatment of "imperfect bodies" (i.e. all substances except gold and silver), the Treatise only comments upon spirits (e.g. mercury and sulfur) and to some extent different kinds of salt; the *Mirror of Lights*' discussion of calces, stones, and similar substances is absent. It is possible that these passages were left out because they were seen as less relevant in the context of the Treatise. While all of the imperfect substances feature prominently in the later practical procedures in the *Mirror of Lights*, the Treatise mentions only a few of them, and so infrequently that discussing them might have been felt to be unwarranted.

Although mentioned in the enumeration, the formal conditions are not discussed at all in the Treatise. The *Mirror of Lights* uses these conditions to introduce different alchemical procedures, including calcination, sublimation, ceration, and fermentation. Perhaps these descriptions, which are fairly basic, were considered

---

[16] See EN 295r: 11–21.

too introductory, or perhaps Lock (or a subsequent copyist) did not agree with the descriptions, and hence they were left out.

The discussion in the Treatise does not seem to derive from any one extant manuscript of the *Mirror of Lights*. However, there are strong indications that the source of the Treatise was written in the vernacular. The English *Mirror of Lights* and the Treatise are identical in many passages in linguistic form and vocabulary;[17] and there are readings in the Treatise that may be the result of a misunderstanding of an earlier English source.[18] Although a specific source manuscript cannot be identified, the Treatise is certainly closer to some manuscripts than to others, most notably to the three closely affiliated manuscripts MS. R. 14. 45, MS. Harley 3542, and MS. Sloane 3580A.[19]

Scoller and Master

One of the most striking examples of the Treatise's reuse of an earlier alchemical text is its adoption of the originally independent dialogue *Scoller and Master*.[20] Presumably inspired by scholastic textual models, this instructional dialogue seems to have been very popular in alchemical circles in the fifteenth century through to the seventeenth century: it survives wholly or in part in at least fifteen English vernacular manuscripts from the period.[21] The setup of *Scoller and*

---

[17] See e.g. the use of *wild* in 295v: 12, which is found in the English versions but not in the Latin; EN 295v: 12.

[18] See EN 295r: 36–295v: 3.

[19] See EN 294v: 13–22.

[20] There is no consensus on the title of this dialogue in the manuscripts: "Interrogaciones discipuli ad magistrum de arte ista" (MS. Sloane 3747, fol. 66r); "Opus de Magistro & Discipulo" (MS. Sloane 3580B, fol. 131r); "Scoller and Master" (MS. Ashmole 1490, fol. 81r); "A Dialogue betwene the Master and the Disciple" (MS. Ashmole 1487, fol. 107r); "A Conference betweene the Maister and the Disciple" (MS. Ashmole 1493, fol. 81r). For the sake of convenience, I will use the title *Scoller and Master* to refer to this text.

[21] Since only a few of these manuscripts have been listed together in previous research, I include them all here: Cambridge, Trinity College MS. O. 2. 22, fols. 40v–41r (15th c.); Glasgow, Glasgow University Library MS. Ferguson 91, fols. 19r–25r (17th c.); London, British Library MS. Harley 2407, fols. 68v–69r (15th c.)(?); London, British Library MS. Harley 6453, fols. 25r–29r (16th–17th c.); London, British Library MS. Sloane 3580B, fols. 131r–134v (16th c.); London, British Library MS. Sloane 3720, fols. 87r–94r (17th c.); London, British Library MS. Sloane 3747, fols. 66r–71v, and 117v–119r (15th c.); Oxford, Bodleian Library MS. Ashmole 759, fols. 42r–45v, 47v–52v (15th c.); Oxford, Bodleian Library MS. Ashmole 1451, fols. 68v–69v (15th c.); Oxford, Bodleian Library MS. Ashmole 1479, fol. 223r (16th c.); Oxford, Bodleian Library MS. Ashmole 1487, fols. 107r–110v (16th–17th c.); Oxford, Bodleian Library MS. Ashmole 1490, fols.

Master is a discussion between a father and his son, a master and a scholar (i.e. a student), or a teacher and a disciple, where the son/scholar/disciple questions his father/master/teacher on matters of alchemical practice and theory.[22] The Treatise reproduces most of this dialogue fairly faithfully in Chapter 2 (296v: 1–299v: 2), but it postpones parts of the text until Chapters 4–9, where they appear in a sometimes substantially reworked form.

As in the case of the first chapter based on the *Mirror of Lights*, strong indications point to the text of the Treatise deriving from a version of *Scoller and Master* written in English. This is suggested by striking similarities in formulation and linguistic form; supporting evidence is also found in some apparent mistakes and misconceptions in the Treatise.[23] Furthermore, I have not been able to identify a Latin source for this dialogue. Although this does not mean that no such source existed, it does suggest that a Latin text may not have circulated as widely as the vernacular *Scoller and Master*.[24]

81r–84r, and 88r (16th c.); Oxford, Bodleian Library MS. Ashmole 1493, fols. 81r–84v (17th c.); Oxford, Corpus Christi College MS. 185, fols. 133r–134v (15th c.); Oxford, Corpus Christi College MS. 226, fols. 31r–33v (15th c.). Braekman, *Studies on Alchemy*, 13, also lists Trinity College, Cambridge, MS. R. 14. 44 (fol. 15r). However, this dialogue is not the same as the one in the other manuscripts and has therefore not been included here; see Keiser, *A Manual*, 3806; Linne R. Mooney, *The Index of Middle English Prose. Handlist XI: The Library of Trinity College, Cambridge* (Cambridge: D. S. Brewer, 1995), 46.

[22] For the use of the dialogue form in scientific writing, see Irma Taavitsainen, "Dialogues in Late Medieval and Early Modern English Medical Writing," in *Historical Dialogue Analysis*, ed. Andreas H. Jucker et al. (Amsterdam: John Benjamins, 1999), 243–68; eadem, "Transferring Classical Discourse Conventions into the Vernacular," in *Medical and Scientific Writing in Late Medieval English*, ed. eadem and Päivi Pahta (Cambridge: Cambridge University Press, 2004), 63–67; Peter Grund, "Albertus Magnus and the Queen of the Elves: A 15th-Century English Verse Dialogue on Alchemy," *Anglia* 122 (2004): 651–52.

[23] For specific examples, see EN 296v: 27, 298v: 17–19, 299r: 19–20, 299r: 34–35, 299r: 37.

[24] *eVK*; Keiser, *A Manual*, 3806; Braekman, *Studies on Alchemy*, 13; Thorndike and Kibre, *A Catalogue*; Singer, *Catalogue*, 324–25. In MS. Ashmole 1490 (fol. 81r), "Egidius de Vadis" has been canceled under the title *Scoller and Master*. A dialogue attributed to Egidius de Vadis exists in a number of manuscripts, entitled "A dialogue of Egidius Devadius, betwene Nature and the Disciple of Philosofie"; see MS. Ashmole 1490, fol. 28r. However, this dialogue is not related to *Scoller and Master*. The *Alchemy Website* contains more than fifty alchemical dialogues (or dialogue-like texts) in its bibliographical component; several of the Latin dialogues seem to have a master/teacher/father and scholar/pupil/son structure. Owing to the large number of manuscripts of potential interest, it has not been within the scope of this study to make an exhaustive investigation into the Latin texts. Some sections in *Scoller and Master* also appear in a Latin treatise entitled *Rosarium philosophorum*; Zetzner, *Theatrum*, 3: 664, 678, 688. The *Rosarium* includes most of the

The overall structure is largely the same in the Treatise and the majority of the copies of *Scoller and Master*.[25] The Treatise includes seven of the eight questions and answers that seem to constitute the core of the dialogue; it excludes only a discussion about mercury. Textually, the two are also close, although differences do occur. For example, in 298r: 4–12, the Treatise presents an elaboration that does not appear in any of the copies of *Scoller and Master*; and in 298v: 40–43, material that is part of an answer in *Scoller and Master* is removed to a question in the Treatise. Chapter 7 contains several passages and phrases that have no parallels in *Scoller and Master*. Many of these statements seem to serve as clarifications of the text, though it is unclear whether these clarifications existed in a version of *Scoller and Master* or were added by Lock (or a subsequent copyist).[26]

A more detailed comparison between the manuscripts of *Scoller and Master* and the individual chapters of the Treatise based on the dialogue reveals slightly different trends in the treatment of the source material in the different chapters. Chapter 2 adheres fairly closely to some of the manuscripts of *Scoller and Master*. At the same time, none of the identified manuscripts can alone have served as the source of the Treatise, since it exhibits varying textual affinities. This suggests that the source of the Treatise must have been a version of the dialogue that contained features which are now found only in different manuscripts. Alternatively, the Treatise conflated several versions of *Scoller and Master*, although this seems less likely. Even though there is no direct relationship between the Treatise and any one manuscript, the Treatise is more closely related to some manuscripts than to others. In particular, the Treatise agrees with MS. Ashmole 759, MS. Harley 6453, the two versions in MS. Sloane 3747,[27] and MS. Ferguson 91. For example, these versions present similar answers to the first question in the dialogue (296v: 23–297r: 6), which deals with heterogeneous and homogeneous substances. The other copies that contain the first question, such as Corpus Christi College, Oxford, MS. 226, MS. Ashmole 1490 (fols. 81r–84r), and MS Sloane 3720 postpone this particular discussion to the very end of the dialogue.

---

discussion on sublimation found in Chapters 6 and 8 of the Treatise, and some statements found at the beginning of Chapter 2. It is unclear, however, if the *Rosarium* is the source for these sections of *Scoller and Master*, or whether they share a common source.

[25] There is some variation in the manuscripts of *Scoller and Master* in inclusion of material, organization of material, and formulation. Some manuscripts, such as Corpus Christi College MS. 185 and MS. Ashmole 1451, contain only fragments of the dialogue, whereas other copies, such as MS. Ashmole 1490 and MS. Harley 6453, are fuller copies, containing eight core questions and answers and even some additional material.

[26] See EN 308r: 34–35, 37–38, 40–42, 308v: 2–9.

[27] MS. Ashmole 759 and MS. Sloane 3747 are written in the same hand. However, there are differences between the manuscripts that suggest that they do not have an exemplar-copy relationship.

In many individual readings, the Treatise also joins the manuscripts of the former group.[28]

In general, the Treatise is closer in wording and structure to *Scoller and Master* in Chapter 2 than in Chapters 4–9. Among these later chapters, Chapter 5 is of special interest. There is only one manuscript of *Scoller and Master* that contains a similar passage: MS. Ashmole 759. However, two features make it highly unlikely that the Treatise is based on this manuscript: the two show many significant textual differences, and, more importantly, MS. Ashmole 759 merges passages that correspond to Chapters 4 and 5 in the Treatise. MS. Ashmole 759 begins with material closely related, though not identical, to Chapter 5 of the Treatise, which is interrupted by material corresponding to Chapter 4. Subsequently, MS. Ashmole 759 reverts back to material that is also found in Chapter 5 of the Treatise (though differences occur). Since other manuscripts of *Scoller and Master*, which contain material that is also found in Chapter 4 of the Treatise, exhibit the same structure and formulations as the Treatise, MS. Ashmole 759 clearly did not serve as the Treatise's source. The textual differences underscore that impression. However, MS. Ashmole 759 is important in that it demonstrates that the material of Chapter 5 indeed derives from a version of *Scoller and Master* and not from elsewhere.

Perfectum magisterium

The passage that follows the extract from *Scoller and Master* in Chapter 2 (299v: 2–300v: 30) draws upon the alchemical tract *Perfectum magisterium*. Manuscripts and printed editions usually ascribe this widely circulated text to the medieval physician Arnold of Villanova (1243–1311).[29] This particular passage is an excellent

---

[28] See EN 297r: 23–26, 298v: 17–19, 299r: 41–299v: 2.

[29] Various titles have been attributed to this treatise, including *Perfectum magisterium*, *Semita semitae*, *Flos florum*, and *Lumen luminum*; for a discussion of these titles, see Lynn Thorndike, *A History of Magic and Experimental Science*, 8 vols. (New York: Columbia University Press, 1923–58), 3: 61, 69–71. I follow Thorndike in referring to this text as *Perfectum magisterium* (although Thorndike uses only the English translation of the title, *Perfect Mastery*, after introducing the title in Latin); *A History*, 3: 53. Payen has established that there is a fundamental difference between texts circulating under the titles *Perfectum magisterium* and *Semita semitae*, since the latter do not include, for example, the enumeration of errors, which is the section used by the Treatise: J. Payen, "*Flos Florum* et *Semita Semite*: Deux traités d'alchimie attribués à Arnaud de Villeneuve," *Revue d'histoire de sciences* 12 (1959): 289–300. I have consulted the following printed and manuscript copies of *Perfectum magisterium*: Arnold of Villanova, *Hec sunt opera Arnaldi de Villa Nova nuperrime recognita ac emendata diligentique opere impressa que in hoc volumine continentur* (Basel, 1509), fols. 302r–304r; *Artis Auriferae, quam alchemiam vocant, volumen secundum* (Basel, 1572), 513–31; Zetzner, *Theatrum*, 3: 128–36; Cambridge, Trinity College MS.

example of how the Treatise is selective in its choice of material:[30] it includes only the first part of *Perfectum magisterium*, which presents a discussion on errors that alchemists have committed in their work; the second part, which is more theoretical in nature, has been left out.[31] The reason for this selection seems obvious: the section culled from *Perfectum magisterium* ties in well with the problem-based question-answer structure of the preceding dialogue. Similarities in content also allow the Treatise to move fairly seamlessly between the two sources. *Perfectum magisterium*'s reader address (given as *scias charissime* 'you should know, my dearest' in some copies) provides a particularly smooth transition. In the Treatise, this reader address becomes a means for the father (of the dialogue) to broach another subject. Subsequently, however, the Treatise follows the monologic stucture of *Perfectum magisterium* rather than the earlier dialogic structure.

The Treatise is much closer in structure to the three manuscript versions that I have seen than the three printed editions.[32] Most significantly, the manuscripts of the *Perfectum magisterium* as well as the Treatise contain a preface that is absent in the printed editions.[33] The manuscripts' formulation of this preface varies, and the Treatise is closer to the English version in MS. R. 14. 44 than to the two Latin manuscripts, MS. Ashmole 1384 and MS. Ashmole 1450, which contain longer discussions. In addition to closeness in structure, the Treatise also adheres to the manuscripts textually, especially the English version in MS. R. 14. 44.[34] At the same time, it is clear that the source of the Treatise must have contained a combination of the features that are now found only in individual manuscripts.

R. 14. 44, Part 4, fols. 1r–8r (15th c. in English); Oxford, Bodleian Library MS. Ashmole 1384, fols. 76r–79r (14th c.); Oxford, Bodleian Library MS. Ashmole 1450, fols. 23v–25r (15th c.).

[30] Whether the fusion of *Scoller and Master* and *Perfectum magisterium* is a result of the compilation of the Treatise or whether an already existing compilation was appropriated is of course difficult to ascertain with certainty. However, it is significant, although not conclusive, that none of the manuscripts of *Scoller and Master* that I have consulted contain a version of the dialogue followed by a version of *Perfectum magisterium*.

[31] Singer, *A Catalogue*, 199, lists the parts given by the Treatise as the preface and the "Practica §1." See also Payen, "*Flos Florum*."

[32] The three editions display a great deal of internal variation in formulation and inclusion of material. *Artis Auriferae* and Arnold of Villanova, *Hec sunt opera*, seem to be more closely related to each other than to Zetzner, *Theatrum*.

[33] 299v: 2–15. For this preface, see *Histoire littéraire de la France* (Paris: Imprimerie Nationale, 1881), 28: 83–84.

[34] See EN 299v: 27–28, 300r: 21–22.

Dicta

Chapter 3 is one of the most intriguing parts of the Treatise, not only because it is based on more sources than any other chapter, but also because it illustrates some striking redactional strategies. The most prominent example appears after the general introduction to Chapter 3, from which point the Treatise presents a series of seemingly unconnected statements (301v: 41–303v: 47). These statements have been lifted from an untitled and anonymous collection of sayings attributed to different philosophers and alchemists (henceforth: *Dicta*). This florilegium survives in at least two vernacular manuscript copies: MS. Ashmole 759 and MS. Gl. Kgl. Saml. 242.[35] Although the different sayings presumably derive ultimately from a large number of Latin sources, I have found no Latin version of this particular collection.[36] The Treatise's close adherence to the text of the *Dicta* in MS. Ashmole 759 even suggests that the Treatise may very well be based on an English text.

In order to highlight the way in which the Treatise seems to transform the *Dicta*, I have included an overview of the correspondences between the two texts in the table below.

TABLE 3.2. COMPARISON OF THE *DICTA* AND THE TREATISE

Dicta (Ashmole 759, fols. 69r–76v)	Treatise
Hermes	301v: 41–44
Alfidius	301v: 44–46
Lylly	302r: 1–6
Geber	302r: 6–9
Lilly	302r: 9–11
Hermes	—
Rosary	302r: 12–21
Rosyne	—
Nouella marita	302r: 21–22
Gregory	—
Zenon / Rosyns / Rasys / Lumen Luminum / Morien	302r: 22–38
Empedicles / Joh*ann*es de ffrancia. . .in his nouella margareta	302r: 39–41
Rosarius	302r: 41–44
Aristotell	—

---

[35] Oxford, Bodleian Library MS. Ashmole 759, fols. 69r–76v (15th c.), and Royal Library of Copenhagen, MS. Gl. Kgl. Saml. 242, fols. 44v–47v (16th c.). I have not been able to inspect Gaml. Kgl. Saml. 242 in person. For information on this manuscript, see Taavitsainen, *The Index*, 4.

[36] Neither Singer, *A Catalogue*, 12–17, nor Thorndike and Kibre, *Catalogue*, lists a related text. I have not tried to trace the ultimate source of the different sayings.

Dicta (Ashmole 759, fols. 69r–76v)	Treatise
Arisleus	—
Horfolius in Turba / Empedocles	302r: 44–302v: 9
Nouella M	302v: 9–13
Lumen Luminum	302v: 13–18
Rasys	302v: 18–20
Zenon	302v: 20–22
Alex*ander*(?)	302v: 22–29
Hermes	—
Lylly	—
Rosynus in Turba / Morienus	302v: 29–30
Phiaras	—
Rasys	—
Lylly / Lumen Luminum	302v: 31–40
Gebar	302v: 42–45
Rasys	302v: 45–46
Gebar	302v: 46–303r: 3
Morienus	303r: 3–8
Rosarius	303r: 8–303v: 3
Lumen	303v: 3–12
Neuella M	303v: 12–21
Reymond	303v: 24–47
Hermes	—
Paganus	—
Reymond	—

The Treatise follows the series of the *Dicta* fairly closely, and it is also, mostly, close to it in formulation. There are only modest attempts at reconciling the passages of the *Dicta* with passages in other parts of the Treatise drawn from other sources, and the Treatise includes very few stretches of text that are not recorded in the *Dicta* as well.[37] But the Treatise is far from slavish in its adherence: some sayings are absent, and others are represented in a condensed format.[38] The most conspicuous feature is that the Treatise has systematically removed all the attributions to different alchemists and alchemical texts. If this was done by Lock (rather than his source text), it may have been a conscious effort to provide a smoother, more coherent text, and perhaps to downplay the contributions of earlier alchemists, to his own benefit. If Lock tried to produce a more logically linked text, his efforts did not extend much further than removing the attributions: most of the time the statements are left unconnected, or they are loosely tied together with linking words such as *wherefore, therefore, for*, and *and*.

---

[37] See EN 302r: 16–21, 302v: 40–42, 303v: 21–24.
[38] See EN 303v: 3–12.

Notabilia Guydonis Montaynor

Notabilia Guydonis Montaynor, a fairly unknown alchemical tract (as it seems), forms the basis for a discussion of the elixir in Chapter 3 (304v: 9–23). Extant in a fifteenth-century manuscript, MS. Sloane 3744, and in a printed edition under the title *De arte chymica*, the *Notabilia Guydonis* is a collection of questions and answers about various alchemical procedures.[39] The identity of the author, Guido Montaynor or Guido de Monte, remains uncertain: Thorndike suggests that he is a fifteenth-century author, whereas Singer identifies him as one Guido de Monte Rocherii (*fl.* 1333).[40]

The Treatise appears to include only the first question and answer of *Notabilia Guydonis*, which deals with the existence of two *materiae primae*.[41] The Treatise is much closer to MS. Sloane 3744 than to the printed text (a full-text comparison is given in the table below).

TABLE 3.3. COMPARISON OF THE TREATISE AND *NOTABILIA GUYDONIS*[42]

Treatise (304v: 9–23)	*Notabilia Guydonis* (MS. Sloane 3744, fol. 27r)
For ther is doble materia prima, and yet but on thinge, that is to say, water from water was before heaven and earth was made, as it is said in the bocke of Genises that the earth was vain & void, and the sprite of the Lord was borne vpon the waters, and God made the firmament & dep*a*rted the waters from water. & forasmoch as water was before heaven & earth, it is moste manifeste that water was the firste matter of all thinges, wherin was a beinge in poware, reduciable to the acte that is of noe quality nor quantitie, of the w*hich* matter and	Quid est materia prima. Dico quod 2^x habetur materia prima, vna viz primordialis materia quae fuit aqua antequam celum esset neque terra vt dictum[?] libro Genesis. Terra erat inanis & vacua & spiritus Domini ferebatur super aquas et dicitur ibidem Deus fecit firmamentum in medio aquarum & diuisit aquas ab aquis. Ergo aqua fuit antequam celum & terra &c. Aqua fuit primordialis materia omnium rerum. Reliqua vero primaeva materia in hac arte dicitur materia prima corporum mineralium, quae est aqua viscosa in visceribus

---

[39] London, British Library, MS. Sloane 3744, fols 27r–31r (15th c.); *Harmoniae inperscrutabilis chymico-philosophiae sive philosophorum antiquorum consentientivm. Decas I* (Frankfurt, 1625), 125–52. See also Singer, *A Catalogue*, 280.

[40] Thorndike, *A History*, 4: 333; Singer, *A Catalogue*, 941. He may also be the same as the Guido drawn upon in *Concordantia Raymundi et Guidonis* (discussed below), and identified as the author of *Scala philosophorum*, found e.g. in Manget, *Bibliotheca*, 2: 134–47. See also Singer, *A Catalogue*, 280; S. J. Ogilvie-Thomson, *The Index of Middle English Prose. Handlist VIII: Oxford College Libraries* (Cambridge: D. S. Brewer, 1991), 21; Linden, *George Ripley's* Compound, 117.

[41] MS. Sloane 3744, fol. 27r; the text extract constitutes chapter four (incorrectly labeled as chapter five) in *Harmoniae inperscrutabilis... Decas*, 131.

[42] MS. Sloane 3744 uses a large number of abbreviations, which I have silently expanded.

substantiall form is broughte forth the second matter that is a beinge in acte of the beinge in poware. Soe likewise the second materia prima, which is the first matter of the bodies of the mine, is ingendred in the bowells of earth, as a viscus water, ro*n*ninge in the vains therof, takinge noe reste but lightly flyeth from the fyer [. . .]

terrae concurrens & vnita tali unione quod humidum teneatur a sicco & siccum ab humido equali dispocistione. Ista aqua viscosa creata fuit ex illa primordiali aqua per potenciam Domini et ex ista aqua viscosa omnia corpora mineralia habuerunt principium. Et nota quod primordialis materia omnium rerum secundum Aristotelem proprie dicitur ens in potencia reducibilis ad actus quae non est actus qualitatis nec quantitatis, ex qua materia & forma substanciali resultat materia secunda, quae est ens in actu ex ente in potencia, & forma substanciali adnecta penitus est composita

[What is prime matter. I say prime matter is considered twofold. There is one matter, i.e. the primordial, which was water before heaven and earth were made as it is said in the book of Genesis. The earth was empty and void and the spirit of the Lord was borne over the waters, and it is said in the same place that God made the firmament in the middle of the waters and separated waters from waters. Therefore, water existed before heaven and earth etc. Water was the primordial matter of all things. But the remaining initial matter in this art is called the prime matter of the mineral bodies, which is viscous water running in the veins of the earth and joined in such a union that the moist is kept by the dry and the dry by the moist in equal disposition. This viscous water was created from the primordial water by God's power and out of this viscous water all mineral bodies have their beginning. And note that the primordial matter of all things according to Aristotle is properly called a potential being reducible to a real being which is not a real being of quality and quantity, from which matter and substantial form comes the second matter, which is a real being of the potential being and it is made when the substantial form has been firmly united with it [?]*A*

A My translation.

The most striking difference between the Treatise and MS. Sloane 3744 appears in the structure of the argument. The *Notabilia Guydonis* introduces the discussion on real and potential beings after discussing the second *materia prima*. The Treatise, on the other hand, presents a smoother transition between the exposition of the two *materiae*. It first exhausts the discussion on the first *materia*, including the real and potential beings. It subsequently proceeds to talk about the second *materia*, taking its cue from the fact that the second *materia* is created from the first.

Concordantia Raymundi et Guidonis

Following the passage based on the *Notabilia Guydonis* in Chapter 3 is a long stretch of text that frequently refers to "Raymond" and "Guido" (305r: 19–306r: 17). This is slightly surprising: the multiple references are uncharacteristic for the Treatise. However, none of the references are in fact direct attributions: they have all been transferred from the text that the Treatise recycles: *Concordantia Raymundi et Guidonis*. Possibly composed in the 1470s, this tract is usually attributed to the famous English alchemist George Ripley (1415?–1490?).[43] It consists of a comparison of views on different points of alchemy allegedly held by Raymond Lull (1232/3–1315/6) and Guido, "a great ph*ilosofe*r of Greace."[44]

The Treatise is selective in its inclusion of material from this text: it adopts only what the *Concordantia* manuscripts refer to as the theoretical part, which is perhaps to be expected since Chapter 3 is concerned with alchemical theory. The discussion in the Treatise follows that of the *Concordantia* fairly closely, though passages do occur where the Treatise presents alternative versions that are not attested in any of the copies of the *Concordantia* that I have consulted.[45] The differences do not seem to mark a certain redactional aim, and it is impossible to determine whether they are the result of a revision by the compiler of the Treatise, or, alternatively, whether the Treatise may be based on a different version of the *Concordantia*. The Treatise is more in line with the English manuscript versions than with the printed Latin version that I have seen.[46] The affiliated MS. Rawlinson B. 306, MS. Ashmole 1424, and MS. Ashmole 1479 are particularly close to the Treatise.[47]

---

[43] For more information on George Ripley, see Linden, *George Ripley's* Compound. Gareth Roberts suggests that the *Concordantia* must have been composed sometime after 1471, but it is unclear what he bases his suggestion on: *The Mirror*, 41.

[44] For Guido, see the discussion on the *Notabilia Guydonis*.

[45] See EN 305r: 25–29, 305v: 11–13, 305v: 22–24.

[46] Oxford, Bodleian Library MS. Ashmole 1424, fol. 100r–v (17th c.); Oxford, Bodleian Library MS. Ashmole 1479, fols. 43r–44r (16th c.); Oxford, Bodleian Library MS. Ashmole 1487, fols. 180v–181v (17th c.); Oxford, Bodleian Library MS. Rawlinson B. 306, fols. 41v–42v (16th–17th c.); Ripley, *Opera omnia*, 323–31.

[47] See EN 305v: 18, 306r: 8–9.

Medulla alkemiae

One of the most important and extensively used sources in the Treatise is *Medulla alkemiae*, 'The Marrow of Alchemy,' usually ascribed to George Ripley. Dedicated to George Neville, Archbishop of York, the *Medulla* was allegedly finished in 1476.[48] It deals primarily with three different forms of the philosophers' stone: the mineral, vegetable, and animal stones. The *Medulla* material appears in a substantially revised form in the Treatise. Chapters 15 to 20 and a section in Chapter 3 present the *Medulla*'s discussion of the mineral stone. Chapter 26 is concerned with the vegetable stone, but it interpolates sections from the *Medulla*'s mineral stone discussion into its description. Chapter 27, finally, derives from the *Medulla*'s treatment of the animal stone. The Treatise omits the *Medulla*'s long introduction. This is hardly surprising, since the introduction is primarily concerned with trying to gain the favor of Archbishop Neville and with presenting some basic alchemical concepts, many of which have already been introduced in Chapters 1 and 2 of the Treatise.

Existing in numerous manuscripts and printed editions, in Latin as well as in English translation, the *Medulla* appears to have been extremely popular in the sixteenth and seventeenth centuries.[49] A comparison between the Treatise and the versions of the *Medulla* reveals that the Treatise must be based on a text that was in parts very different from all of the *Medulla* copies that I have consulted. Alternatively, Lock (or a subsequent copyist) substantially revised the *Medulla* text. If indeed Lock was behind this revision, it is difficult to discern a clear exegetical strategy. Although the Treatise does not correspond to any one copy in particular, it does seem to be more in line with the manuscript versions than with the printed editions. In a few instances, the Treatise exclusively agrees with MS. Ashmole 1479 and MS. Rawlinson B. 306,[50] and, in a few other cases, with MS. Ashmole 1507.

The complex relationship between the Treatise and the *Medulla* warrants attention to the structure and content of the individual chapters of the Treatise

---

[48] Thorndike, *A History*, 4: 351; Linden, *George Ripley's* Compound, xii.

[49] I have consulted the following manuscripts and printed editions: Oxford, Bodleian Library MS. Ashmole 1479, fols. 35r–42v (16th c. in Eng.); Oxford, Bodleian Library MS. Ashmole 1480, Part 5, fols. 9r–15v (16th c. in Eng.); Oxford, Bodleian Library MS. Ashmole 1485, Part 3, fols. 69r–88r (16th c. in Eng.); Oxford, Bodleian Library MS. Ashmole 1487, fols. 172r–180r (16th c. in Eng.); Oxford, Bodleian Library MS. Ashmole 1490, fols. 66v–74v (16th c. in Lat.); Oxford, Bodleian Library MS. Ashmole 1492, pp. 157–175 (16th c. in Eng.); Oxford, Bodleian Library MS. Ashmole 1507, fols. 109v–120v (17th c. in Eng.); Oxford, Bodleian Library MS. Rawlinson B. 306, fols. 55r–65v (16th c. in Eng.); Ripley, *Opera omnia*, 123–79 (in Lat.); William Salmon, ed., *Medicina Practica, or Practical Physick* (London, 1692), 643–87 (in Eng.).

[50] See EN 322r: 33, 322v: 10–11, 323r: 24–26.

that draw upon it. The section in Chapter 3 (304r: 18–30) presents only a brief extract from the *Medulla*'s introduction to the mineral stone,[51] and Chapter 15, which constitutes the Treatise's introduction to the same stone, seems to summarize various aspects of the *Medulla*'s treatment. Chapter 16, on the other hand, is a more complicated case. The chapter begins with a passage from the *Medulla* (312v: 34–39), but then proceeds with material that does not seem to derive from the same source text. This material incorporates a recipe for "the fier againste nature." Although this substance is repeatedly mentioned and employed in the *Medulla*, the text never makes clear how to produce it. The recipe may thus have been included to remedy this silence. A similar recipe, which employs vitriol and vermilion together with saltpeter, is found in MS. Ashmole 1479 (fol. 52v), attributed to Raymond Lull; I have not been able to identify a similar recipe in (pseudo-)Lull's writings. The passage following the recipe (313r: 33–313v: 3) appears to revert back to *Medulla* material, though there are many dissimilarities between the two texts. The differences again point to substantial revisions of the *Medulla* text, or to a very different version of the underlying source.[52]

Chapter 17 presents a heavily abbreviated version of what the *Medulla* refers to as the second procedure of the mineral stone.[53] The first three sentences of the chapter exemplify the type of transformation that the *Medulla* text has undergone. The sentences have been lifted from widely separated passages in the *Medulla*'s discussion, and the intervening passages have been excluded, as is shown in the comparison below.

Table 3.4. Comparison of the Treatise and the *Medulla*

Treatise (313v: 19–25)	*Medulla* (Ripley, *Opera omnia*, 143–45) [Emphasis and translation added]
In the second parte of the minerall stone, gould is veri excellently elixeratid with the fier againste nature compounded with the fier of nature, beinge distilled both together;	Per secundum modum fit aurum magis miraculosum, elixiratum per ignem contra naturam, compositum cum igne naturali, [By the second method gold becomes more miraculous, elixerated by the fire against nature, compounded with the fire of nature] & modus operandi est talis. R.^{A} Vitriolum factum de acutissimo humore uvarum (cum igne naturae) & Sericone, conjunctis in una massa cum Vitriolo naturali aliquantulum exiccato simul cum salenitri: atque ex his destilletur aqua, quae primo erit debilis phlegmatica sine coloratione vasis, quam

---

[51] Ripley, *Opera omnia*, 138–39.
[52] EN 312v: 39–313v: 16.
[53] Ripley, *Opera omnia*, 143–47.

and this water beinge minerall & vegitall, it worketh co*n*trary thing*es*.

For so moch as it dissolueth, soe moch again it congealeth and lifteth in glorious earth, w*hich* of phi*loso*fe*r*s is called the secrete dissolution of the stone, w*hich* is not without the co*n*gelation of his water.

projice. Tunc ascendet fumus albus, qui faciet apparere vas quasi citrinum (al. quasi lac.) qui colligendus est, quousque cesset, & vas revertatur ad suum pristinum colorem. Nam illa aqua est menstruum foetens, in quo est quinta essentia nostra, id est fumus albus, qui vocatur ignis contra naturam: qui si esset absens, esset ignis noster naturalis, de quo plura dicemus suo loco, h. e. capite sequenti. **Et hae aquae, scilicet mineralis & vegetabilis, commixtae & aqua una (al. viva) factae, operantur contraria, [And these waters, i.e. the mineral and vegetable, mixed and made one (alternatively living) water, work contrary things]** quod est mirandum. Nam haec una dissolvit & congelat, humectat & exiccat, putrefacit & purificat, disgregat & conjungit, separat & componit (al. restaurat,) mortificat & vivificat, de-destruit [sic] & redintegrat, attenuat & inspissat, nigrificat & albificat, comburit & frigefacit, incipit & perficit. Hic sunt duo dracones pugnantes in gurgite Sathaliae, hic est fumus albus & rubeus, quorum unus devorabit alterum. Atque hic vasa resolutoria non debent lutari, sed solummodo leviter sint obstructa cum panno lineo & mastiche aut cera communi. Nam haec aqua est ignis & Balneum intra vasa & non extra, quae si alium ignem fortem senserit, statim elevabitur ad summum vasis, & si ibi requiem non invenerit, vasa rumpe*n*tur, atque ita compositum relinquetur frustratum. **Haec aqua composita in quantum solvit, in tantum congelat & elevat in terram gloriosam. Atque ita est secreta dissolutio nostri lapidis, quae semper fit cum congelatione suae aquae. [This composite water dissolves into so much as it congeals and elevates into glorious earth. And this is the secret dissolution of our stone, which always takes place together with the coagulation of its water.]**

A ℞ = *recipe*/*recipite* ('take').

The exact reason for the reworking is unclear. Perhaps the intervening passages were less relevant for the discussion, since the three remaining sentences were felt to sum up the argument. Another significant example of omission in Chapter 17 concerns the procedure on a compound water. This procedure is only abbreviated here and presented more fully in Chapter 26, which actually deals with the vegetable stone and not the mineral. This choice seems to have been made to provide a smoother, more accessible text in the later chapter (see below).

Chapters 18 and 19 are fairly close renderings of the *Medulla* discussion of the third and fourth procedures of the mineral stone,[54] while Chapter 20's relationship to the *Medulla* is less straightforward. It may be a heavily abbreviated version of a procedure that the *Medulla* labels as a mercury-only process.[55] The brevity of Chapter 20 and the general nature of its content, however, make the exact connection between the two texts less than certain.

With the possible exception of Chapter 3, Chapter 26 is the most structurally complex Treatise chapter in terms of sources. To illustrate the complexity, I give a summary of the sources used in the chapter in Table 3.5.

Table 3.5. Source structure of Chapter 26

Passage	Source
320v: 3–15	*Medulla* (Vegetable Stone)
320v: 15–321r: 10	Unidentified
321r: 10–31	*Medulla* (Vegetable Stone)
321r: 31–36	*Notabiliora philosophorum*
321r: 37–321v: 2	*Thesaurus philosophorum*
321v: 3–5	*Declaratio lapidis physici Avicennae filio suo Aboali*
321v: 5–18	Unidentified
321v: 18–322r: 4	*Medulla* (Vegetable Stone)
322r: 5–322v: 18	*Medulla* (Mineral Stone)
322v: 18–25	*Medulla* (Vegetable Stone)

Devoted to the vegetable stone, the chapter begins with a section (320v: 3–15) from the *Medulla*'s discussion of this particular stone, though not a very close rendering of the *Medulla* text as preserved in the copies consulted. The subsequent passage (320v: 15–321r: 10) is not found in the *Medulla* copies, and I have not been able to identify a possible source for this passage.[56] It deals primarily with two "menstrues," especially the "menstrue resoluble," which is said to bring about the conjunction of the two opposite principles, mercury and sulfur. Continuing the discussion of the two menstrues, the section that follows (321r: 10–31) reverts

---

[54] For some differences between the Treatise and the *Medulla*, see EN 314r: 37–40, 315v: 2–4.

[55] Ripley, *Opera omnia*, 157–60.

[56] Cf., however, 304v: 43–305r: 14.

back to the *Medulla* text. However, once again, the discussion is interrupted, this time by a passage referring to "Antazagurras," which has probably been imported from the text *Notabiliora philosophorum*.[57] After this insertion, the Treatise continues to draw on a source other than the *Medulla*. For the passage in 321r: 37–321v: 2, the Treatise subsumes material from *Thesaurus philosophorum*, the source of Chapters 28–33, and follows it up by a quotation from (pseudo-)Avicenna's *Declaratio lapidis physici Avicennae filio suo Aboali* in 321v: 3–5.[58] Not until 321v: 18 does the Treatise return to the *Medulla*. However, although the Treatise contains exclusively *Medulla* material from here on, it does not simply reproduce the *Medulla* discussion. Instead, in 322r: 5, the Treatise interpolates a long description of a compound water that it previously presented only partly in Chapter 17. The reason for this interpolation is clear. At the corresponding point (right before the interpolation), the *Medulla* states "[e]t deinceps ordine sicut in op*ere* aquae compositae vsque ad eius totale co*m*pleme*ntum* per *om*nia procedatur" 'and after that proceed in everything in due order as in the work on the composite water until its complete accomplishment' (MS. Ashmole 1490, fol. 73r). The passage referred to here is indeed the one that the Treatise inserts. So instead of sending the reader back to Chapter 17, the Treatise makes the proper discussion conveniently available to the reader. Finally, in 322v: 18, the discussion switches back to the *Medulla* discussion of the vegetable stone, which continues to the end of the chapter.

Finally, Chapter 27 is based on the *Medulla* discussion of the animal stone. Although mostly close to the *Medulla*, the Treatise exhibits some variant passages and readings.[59] In four places, the Treatise agrees with MS. Ashmole 1507, copied by Elias Ashmole, against all the other manuscripts.[60] However, MS. Ashmole 1507 clearly marks these passages as somehow special by enclosing them within square brackets. This may indicate that they are not part of the text proper, and have been added from elsewhere. Even if they are not a natural part of MS. Ashmole 1507, the fact that the passages exist in the manuscript all the same suggests that they are not idiosyncratic features of the Treatise, but rather belong to another version of the *Medulla* or to another source text.[61]

---

[57] See p. 51.

[58] Zetzner, *Theatrum*, 4: 876. I have not treated this text in a separate section since only one sentence seems to be quoted from it. The same text is quoted in Chapter 33 (325r: 13–15), where a slightly longer extract is given. In Chapter 33, Avicenna's text constitutes a secondary source taken over from *Thesaurus philosophorum*, and it is quite possible that the quotation in Chapter 26 is secondary as well, similarly originating in *Thesaurus philosophorum*.

[59] EN 323r: 4–8, 323r: 31–39.

[60] EN 323r: 27, 323r: 27–28, 323r: 29–31, 323v: 3–4.

[61] Although it is theoretically possible that MS. Ashmole 1507 could be based on the Treatise in the passages where only the two agree, this seems highly unlikely, since the Treatise and MS. Ashmole 1507 are very different in other parts of the *Medulla*.

Notabiliora philosophorum

The *Dicta*, discussed earlier, is not the only florilegium that the Treatise seems to draw upon: Chapters 25 and 26 feature brief passages that are ascribed to Morien and Antazagurras, respectively. These same statements appear in an anonymous collection of sayings, entitled *Notabiliora philosophorum*, which exists in at least three English vernacular manuscripts from the fifteenth and sixteenth centuries.[62] The full-text comparison below reveals an undeniable relationship between the Treatise and the *Notabiliora*.

TABLE 3.6. COMPARISON OF THE TREATISE AND THE *NOTABILIORA PHILOSOPHORUM*

Treatise	*Notabiliora philosophorum*
Morien the ph*ilo*so*fe*r said in Turba Ph*ilo*so*fo*rum: Honor our kinge com*m*inge from battaille, vz., from the fier, crowned with a dyadem and clothed with a king*es* clothinge. And he saith moreover: Let him be norished vntill he com to p*er*fecte age, whose father is ☉ & whose mother is ☽ [319r: 43–319v: 3]	Morien*us* saythe: Honour o*ur* kyng commyng frome batell, þat is, frome fyer, crowned wi*th* a dyadem & norysche hym vnto he come to a p*er*fy3te age, whos father is þe sonne & hys mother þe moone. [MS. Ashmole 1486, Part 3, fol. 29v]
[. . .] as Antazagurras said. On time, said he, I soughte in bock*es* that I might com to the knowledg of this science. Afterward I praid vnto God that he should teach it me, and*ᴬ* my prayer was hard that ther was showed me a cleane water, wh*i*ch I knewe to be a clean viniger, and howe moch more I red in bockes, soe moch more they weare made open vnto me. [321r: 31–36]	Antazaguarras saythe in on tyme I sowghte in bokys þat I xulde cu*m*e to þe sentens of on of þis. Afterwarde I prayed vnto Gode þ*at* he shulde teche me. Forsothe my prayer was harde & þ*er* was schewed vnto me clene water þe wyche I knewe to be o*ur* clene venyger, & howe myche more þ*at* I rede in bokys so myche more were þey ly3tenede & when þe male & þe female ar ioyned þ*er* is made a woman noght fugytyffe. [MS. Ashmole 1486, Part 3, fol. 29r]

ᴬ The reading *forsooth* is found in the following copies of the Treatise, corresponding to the source: G, S1, S2 and S4.

---

[62] Cambridge, Trinity College MS. O. 5. 31, fols. 9v–10v (15th c.); Oxford, Bodleian Library MS. Ashmole 1486, Part 3, fols. 28v–29v (15th–16th c.); Copenhagen, Royal Library of Copenhagen, MS. Gamle Konglige Samling 1727, fols. 97r–98v (15th–16th c.). I have not been able to inspect the Copenhagen manuscript in person; see Taavitsainen, *The Index*, 10. A fourth copy might exist in Corpus Christi College, Oxford, MS. 244, fol. 84r, which I have not been able to consult; see Michela Pereira and Barbara Spaggiari, eds., *Il Testamentum alchemico attribuito a Raimondo Lullo: Edizione del testo latino e catalano dal manoscritto Oxford, Corpus Christi College, 244* (Florence: Sismel, Edizione del Galluzzo, 1999), 596–97. The Latin title of the text is found in MS. Ashmole 1486, whereas MS. O. 5. 31 (fol. 9v), gives the English title "Notable of þe phelosophers": Mooney, *The Index*, 122.

Whether the Treatise derives directly from an English version or a Latin one is uncertain, though the striking linguistic similarities to the English versions of the *Notabiliora* point to a possible vernacular source. Like the *Dicta*, the *Notabiliora* probably goes back to a large number of Latin sources, but I have not found a Latin version of this particular collection. An interesting and at the same time peculiar reference appears in the Treatise's version of Morien's saying. Here the saying is attributed to the seminal Latin tract *Turba philosophorum*.[63] However, this cannot be the ultimate source for the Treatise (or the *Notabiliora*), since Morien's saying does not appear in the *Turba* (and neither does Antazagurras's saying). The Treatise's attribution to the *Turba* may in fact have been influenced by a different source. The very same statement as in the Treatise appears in some versions of (pseudo-)Arnold of Villanova's *Perfectum magisterium*, which, as we saw earlier, is excerpted in Chapter 2. Like the Treatise, *Perfectum magisterium* ascribes the saying to the *Turba*, but, unlike the Treatise, it claims that it is by an anonymous "philosophus."[64] It may be speculated, then, that the statement in the Treatise derives directly from the *Notabiliora*, but that the *Turba* attribution is secondary (fueled by *Perfectum magisterium* or another source text).

Thesaurus philosophorum

Sometimes ascribed to George Ripley, the source of Chapters 28 to 33, as well as a passage in Chapter 26, exists in various versions in manuscripts and editions of the sixteenth and seventeenth centuries.[65] The printed collection of Ripley's works presents it as two chapters of the *Concordantia*, under the headings *Compositio aceti acerrimi vegetabilis* and *Compositio mercurii alchymici*.[66] In some alchemical manuscripts, on the other hand, the text is unattributed and split up.[67] Although the exact relationship between the widely different versions is unclear, the Treatise's source is undeniable: it corresponds very closely to the version

---

[63] Julius Ruska, ed., *Turba Philosophorum: Ein Beitrag zur Geschichte der Alchemie* (Berlin: Julius Springer, 1931).

[64] Zetzner, *Theatrum*, 3: 136.

[65] The passage in Chapter 26 (321r: 37–321v: 2) differs only slightly from the beginning of Chapter 28 (324r: 2–16).

[66] Ripley, *Opera omnia*, 331–36. It is notable that none of the manuscript versions of the *Concordantia* consulted contain these two chapters.

[67] I have consulted the following manuscripts and printed edition: Oxford, Bodleian Library MS. Ashmole 1407, Part 4, pp. 24–26 (17th c.); Oxford, Bodleian Library MS. Ashmole 1441, pp. 197–199 (17th c.); Oxford, Bodleian Library MS. Ashmole 1459, pp. 186–191 (16th c.); Oxford, Bodleian Library MS. Rawlinson D. 1217, fols. 2r–4r (16th c.); Ripley, *Opera omnia*, 331–36. MSS. Ashmole 1407 and 1441 are both written in the hand of Thomas Robson (see p. 115), but the two versions differ significantly in some parts.

presented in the clearly related copies MS. Ashmole 1459 and MS. Rawlinson D. 1217. Both these manuscripts present a single text under the title *Thesaurus philosophorum*, consisting of four chapters.[68]

That the Treatise is closer to the *Thesaurus* than the other versions is clearly illustrated in a number of passages. In Chapter 33, the Treatise cryptically states that "the salte is made in the first chapter from the w*hich* our vegitable ☿ was drawen" (324v: 36–37). The reference to "the first chapter" is slightly peculiar since Chapter 1 of the Treatise obviously does not discuss this salt. MS. Ashmole 1407, MS. Ashmole 1441, and the printed edition provide no clues to this cryptic statement since they only have a very vague reference: "salis vegetabilis superius reservati" 'vegetable salt reserved earlier' (or a similar expression).[69] But the *Thesaurus* makes clear what the Treatise refers to. In the margin of chapter four, the *Thesaurus* contains the following note: "As is shewed in the first chap*ter* out of w*hi*ch our vegetable ☿ is drawne" (MS. Ashmole 1459, p. 190). The chapter referred to, then, is chapter one in the *Thesaurus*, which is Chapter 28 in the Treatise. The reference has obviously been adopted uncritically in the Treatise from the source. In addition to this, there is another striking similarity between the Treatise and the *Thesaurus*. The *Thesaurus* alone contains a passage corresponding to Chapter 29 of the Treatise. Somewhat surprisingly, it is not an independent chapter in the *Thesaurus* but the heading of chapter two, which has been transformed into Chapter 30 in the Treatise.

Although the Treatise is close to the *Thesaurus*, it does not follow its source slavishly. For example, the chapter structure of the two differs, since the Treatise divides up chapter three of the *Thesaurus* into Chapters 31 and 32. The Treatise also seems to significantly abbreviate the *Thesaurus* discussion in a number of passages, although it rarely adds material that is not also present in the *Thesaurus* in one form or another.[70] One of these rare additions is particularly interesting. At the beginning of Chapter 32, the Treatise ascribes the invention of the "lesse worke of alkamy" to "Alkamus the ph*ilos*ofer" (324v: 25). The *Thesaurus*, by contrast, is less specific and simply attributes this procedure to anonymous alchemists. The Treatise attribution echoes a line in one of the poems following the Treatise, "The way to calcen perfectly," where the same Alkamus is mentioned,

---

[68] The *Thesaurus* also contains a prologue, which is not included in the Treatise; cf. MS. Ashmole 1459, pp. 186–187. The headings of chapter one, "The composition of ye ph*ilosoph*ers vineger in w*hich* are the p*ro*perties of our ☿" (MS Ashmole 1459, p. 187), and of chapter four, "Compositio Mercurii Alchymista*rum*" (MS Ashmole 1459, p. 190), in the *Thesaurus* resemble closely the chapter headings that are found in the printed edition of the *Concordantia*. The exact relation between the *Thesaurus* and the texts ascribed to Ripley is uncertain. Perhaps the *Thesaurus* is a substantial reworking of a tract originally attributed to Ripley.

[69] Ripley, *Opera omnia*, 333.

[70] See EN 324r: 3, 324r: 20, 324v: 42–325r: 1, 325r: 29–31.

but there he is said to have invented alchemy as a whole, and not only one procedure.[71] The reference in the poem indicates that Alkamus is probably the same as *Alchymus*, who appears in other alchemical texts as the mythical inventor of alchemy, most notably in Thomas Norton's *Ordinal of Alchemy*.[72] The exact reason why the Treatise invokes Alkamus in Chapter 32 is not clear, and it is of course possible that the attribution was inherited from a copy of the *Thesaurus* that has not been identified. However, the fact that he appears both in the Treatise and in one of the subsequent poems (which was probably written by Lock) illustrates an interest in this mythical authority. It also seems significant that, in all copies except A, Alkamus speaks in the first person, pronouncing the veracity of the procedure. Mentioning Alkamus may thus have been part of a strategy to lend credibility and authority to the procedure under discussion.

Unidentified Sources and Parallel Texts

Although I have been able to identify a substantial number of the Treatise's sources, there are still chapters or sections of chapters whose sources remain unidentified, most notably the end of Chapter 2, Chapters 10–14, and Chapters 21–25. Of course, it cannot be ruled out that Lock or a subsequent copyist partly or wholly composed some sections. However, the structure and characteristics of these chapters and sections indicate that a majority, if not all, of them are based on previous alchemical tracts. The existence of texts that discuss more or less closely related procedures strongly supports the idea of underlying, though unidentified, sources. In what follows, I will address the relationship between the Treatise and some of these parallel texts in order to illustrate that most of the Treatise is probably indebted to earlier sources, either specifically or more generally.

As we saw above, Chapter 2 is by and large based on two earlier alchemical texts: *Scoller and Master* and *Perfectum magisterium*. But when the extract from *Perfectum magisterium* ends in 300v: 30, the Treatise carries on with the same structure of presenting errors by previous alchemists until the end of Chapter 2. Although they do not appear to be direct sources, several parallel texts discuss procedures similar to the ones criticized in this section of the Treatise. This is not very surprising. The fact that the Treatise includes certain procedures and errors for discussion suggests that they were advocated by some alchemists and needed to be refuted. For example, a recipe in Cambridge, Trinity College MS. R. 14. 45 (fol. 97v) records a procedure that resembles one that is rejected by the Treatise. The recipe is more detailed, as may be expected, since the point of the Treatise is not to develop the argument but rather to refute the general strategy. I include parts of the discussion for illustration.

---

[71] See p. 103.
[72] Reidy, *Thomas Norton's* Ordinal, 18, 122.

TABLE 3.7. COMPARISON OF THE TREATISE AND A RECIPE IN MS. R. 14. 45

Treatise (300v: 31–38)	MS. R. 14. 45 (fol. 97v)
Some I haue kowen [sic] that haue drawen a stronge water of vitriall, salpeter, and allom, and in the said water they dissolued siluer, clensed it, and distilled it from his flegma; and afterward they drewe another strong water of vitriall, sal armoniak, & allom, and in the same water they dissolued gould, clensed it, and distillud it from his flegma as in the other water, and then they ioyned the too waters together in on glasse, and closed vp the glasse, and set him in fier of the first degre [. . .]	Make a coresif of wetriall sat [sic] petyr and of alym a*na* and dyssolue therin an ownce of syluyr and iii z m*er*cury crew and whan they be dissoluyd put vnto ham di li m*er*cury sublymyd iiii z of salt tartyr iiii z of salt arm*oni*ac fyxyd iiii of salt petyr fyxyd and ii z of boras and grynd all thyse togedyrys & put ham all into the coresyfe wit*h* yowre forseyd matyrys and set ham in aschys and set on an hede and draw the watyr from*e* ham and wan*e* all the watyr ys com*e* take of your*e* hede and stop your*e* glas and make good fyr*e* vndyr your*e* glas tyll all your*e* mater*e* be sublymyd þat woll sublyme [. . .]

Following this rejected procedure is an anecdotal passage which claims that the product resulting from the procedure was a miraculous soil fertilizer (300v: 38–45). Similar stories appear elsewhere in alchemical literature. For example, Taylor cites the following note written by Elias Ashmole in his annotation of an alchemical text:

> [. . .] there was a Glasse fond in a Wall full of Red Tincture, which being flung away to a dunghill, forthwith it coloured it, exceeding red. This dunghill (or Rubish) was after fetched away by Boate by Bathwicke men, and layd in Bathwicke field, and in the places where it was spread, for a long tyme after, the Corne grew wonderfully ranke, thick, and high [. . .][73]

What these two comparisons make clear is that there are parallels, but also that the parallels are very general. The question is how the Treatise relates to these texts. The simplest explanation is that the discussion in the Treatise is a product of Lock's compilation. Finding additional lists of alleged errors in other texts, Lock supplied what had not already been covered by *Scoller and Master* and *Perfectum magisterium*.

Chapters 10–14 in the Treatise are slightly different from the section in Chapter 2 in that there is stronger evidence suggesting that they all derive from a single source. Cambridge, Trinity College MS. O. 4. 39 (fols. 244v–250r) and Oxford, Bodleian MS. Ashmole 1507 (fols. 121r–128v) contain a text that exhibits several affinities to these chapters. The untitled and anonymous text of

---

[73] Frank S. Taylor, "Thomas Charnock," *Ambix* 2 (1946): 148–76, here 160.

the two manuscripts appears to be a compilation, consisting of a main text with several additions or annotations mentioning texts attributed to Raymund Lull.[74] Though there are similarities between this compilation and Chapters 10–14, it can be said with some certainty that the compilation is not the direct source of the Treatise. At some points, the relationship is unclear or only very general. Chapter 10 of the Treatise is closely related to the second chapter of the compilation, as seen in the table. At the same time, there are also several dissimilarities, and the procedure described in the Treatise is much more detailed.

TABLE 3.8. COMPARISON OF TREATISE CHAPTER 10 AND MS. ASHMOLE 1507

Treatise Chapter 10 (310r: 3–42)	MS. Ashmole 1507, fols. 122r–22v
First make a malgam of ii^{li} of quicksiluer and of ii^{li} of puer tyne in this manner: Melte your tyne in a meltynge pot that shalbe large and greate ynoughe for yo*ur* malgam. Then haue another crucible, wherin youe muste make your quicksiluer hote, and when he is hote, put him into the tynne that is moulton, and stur them strongly together. Then take the crucible from the fier, & stir it co*n*tynually vntill it be somwhat could. Then caste the malgam into a pan of clean water, and washe the malgam vntill the water becom all blacke. Then pour out the same water, and put to newe, and soe contynue washinge and powringe out the water vntill it com cleare from the malgam w*ith*out any more blacknes. Then dry the said malgam, and grind yo*ur* ☿ sublimed and your	Nowe to dissolue this mercury sublimid into oyle, first to euery pound of the said mercury, yee must haue a {pound} of tynne & a^{li} of rawe quicksiluer, and of theis two you shall make into a malgama as the goldsmithes doe. First ye shall melt yo*ur* tynne, and then put in yo*ur* quicksiluer to the tynne, the mercury beeinge hoote. Then beate it when it is colde in a yron morter small that noe knobbes may be felt betwixt yo*ur* fingers, then it is well beaten. Then wash it cleane, but or it be cleane it woll make many a foule water, but wash it till the water come clere from it. Then drye it at the sonne, or els a very softe fyer. Now when it is drye take a^{lb} of that tynne and mercury. Then mingle it with a^{li} of the mercury sublymed and the said malgama togedre grindeing it toguether vpon a plate

---

[74] The origin of the compilation and the relationship between MS. O. 4. 39 and MS. Ashmole 1507 are unclear. Chapter 8 of the text ends with the following statement: "Here endeth th'ole practise plaine set forth w*i*thout intricate word*es* of all th'ole science. Lawde be to God, vnto the moost noble king Henry the viiith by me Thomas Petre" (Ashmole 1507, fol. 125r). I have not been able to identify this person, who may be the original compiler or translator. MS. O. 4. 39 is a sixteenth-century copy, whereas MS. Ashmole 1507 was copied sometime in the seventeenth century by Elias Ashmole. The copies are almost identical textually, sharing a number of idiosyncratic spellings and obvious errors. In MS. Ashmole 1507 (fol. 121r), a note has been added in the upper left-hand corner stating: "MS.: penes do*ct*orem Flood N: 2: fol." I have not been able to ascertain whether the exemplar referred to is possibly MS. O. 4. 39, but my cursory collation of the two manuscripts suggests that MS. O. 4. 39 could well be the exemplar of MS. Ashmole 1507. Although referring to several pseudo-Lullian texts, the compilation does not seem to be drawing on these texts but simply referring to parallel procedures.

malgam together. And her note that to your malgam of ii^{li} of quicksiluer and ii^{li} of tynn, youe muste put ii^{li} of sublimatu*m*, and grind them well together vpon a marble stone, and when they ar well ground, lay them vpon a table of glasse that muste be made for the purpose, hauinge a phillet about it to kepe in the water that it haue noe way out but at the spoute, w*hich* muste be made for the yssue of the said water as it dissolueth, w*hich* table muste be set in a cold place or in som seller, for otherwise it will not dissolue to any good purpose. And when it is wholly dissolued and distilled from the plate of glasse into som great still body or other vessell of earth, well leaded, then let it be filtered dyuers tymes, for soe shall ye filth remain at every filtringe behind in the bottom of the vessell. Then take the malgam and the quicksiluer that remayneth after the filteringe, and put them together in a pan of clean water, and washe them soe longe as any filth will com from it and that the matter be clean and clear. Then sep*a*rate your quicksiluer from the tynn by strayninge it through a stronge lynnen cloth, w*hich* quicksiluer will serue youe again for to make another malgam. Afterward*es* take youre tyn that remayneth after the straninge, and lay it vpon the plate of glasse somwhat thine. And yf ther be any quiksiluer remani*n*ge amonge the tyne, it will com ron from it. Then enkitch yo*u*r malgam again, and soe dissolue from tyme to tyme at yo*u*r pleasure vntill youe haue soe moch as will suffice for youre worke &c. Finis.

of yron tynned ouer, which you may haue at the yronmonger. But let the plate be turned vp by the sides an ynche of height. Then put theis into a colde seller to dissolue into a faire browne oyle, which wolbe within xx dayes and sooner yf yo*u*r mercu*ry* be well sublimed with vitriol and saltpetre as is aforesaid. Then haue you a preciouse oyle which few men can make.

Chapters 11, 12, and 13 of the Treatise are related only on a general level to the fourth, fifth, and sixth chapters of the compilation in MS. O. 4. 39 and MS. Ashmole 1507: they describe similar procedures that involve similar processes, but the language and the specifics of the two texts are miles apart. Chapter 14, on the other hand, exhibits a closer correspondence to chapter eight in the compilation, but there are still many dissimilarities. Instead, for Chapter 14, there is a much closer parallel. In fact, MS. Sloane 3580B (fols. 53v–54r) contains a text that is nearly identical to Chapter 14, even in formulation. This demonstrates that the Treatise is clearly based on and very faithful to a source. The text in MS. Sloane 3580B appears to be part of a longer text with the general title "for Sol & Luna" (fol. 52r), but no other section of this text is related to the Treatise. I have not been able to identify this text elsewhere, and it is unclear whether it is an independent tract or whether it has been excerpted from a longer composition.[75] Although not clearly identifiable, this source text is significant in that it demonstrates that the Treatise and the compilation in MS. O. 4. 39 and MS. Ashmole 1507 are only parallel texts that may share one or more sources. Like Chapter 14, Chapters 10–13 most probably also have more immediate sources, reflected indirectly in the MS. O. 4. 39 and MS. Ashmole 1507 compilation.

Like Chapters 10–14, Chapters 21–25 seem to constitute a unit and may be based on a single source. However, I have not been able to find any clear parallels to these chapters. Although the chapters have been sprinkled with references to Morien the philosopher, I have not in most cases been able to find a direct source in writings that have been ascribed to Morien, such as *Liber de compositione alchemiae quem edidit Morienus Romanus, Calid Regi Aegyptiorum*.[76] It is more likely that the references are secondary, that is, they were taken over from the direct source.

Chapter 21 contains a fascinating, but also slightly puzzling reference. The Treatise here refers the reader to George Ripley's famous *Twelve Gates*, more commonly known as the *Compound of Alchymy*, which is said to have been instrumental for the narrator's (Lock's?) "p*er*fecte vnderstandinge of this science":[77]

> Therfore, such as shall year after be p*ar*takers of this my simple labour, I farther co*m*mit them to y*e* p*er*vsinge of such bockes as ar writen of this science, and specially the worke of George Reply in English miter called his XII Gates. For I haue found more fruite in that on bocke for the p*er*fecte vnderstandinge of this science then in all other bock*es* that haue com to my hand*es*. [317r: 14–20]

The *Compound of Alchymy* was one of the most influential vernacular texts in the sixteenth and seventeenth centuries, and, in this respect, a reference to it is hardly surprising. But there are other aspects of this passage that makes the statement particularly interesting. The narrator obviously attributes great importance to the *Compound*. At the same time, there is no evidence in Chapters 21–25 of any direct influence from this text. Even if we consider the possibility that Lock added this passage, and hence that other parts of the Treatise may have been influenced

---

[75] It would not be surprising if it were an excerpt: the compiler of MS. Sloane 3580B (and its sister volume 3580A), Thomas Potter, includes extracts from numerous tracts; see Keiser, "Preserving the Heritage."

[76] Manget, *Bibliotheca*, 1: 509–19; for a possible exception, see EN 316v: 13–16.

[77] For Ripley, see Linden, *George Ripley's* Compound, and previous discussions on the *Medulla* and the *Concordantia* on pp. 45 and 46. See also p. 20.

by the *Compound*, the evidence is still negative: I have found no clear parallels between the two texts. On the other hand, the passage never alleges that the *Compound* has been used as a direct source, but only attests to its general usefulness, and directs the reader to peruse it. This admonition may even have led to the addition of the *Compound* to the collection of poems that follows the Treatise in one copy of the Treatise: S2. Since the *Compound* occurs only in this one manuscript, it seems easier and perhaps more logical to ascribe the addition to the S2 copyist or an ancestor of this copy. But it cannot be ruled out completely that Lock himself appended a version of Ripley's *Compound* to his collection of alchemical texts in response to the comment in Chapter 21 (whether added by himself or by his source), since the comment so emphatically underscores the importance of the *Compound*.[78]

Despite several references to alchemical authorities, a clear source for Chapters 21–25 can thus not be established. Nevertheless, the unity of content and formulation, with ample cross-referencing between the chapters, does suggest that they derive from a single source. The very end of Chapter 20 provides some support for this assumption:

> And forasmoch as the 4 chapters hearafter folowynge be not wroughte after the manner of the minerall stone, nor alltogether after the manner of the vegitall and animall stone, I haue therfore set them as a mean between both to a farther relation, as shalbe showed hearafter, not terminge them the on nor the other, because of their diuersitie in workinge &c. [315v: 25–30]

This comment helps to account for why the descriptions based on the *Medulla* are split up into Chapters 15–20 (which deal with the mineral stone), and Chapters 26–27 (which deal with the vegetable and animal stones). The person adding the comment (presumably Lock) saw the content of Chapters 21–25 as intermediary. But it is slightly peculiar that the note refers to four chapters instead of the five that are now found between the *Medulla*-based chapters in the Treatise. A possible explanation for this is that Chapter 25 actually deals with the animal stone, which is again discussed in Chapter 27, and is thus not an intermediary procedure. At the same time, Chapter 25 is clearly related to Chapters 21–24 in that it uses the substances produced in these chapters as the basis for its procedures. Another notable feature of Chapter 25 is that it contains seemingly incongruous material, which may suggest that several sources were used for the compilation

---

[78] See also p. 16. S2 introduces the *Compound* with a note that echoes both the Treatise passage and the note which prefaces all the additional texts (see p. 17): "I haue heare also ioyned vnto this my booke the work of G. Ripply called his 12 Gates in English meeter, whose excellencie for the profound document that is taught therin is worthy to be regestred with *lett*res of goulde; and for that the said booke is so necessary for euery man that shall woorke in this science, I haue added vnto this booke all & so many chapters as I had of the same &c" (fol. 43v).

of this chapter. It begins with a statement that the chapter will include what has been described in the four previous chapters; it will be divided into two parts, but it will be "by one order of workinge" (318v: 31–50). But before the text launches into the technical procedures, it elaborates on the virtues of the animal stone. It presents the stone as the gift of God, and the chapter catalogues a number of ancients who have allegedly possessed the secret, including Noah, Moses, and Solomon (319r: 12–27). After the catalogue follows a description of the "less world" or microcosm (319r: 30–37), a discussion that is paralleled in Chapter 27, which also deals with the animal stone (though this chapter is based on the *Medulla*). In 319v: 7, after a quote attributed to Morien, the Treatise finally presents the procedure that the introduction promised. Continuing to the end of the chapter, the discussion revolves around the substances produced in the previous chapters. Perhaps the different parts that are now found in Chapter 25 were already joined in a source, or the current structure may be the result of the compilation of the Treatise. The incongruity of the material may even stem from sections of the source text being left out in the Treatise.

The Adaptation of the Sources

The descriptions of the individual sources have clearly demonstrated that, although some of the sources have been integrated into the Treatise with minimal changes in content and formulation, numerous passages exist where the Treatise and the source text exhibit substantial differences. Naturally, as I have emphasized, several reasons may lie behind these differences: the passages may only appear to be dissimilar since the exact source or version of the source has not been identified. More complicated patterns such as the conflation of two versions of a source may also have occurred, although there is no direct evidence for such a process. And differences may also be the result of a misconception of the source. However, Lock (or a subsequent copyist) was undoubtedly also responsible for some of the differences: although seemingly limited in scope, adaptations do occur, reflecting attempts to reconcile incongruous sources. Chapters 1 and 2, in particular, feature a number of such adaptations of the source texts, which seem to aim at making the Treatise more coherent. The following passages are illustrative examples of this strategy of reworking. In Chapter 2 based on *Scoller and Master*, the father of the dialogue stresses that salts and alums are unnecessary in most procedures and even detrimental to some expermients:

> And as towching all thes salltes and allames and all maner of
> strange thinges that be not of the kinde of mettelles, thay be

not helping to this crafte, but in the beginninge of preperatione, so*m*me be nedfull, as vitryall and commane sault, or sault armoniack; and these be all that we ned in this worke. For if thaye be put in any manner of degre or otherwise then in sublimac*i*on of quicsilver, all is lost without remedye. [297r: 7–15]

However, in Chapter 16 (based on the *Medulla*), the Treatise asserts that the production of the mineral stone requires vitriol and vermilion (313r: 12ll.). To accommodate this idea, the Treatise adds the following passage to Chapter 2, right after the section cited above:

But here thow must vnderstand that I speake now of the a{n}i_{m}all and vigitable stone and not of the mynerall. For in the minerall stone, the ardent water, being ioyned with the spirit of vit{riall} and vermelon, is ingrossed and made thicke, and more cleuinge vnto the bodye, with a swifter fixac*i*on then before, by the ress{o}n whereof his fixac*i*on is complete in a shorter time then in the other stones. [297r: 15–23]

The intention here and in other instances seems to be to reconcile diverse theoretical and practical frameworks.[79]

The nature and order of some of the chapters also indicate that the Treatise was not the result of a completely random and disorganized incorporation of texts. For example, as pointed out earlier, the Treatise joins *Scoller and Master* and *Perfectum magisterium* simply by letting the reader address in the latter text become part of the father's response to his son in the former. The transition from one text to the other is thus hardly noticeable. Furthermore, the Treatise postpones the discussion on sublimation, which originally belongs to the dialogue in *Scoller and Master*, to after Chapter 3. This decision is logical since, though dealing with some theoretical issues, the discussion is more practical in nature. The intermediate chapter, Chapter 3, by contrast, deals exclusively with matters of alchemical theory. Yet another example of deliberate restructuring is Chapter 26 (as we have seen). By bringing sections belonging to the *Medulla*'s discussion of the mineral stone into its treatment of the vegetable stone, the Treatise helps the reader follow the text. Instead of being referred back to an earlier part, the reader is thus conveniently presented with the necessary information at the time when it is required.

Although the chapters mentioned above provide illustrative examples of a conscious compilation strategy, the most striking evidence is found in Chapter 3. I have shown earlier that Chapter 3, which is the longest chapter in the Treatise,

---

[79] For further examples, see e.g. EN 294v: 30–35, 295r: 31–34, 295v: 3–9, 298r: 17–21, 300v: 21–24.

is based on more sources than any other chapter. The structure of Chapter 3 is illustrated in the table below.

TABLE 3.9. STRUCTURAL COMPONENTS IN CHAPTER 3

Line	Content
301v: 3–40	General introduction
301v: 41–303v: 47	Alchemical theory (*Dicta*)
303v: 47–304r: 32	Outline of the Treatise (304r: 18–30: *Medulla*)
304r: 32–304v: 4	Introduction to the discussion about the elixir and the ferment
304v: 5–305r: 14	The subject of the elixir (304v: 9–23: *Notabilia Guydonis*)
305r: 15–306r: 17	The discussion about the ferment (305r: 19–306r: 17: *Concordantia*)
306r: 18–39	Conclusion

The four major sources of Chapter 3 are the *Dicta*, the *Medulla*, the *Notabilia Guydonis*, and the *Concordantia*. In addition to passages based on these sources, Chapter 3 also contains sections that cannot be, and probably should not be, attributed to a source. These sections include an introduction and a conclusion to the chapter as well as transitional passages between the major sources. It is reasonable to assume that the transitions have been added by Lock (or less probably by a subsequent copyist), since some of them refer to other parts of the Treatise as it now stands. Similarly, the introduction and conclusion are perhaps best ascribed to Lock since they frame the alchemical discussion provided by the sources; also, the end of Chapter 3 refers to the subsequent chapters. Introducing the topic of discussion, the metatext thus logically links the different sources. Such structural "glue" shows that at least some attention was paid to making the text coherent and smooth.

Although Lock (sometimes perhaps with the help of later copyists) reworked some of his sources, was he in essence merely a plagiarist, putting his name on a composition mostly culled from the work of others? From a modern perspective, the answer would undoubtedly be yes, but from the perspective of early modern writers on alchemy, almost certainly no.[80] Lock followed a centuries-long tradition of recycling and re-adaptation of earlier texts, a tradition that also allowed leaving out attributions and indebtedness to the sources. Chapter 1 and its source provide a particularly striking example of this tradition and the sometimes multilayered reworking of alchemical texts. As demonstrated earlier, Chapter 1 substantially revises the earlier pseudo-Albertan tract, the *Mirror of Lights*, without any recognition of the source. The *Mirror of Lights* is in turn an even more

---

[80] The *OED* first records the word *plagiarism* in 1621, although the concept of literary theft naturally existed long before that; *OED* s.v. *plagiarism*.

comprehensive redaction of the *Semita recta*, another pseudo-Albertan text. The redactor seems to have felt that the *Semita recta* was too loosely structured, and wanted a clearer division between theory and practice. Although adding very little, if anything at all, of his own, the redactor does not acknowledge the source. But the trail of reworking does not stop here: William Newman has convincingly argued that the *Semita recta* borrows liberally from the *Summa perfectionis*, usually ascribed to Geber.[81] Despite long verbatim citations, the *Semita recta* is silent about its source. A fifth step is represented by one of the extracts of the Treatise, E1. Although this extract does not appear to fashion itself as an independent text, it does rework the Treatise (for example, by renaming some of the conditions). As its predecessors, E1 foregoes a reference to its source, that is, the extract is not explicitly labeled as an excerpt from the Treatise. Lock's omission of attributions and references is thus hardly unique or even peculiar; it is more likely conventional.

The Alchemy of the Treatise

The earlier discussion of sources reveals that the Treatise is steeped in several traditions of alchemical theory and practice current in the late Middle Ages and early modern period. It is clear from the content that Lock was not a particularly innovative alchemical thinker. As we have seen, he does not appear to have added much to his sources from his own experience (as far as we can tell). He even seems to have made only modest attempts at reconciling the often contradictory information in his sources. He also hints that he made the Treatise intentionally obscure in the hope that Cecil and Elizabeth would bring him back to England to perform the procedures himself.[82] His purpose is greatly helped by the opaque metaphorical language that is sometimes employed in the Treatise.[83] All of these circumstances make it difficult to give a coherent picture of the alchemical ideas in the Treatise. However, I will attempt to outline the main points of the Treatise's alchemy, and how it relates to other alchemical frameworks of the time.[84]

---

[81] William R. Newman, "The Genesis of *Summa Perfectionis*," *Archives internationales d'histoire des sciences* 35 (1985): 246–59.

[82] For this statement, see "the Russia note" in 316v: 27.

[83] See further the discussion on alchemical language in the Treatise.

[84] In the following description, I rely heavily on surveys and general histories of alchemy: John Read, *Prelude to Chemistry: An Outline of Alchemy, its Literature and Relationships* (London: G. Bell, 1939); Frank S. Taylor, *The Alchemists: Founders of Modern Chemistry* (New York: Henry Schuman, 1949); Eric J. Holmyard, *Alchemy* (Harmondsworth: Penguin, 1957; repr. New York: Dover, 1990); Robert P. Multhauf, *The Origins of Chemistry* (London: Oldbourne, 1966); Allen G. Debus, *The Chemical Philosophy: Paracelsian Science and Medicine in the Sixteenth and Seventeenth Centuries*, 2 vols. (New York:

The Treatise is primarily concerned with the production of the philosophers' stone or elixir, which is allegedly able to turn base metals into silver or gold. The idea that metals could be perfected has its origin in Aristotelian cosmology, which dictated that all substances consisted of the four elements, fire, water, air, and earth, in different proportions. To alchemical practitioners, it thus followed as a logical conclusion that if the proportions of the different elements in a substance could be rearranged, one substance could be transformed into another.[85] The Treatise frequently alludes to this fundamental concept, suggesting that this transformation can be accomplished by a change in the defining qualities of the different elements, or by the "turning of the elements."[86] Again, underpinning this suggestion is Aristotelian cosmology: each of the elements was believed to consist of a pair of the defining characteristics: hot, cold, moist, and dry. For example, air was considered to be made up of the qualities hot and moist, and water of the qualities moist and cold. If the balance of these qualities could be altered, it would entail a change in the element, which in turn would bring about a change of substance.[87]

However, although the elements play a prominent part in the Treatise, there are two concepts or principles that have an even more conspicuous place in the Treatise's theoretical framework: mercury and sulfur. These are not to be confused with the substances known as sulfur and mercury today; rather, they were constituent parts or principles making up all metals. This intermediate stage between elements and metals was first proposed by the medieval alchemist Geber (often incorrectly equated with the Arabic scholar Abu Musa Jabir ibn Hayyan [*fl.* 760]).[88] Building on the Aristotelian conception of nature, he suggested that if the balance of the two constituents mercury and sulfur could be changed in an appropriate way, the alchemist could effect transmutation. Geber's mercury and sulfur theory gained widespread currency in the Middle Ages, and was commonplace long before the Treatise was compiled and most of its sources written. Although the Treatise subscribes to the mercury-sulfur theory, its adherence to it

---

Science History Publications, 1977); Bernhard D. Haage, *Alchemie im Mittelalter: Ideen und Bilder, von Zosimos bis Paracelsus* (Düsseldorf: Artemis & Winkler, 1996); Roberts, *The Mirror*. Many studies, including some of the works listed here, claim that many alchemical texts describe metaphysical, spiritual, and religious experiences, in addition to or instead of discussing practical laboratory procedures and their theoretical foundation. While I acknowledge that this might be true for some texts, interpreting the Treatise as having a deeper spiritual or experiential meaning would be misleading. I will return to this issue in my discussion of the language of the Treatise.

[85] Read, *Prelude*, 9–10; Roberts, *The Mirror*, 45, 47–48.
[86] E.g. 299v: 18–21, 302r: 44–46, 303r: 31–35, 305v: 22–25.
[87] Read, *Prelude*, 9–10; Roberts, *The Mirror*, 47–50.
[88] Read, *Prelude*, 25; Taylor, *The Alchemists*, 80–81; Roberts, *The Mirror*, 50–51. For Geber, see William R. Newman, ed., *The Summa Perfectionis of Pseudo-Geber: A Critical Edition, Translation and Study* (Leiden: Brill, 1991).

is not completely straightforward or whole-hearted. In fact, the first two chapters maintain that mercury, containing its own internal sulfur, is the sole constituent of metals.[89] This variant theory was particularly popular in the fourteenth and fifteenth centuries, and is especially common in texts that are ascribed to the thirteenth-century physician Arnold of Villanova.[90] It comes as no surprise then that the source for a large part of Chapter 2, where the mercury-alone theory is advocated most strongly, is indeed a text that commonly appears under Arnold's name: *Perfectum magisterium*. Yet, in other parts (which are based on other sources), the Treatise is firmly grounded in the more traditional theory, stressing the importance of sulfur as a complement to mercury or as a product of the manipulation of mercury.[91]

Since all metals consisted of the same *ur*-materials (mercury and sulfur, or mercury alone), base metals were considered perfectible, that is, lead and tin could theoretically be turned into silver or gold. The problem that alchemists struggled with was how to translate this into practice. Throughout the history of alchemy, the idea of a powerful transmutative agent, the elixir or the philosophers' stone, was the most popular. If projected onto imperfect metals, this substance, which was often conceived of as a powder, would transform the metals into a higher perfection, i.e. silver or gold.[92] The theoretical basis for this assumption was that the elixir would restore the proper balance between mercury and sulfur by changing their proportions.[93] Some alchemists even saw the elixir as a medicine that cured diseased or corrupt metals into perfection.[94] The Treatise distinguishes between three different stones or elixirs: the mineral, vegetable, and animal stones. Although this distinction is commonplace in alchemical texts, the Treatise has inherited its particular version of the three stones from two sources that it draws upon heavily, the *Concordantia Raymundi et Guidonis* and *Medulla alkemiae*, both ascribed to the English alchemist George Ripley.[95] The least potent stone, which is primarily discussed in Chapters 15–20, is the mineral stone, which transmutes base metals into silver and gold. The vegetable stone (Chapter 26) and the animal stone (Chapter 25 and 27), on the other hand, are more powerful transmutative agents, and they can also cure diseases. Some parts of the Treatise make an additional distinction between a white elixir and a red elixir of each of the three superordinate elixirs: the white variety transmutes

---

[89] E.g. 295v: 31–296r: 1, 296r: 11–13, 296v: 11, 296v: 33–297r: 2, 298v: 1–2, 300r: 1–7, 300r: 21–23, 302v: 42ll.

[90] Thorndike, *A History*, 3: 58; Roberts, *The Mirror*, 62.

[91] E.g. 298v: 15–20, 300v: 22–23, 300v: 29–30, 302v: 12–20, 303v: 12–17, 303v: 40–41, 304r: 24–25, 304r: 45.

[92] E.g. 309r: 18–20, 312r: 22–23, 315v: 10–13, 317r: 8–12, 322v: 21–22.

[93] Taylor, *The Alchemists*, 82; Roberts, *The Mirror*, 54.

[94] See *medcin* in the Glossary.

[95] See pp. 45 and 46. See also Roberts, *The Mirror*, 54.

base metals into silver and the red turns them into gold.[96] In addition to the three elixirs, the Treatise also recognizes an elixir which is simply named "the less work of alchemy." Chapters 28–33 clarify that it is different from the other three, although it shares with them the power to turn substances into silver and gold.[97]

The production of the elixir involved long and complicated procedures that were known by such names as sublimation, calcination, coagulation, and projection. But before these procedures could be carried out, the alchemist had to decide upon the base material to be used for the stone. The Treatise stresses throughout that since mercury is the prime constituent of metals, it should also be the prime matter of the stone.[98] This mercury should be extracted from metals, and then be manipulated in different ways to produce the stone. The second chapter in particular is concerned with justifying the use of mercury. Organic substances such as blood, hair, and urine are unusable; metals, and hence the stone, must be made of a substance of metallic origin.[99] A later passage (304v: 27–31), which is based on a different source, offers a refinement of this idea. Here it is suggested that there is a difference between the primary matter of metals, said to be quicksilver, and the first constituent of the stone, named mercury. Still other passages advance sulfur as an integral part of the production of the elixir.[100]

Alchemical texts vary greatly in the order and number of practical procedures that they advocate for producing the elixir.[101] Despite disagreement about the actual, specific procedures, some consensus can be distilled from alchemical texts concerning a general sequence of alchemical practice:[102]

1. purification of a base material, which involved the reduction of the base material into the prime matter (i.e. the basic constituent[s] of the substance);

2. preparation of the prime matter, involving processes such as calcination, sublimation, fixation, etc.;

3. the production of the white stone;

---

[96] E.g. 317r: 8–12, 319v: 43–51.

[97] 323v: 26–32. In alchemical literature, the *less* or *little work* usually refers to the production of a white elixir that has the power to transform metals into silver only: Read, *Prelude*, 130.

[98] E.g. 295v: 11–12, 296r: 1–2, 300r: 5–7, 301v: 45, 302v: 42ll., 303v: 34–35, 304v: 27–40.

[99] E.g. 299v: 43–300r: 7; The rejection of organic substances appears in many alchemical texts, and may even have been a commonplace: Roberts, *The Mirror*, 62–63.

[100] E.g. 296r: 11–13, 300v: 21–23.

[101] See e.g. Read, *Prelude*, 135; Roberts, *The Mirror*, 55–60.

[102] The list has been adapted from Read, *Prelude*, 135, and Roberts, *The Mirror*, 57.

4. the production of the red stone;

5. augmentation or multiplication of the potency of the elixir;

6. transmutation.

The Treatise follows this general sequence of processes fairly closely in Chapter 21. The initial passage describes how silver, which is the chosen base material, is reduced to the prime matter (316r: 6–34). Subsequently, the practitioner is instructed on how to turn the resulting prime matter into a white "sulfur," which is the base for the elixir (316r: 34–316v: 25). This involves subjecting the prime matter to a number of procedures including distillation, rectification, putrefaction, and conjunction. Once the white sulfur has been obtained, new procedures follow (316v: 26–40) before the sulfur is fermented (316v: 41–317r: 7) and hence turned into a white elixir with the power to turn mercury or imperfect metals into silver (317r: 8–10). This elixir can be further refined into a red elixir "with longer contrytion of the fier, the goulden oyll, and ferment of red mercury" (317r: 10–11). The Treatise is in this case silent on how to multiply the elixir and on how to perform the transmutation.

Far from all of the processes in the Treatise are as straightforward and transparent as those in Chapter 21, and few experiments appear to be guided by a general sequence of processes. The Treatise stresses the reduction of the base material into the prime matter in the introduction and in the theoretical parts, but it is rarely mentioned explicitly in the later, practical chapters.[103] The preparation of the prime matter, on the other hand, features prominently throughout the Treatise, but the procedures and their sequence vary greatly. Chapter 1 firmly establishes ten processes, underscoring that their order must be kept: preparation, sublimation, calcination, conjunction, putrefaction, fixation, imbibition, fermentation, liquefaction, and projection.[104] But some of these procedures and their order are ignored in later chapters, most likely because these chapters derive from different sources than Chapter 1.

The third and fourth steps, the creation of the white and the red elixirs, are important in some chapters of the Treatise, whereas others make no explicit distinction between the two; there is simply one elixir (e.g. Chapters 22 and 23). In Chapter 25, they are even described as separate processes and not as sequential. Similarly, the augmentation or multiplication appears to be more of an optional nature, since it is mentioned only intermittently in the Treatise.[105] Although the

---

[103] E.g. 296v: 30–297r: 2, 298v: 29–33, 299r: 43–299v: 2, 300v: 22–23, 301r: 28–31, 304v: 40–43, 305v: 33–35.

[104] 294v: 35–295r: 5. See the Glossary for the explanation of these terms.

[105] E.g. 313v: 12–16, 314r: 10–16, 315v: 5–10, 23–24, 322v: 13–18.

Treatise frequently records what effect to look for after an elixir has been applied to an imperfect metal or mercury, it is mostly silent on how to apply the elixir or indeed how transmutation occurs.[106]

Some of the elixirs are also said to possess medicinal properties in addition to being transmutative agents. The Treatise points out the usefulness of both the vegetable and animal stones for therapeutic purposes (while it emphasizes that the mineral stone pertains only to metallurgical alchemy and the transmutation of metals).[107] In off-hand remarks, it also mentions the elixir of life and *aurum potabile*, i.e. 'drinkable gold,' which will allegedly cure all diseases and prolong life.[108] Comments about the medicinal properties of elixirs are quite common in alchemical texts from this period, and the idea of the elixir as a panacea appeared in alchemy in the Western world as early as the beginning of the thirteenth century.[109] Treatises attributed to Roger Bacon (*c.* 1220–1292) and Rupescissa's (*c.* 1310–1365) *De consideratione quintae essentiae*, which advanced alchemically prepared medicines, were highly popular, circulating in numerous manuscripts.[110] The fifteenth and sixteenth centuries witnessed a surging interest in medico-alchemy in England, an interest that was spurred on in the late sixteenth and early seventeenth centuries by the manuscript circulation and printing of texts written by Paracelsus or his followers.[111] Although the Treatise attests to the medical potential of alchemical procedure, it does not advance a medical agenda for alchemy or fashion itself as a medical text. It instead seems to consider the therapeutic properties of the elixirs secondary to their transmutative powers. It never places the elixirs in a medical framework; nor does it describe how they may be applied.

---

[106] E.g. 309r: 18–20, 312r: 22–23, 315v: 10–13, 317r: 8–12, 322v: 21–23.

[107] E.g. 312v: 17–25, 314r: 3–5, 315r: 8–10.

[108] E.g. 295v: 3–7, 309v: 19–21, 312r: 24–27, 322v: 21–27, 324r: 28–29. See also dedication, ll. 146–61.

[109] Chiara Crisciani, "Alchemy and Medicine in the Middle Ages," *Bulletin de philosophie médiévale* 38 (1996): 9–21; Michela Pereira, "*Mater Medicinarum*: English Physicians and the Alchemical Elixir in the Fifteenth Century," in *Medicine from the Black Death to the French Disease*, ed. Roger French et al. (Aldershot: Ashgate, 1998), 26–52, here 28.

[110] Marguerite A. Halversen, ed., "*The Consideration of Quintessence*: An Edition of a Middle English Translation of John of Rupescissa's *Liber de Consideratione de Quintae Essentiae* [sic] *Omnium Rerum* with Introduction, Notes, and Commentary" (Ph. D. diss., Michigan State University, 1998), 34–37; Webster, "Alchemical and Paracelsian Medicine," 302.

[111] Webster, "Alchemical and Paracelsian Medicine," 302–3, 312–14, 317; Pereira, "*Mater*," 26–27; Linda E. Voigts, "The Master of the King's Stillatories," in *The Lancastrian Court: Proceedings of the 2001 Harlaxton Symposium*, ed. Jenny Stratford, Harlaxton Medieval Studies 13 (Donington: Shaun Tyas, 2003), 233–52; Kassell, *Medicine*, 105.

Underscoring the elixirs' metallurgical, transmutative powers may have been a conscious choice. Lock presumably knew that Elizabeth's and Cecil's main interest in alchemy lay in the production of valuable metals, and less in medicine. His Treatise would thus undoubtedly have been more appealing to them if this aspect of alchemy were stressed.[112]

---

[112] Note, however, that Lock devotes four stanzas of his dedication (ll. 146–61) to extolling the *aurum potabile*.

Language and Alchemy: The Case of the Treatise

Before expounding upon the theoretical framework of the text in Chapter 3, the Treatise declares that

> when by great laboure and moste painfull travaill the aunciente phi*loso*fers had attayned to the pe*r*fecte mastery therof [i.e., of alchemy] and then not willinge that their labours should be buried in the bottomlesse lake of oblyuion, haue co*m*piled great volums of the same worke, hydinge it vnder shadowe of misticall word*es* and names infinite, and haue sealed it in their writinge from vnworthy people vnder the mistie mantle of phi*loso*fie [301v: 11–18]

Comments about the obscurity and intentional ambiguity of alchemical language such as those found in the passage above abound in alchemical writings. These comments are often accompanied by claims that the text at hand is more transparent than any other text on alchemy. The Treatise is no exception:

> althoughe I haue not in this my simple labour set forth all thing*es* soe plainly as I haue knowen them, yet I haue more largly opened the hid secret*es* of this science then I haue seen them by any other set forth [306r: 26–30]

The alchemists' use of language was also one of the prime targets for early critics of alchemy. The thirteenth-century scholar Albertus Magnus considered the obscure, metaphorical language found in alchemical texts unsuitable for the exposition of science.[1] Writing four hundred years later, Robert Boyle (1627–1691) in his *Sceptical Chymist* (1661) heavily criticized the alchemists' use of unclear and confusing language, which he thought created unnecessary ambiguity.[2]

---

[1] Kibre, "Albertus Magnus," 190.

[2] Lawrence M. Principe, *The Aspiring Adept: Robert Boyle and his Alchemical Quest* (Princeton: Princeton University Press, 1998), 48–49; Maurice P. Crosland, *Historical Studies in the Language of Chemistry* (Cambridge, MA: Harvard University Press, 1962), 59–62; Maurizio Gotti, *Robert Boyle and the Language of Science* (Milan: Guerini scientifica, 1996), 33, 39, 90–92. Cf. also the opinions of Andreas Libavius, the sixteenth-century German writer of the textbook *Alchemia* (1597), as given in Owen Hannaway,

Serious attempts have been made in modern scholarship to account for and explicate the alchemists' use of language, focusing on the codes, sigils,[3] similes, metaphors, and other linguistic or semi-linguistic strategies frequently employed in alchemical texts.[4] The obscurity and ambiguity of the language has led to a longstanding debate about what the symbols and metaphors actually signify, a debate which has recently been invigorated. Although many researchers take the middle road between the two extremes, some see exclusively straightforward laboratory procedures encoded in the language. Others declare that the alchemists' symbolic descriptions carry other significances—psychological, experiential, religious, or mystic.[5] While the latter approach has recently been heavily criticized by William R. Newman and Lawrence M. Principe,[6] their own emphasis on chemical processes has come under fire as well.[7]

Although I acknowledge that some alchemical texts may perhaps be interpreted in light of esotericism, religion, or even the psychology of the author, I will account for the language of the Treatise exclusively in terms of basic alchemical practice. The Treatise uses only a limited set of commonplace alchemical symbols, and there is no indication that these would encode Lock's mystical or religious experiences. In fact, Lock himself reveals his practical engagement with alchemy and the practical goal of his Treatise: in the "Russia note," he makes clear that he made the text obscure with the intention that Cecil and Elizabeth should bring him back to England so that he could carry out the procedures personally.[8]

The Chemists and the Word: The Didactic Origins of Chemistry (Baltimore: Johns Hopkins University Press, 1975), 78, 109–10, 119–20.

[3] For the use of the term *sigil* to refer to representations such as ☉ for *gold*, see the discussion below p. 83.

[4] See e.g. Robert Halleux, *Les textes alchimiques*, Typologie des sources du Moyen Âge occidental 32 (Turnhout: Brepols, 1979); Raimond Reiter, "Die 'Dunkelheit' der Sprache der Alchemisten," *Muttersprache* 97 (1987): 323–26; Crosland, *Historical Studies*; Roberts, *The Mirror*; Abraham, *A Dictionary*; William R. Newman, "*Decknamen* or Pseudochemical Language? Eirenaeus Philalethes and Carl Jung," *Revue d'histoire des sciences* 49 (1996): 159–88.

[5] For an overview, see Taylor, *The Alchemists*, 159–60; Crosland, *Historical Studies*, 4–5; Roberts, *The Mirror*, 66. For a recent attempt to interpret late fifteenth-century alchemical poems as political propaganda, see Jonathan Hughes, *Arthurian Myths and Alchemy: The Kingship of Edward IV* (Phoenix Mill, UK: Sutton, 2002).

[6] Lawrence M. Principe and William R. Newman, "Some Problems with the Historiography of Alchemy," in *Secrets of Nature: Astrology and Alchemy in Early Modern Europe*, ed. Newman and Grafton, 385–431. See also Newman, "*Decknamen*"; Lawrence M. Principe, "Decknamen," in *Alchemie: Lexikon einer hermetischen Wissenschaft*, ed. Claus Priesner and Karin Figala (Munich: C. H. Beck, 1998), 104–6.

[7] See Hereward Tilton, *The Quest for the Phoenix: Spiritual Alchemy and Rosicrucianism in the Work of Count Michael Maier (1569–1622)* (Berlin: Walter de Gruyter, 2003).

[8] 316v: 27.

In what follows, I will first situate vernacular alchemical writings, and the Treatise in particular, in the context of other scientific texts in the transitional period between the Middle Ages and early modern period. My aim in the subsequent sections is to outline some of the most notable characteristics of the language of the Treatise, using it as an illustrative case study of the linguistic features of early vernacular writings. Since a full description of genre conventions and characteristics of the language of these texts will have to await the availability of a larger body of edited writings, the discussion will to some degree be programmatic, pointing to important areas for future exploration. I will suggest, for example, that, in concentrating on the use of symbols, metaphors, and imagery, many previous studies have failed to consider the full scope of alchemical language. The procedural passages that are formulated as recipes, in particular, receive short shrift in most descriptions of alchemical language, if they are mentioned at all. This is regrettable, since the recipe structure connects alchemical texts with other types of contemporaneous scientific literature and with instructive texts more generally.[9]

English Alchemical Texts in Context

The period of about 1100 to 1375 witnessed scientific writing in England as the near exclusive domain of Latin or, in some instances, French.[10] Not until the late fourteenth century and increasingly in the fifteenth century did scientific texts appear in English. Several factors may have fueled this vernacularization process.

---

[9] A caveat is necessary here. There is a great range of alchemical texts, written in prose as well as verse, whose genre conventions have yet to be determined: some alchemical texts are exclusively theoretical, and hence do not contain practical sections in the form of embedded recipes. Other texts consist primarily or solely of alchemical recipes. Some use nothing but symbolic language; other use none (or at least they do not appear to). With this caveat in mind, my objection to the equation of alchemical language with symbolic language is still valid, since most studies address alchemical texts as a homogeneous group. For a fruitful suggestion of genre classification of alchemical texts, see Halleux, *Les textes*, 73–86.

[10] By contrast, a sizeable body of vernacular scientific writings have survived in Old English, primarily dealing with medicine. The texts range from single recipes to collections of recipes and learned herbals; see Linda E. Voigts, "Anglo-Saxon Plant Remedies and the Anglo-Saxons," *Isis* 70 (1979): 250–68, here 250. See also the *eVK*. Tony Hunt (with the assistance of Michael Benskin) has recently edited a collection of Middle English recipes from the first half of the fourteenth century: Tony Hunt, ed., *Three Receptaria from Medieval England: The Languages of Medicine in the Fourteenth Century*, Medium Aevum Monographs n. s. 21 (Oxford: Society for the Study of Medieval Languages and Literature, 2001). However, this collection seems to be an exception; most scientific texts in Middle English appear to stem from after 1375.

First and foremost, this trend was not peculiar to writings on science, but formed a small part of a more general move toward using the English language in all kinds of writings.[11] Irma Taavitsainen suggests that the Englishing of scientific texts was yet another step in the Lancastrian monarchs' program of establishing and promoting the English language. This policy was instigated by nationalist strivings, as the monarchs sought to distance themselves and England from France and French culture.[12] However, other factors, such as growing vernacular literacy and the intended audience, undoubtedly also played major roles.[13] Whatever the precise mix of influences, English seems to have been firmly established as a language of science as early as the late fifteenth century. According to Linda Voigts, the "process of using English for learned science and medicine appears to have come to something like completion by 1475, for in the last quarter of the fifteenth century we can find a full range of English-language scientific and medical texts of university origin."[14] This does not imply that English replaced Latin in the fifteenth century; rather, the two languages were used as (near) equals.[15] Publication figures for medical works show that by the mid-seventeenth century English had taken over as the language of choice in medical texts: of 238 printed books on medicine in England, 207 (or eighty-seven percent) were in English, and only thirty-one in Latin.[16] Needless to say, for authors writing for an international audience Latin would still have been the premier language, very much like English is today.

Vernacular medical texts in particular have received a great deal of attention in the last fifteen years from historical linguists interested in the dynamics of the vernacularization of scientific texts. Much of this groundbreaking work has been based on a computer-readable collection of medical writings spanning the period 1375 to 1750, which has been compiled under the direction of Irma Taavitsainen

---

[11] Juhani Norri, *Names of Sicknesses in English, 1400–1550: An Exploration of the Lexical Field*, Annales Academiae Scientiarum Fennicae: Dissertationes Humanarum Litterarum 63 (Helsinki: Academia Scientiarum Fennica, 1992), 29–30.

[12] Irma Taavitsainen, "Scientific Language and Spelling Standardisation 1375–1550," in *The Development of Standard English 1300–1800: Theories, Descriptions, Conflicts*, ed. Laura Wright (Cambridge: Cambridge University Press, 2000), 131–54, here 132–33; see also Päivi Pahta, ed., *Medieval Embryology in the Vernacular: The Case of* De Spermate, Mémoires de la Société Néophilologique de Helsinki 53 (Helsinki: Société Néophilologique, 1998), 54.

[13] Norri, *Names of Sicknesses*, 30; Pahta, *Medieval Embryology*, 59–61.

[14] Linda E. Voigts, "What's the Word? Bilingualism in Late-Medieval England," *Speculum* 71 (1996): 813–26, here 816.

[15] Linda E. Voigts, "The Character of the *Carecter*: Ambiguous Sigils in Scientific and Medical Texts," in *Latin and Vernacular: Studies in Late-Medieval Texts and Manuscripts*, ed. Alastair J. Minnis (Cambridge: D. S. Brewer, 1989), 91–109, here 95.

[16] Charles Webster, *The Great Instauration: Science, Medicine and Reform 1626–1660* (New York: Holmes and Meier, 1976), 267.

and Päivi Pahta at the University of Helsinki. Studies that use this corpus of texts have dealt with topics from metatextual expressions and prescriptive formulations to code-switching and noun phrase structure.[17] Their results point to many interesting developments in the establishment of English medical prose, which are linked to different underlying traditions of medical writing and changing attitudes toward science.[18] The early English medical lexicon has also received a great deal of attention in recent years, and studies have shown that translators and writers created a sea of new terms for concepts and procedures that had previously only been written about in Latin or French.[19]

Although medical texts appear to have been in the forefront of the vernacularization movement, texts in other disciplines were not far behind. The fifteenth and sixteenth centuries witnessed a virtual explosion of vernacular texts on topics such as alchemy, astronomy/astrology, botany, chiromancy, geomancy, geometry, and many more.[20] Next to texts on medicine, alchemical writings are the most numerous, surviving in the form of alchemical tracts, treatises, notes, and recipes, written in prose as well as verse.[21] Despite their frequency, historical linguists have paid scant attention to alchemical texts. Part of the reason for this

---

[17] Part of the collection is published in Irma Taavitsainen, Päivi Pahta, and Martti Mäkinen, *Middle English Medical Texts*. CD-ROM (Amsterdam: John Benjamins, 2005). For the studies, see Irma Taavitsainen, "Metadiscursive Practices and the Evolution of Early English Medical Writing 1375–1550," in *Corpora Galore: Analyses and Techniques in Describing English*, ed. John M. Kirk (Amsterdam: Rodopi, 2000), 191–207; eadem and Päivi Pahta, "Vernacularisation of Medical Writing in English: A Corpus-Based Study of Scholasticism," *Early Science and Medicine* 3 (1998): 157–85; Päivi Pahta, "Code-Switching in Medieval Medical Writing," in *Medical and Scientific Writing in Late Medieval English*, ed. Taavitsainen and eadem, 73–99; eadem and Irma Taavitsainen, "Noun Phrase Structures in Early Modern English Medical Writing," paper presented at the Twelfth International Conference on English Historical Linguistics, Glasgow, 21–26 August 2002.

[18] Irma Taavitsainen et al., "Analysing Scientific Thought-Styles: What Can Linguistic Research Reveal about the History of Science?" in *Variation Past and Present: VARIENG Studies on English for Terttu Nevalainen*, ed. Helena Raumolin-Brunberg et al., Mémoires de la Société Néophilologique de Helsinki 61 (Helsinki: Société Néophilologique, 2002), 251–70.

[19] Norri, *Names of Sicknesses*; idem, *Names of Body Parts in English, 1400–1550*, Annales Academiae Scientiarum Fennicae 291 (Helsinki: Academia Scientiarum Fennica, 1998); Roderick W. McConchie, *Lexicography and Physicke: The Record of Sixteenth-Century English Medical Terminology* (Oxford: Clarendon Press, 1997).

[20] Keiser, *A Manual*; *eVK*; Lister M. Matheson, ed., *Popular and Practical Science of Medieval England* (East Lansing: Colleagues Press, 1994).

[21] Keiser, *A Manual*, 3627–37, 3788–3808; *eVK*. For the pan-European vernacularization of alchemical texts, see Michela Pereira, "Alchemy and the Use of Vernacular Languages in the Late Middle Ages," *Speculum* 74 (1999): 336–56.

probably lies in the lack of editions of English-language writings.[22] However, a more important issue is perhaps alchemy's ambiguous status as a science. Today alchemy is often viewed as an occult discipline or at best as an area of pseudo-science. But in the Middle Ages, many contemporary scholars, including Albertus Magnus and Roger Bacon, clearly considered it on a par with other scientific disciplines, although they also expressed reservations.[23] Medieval codices also reveal the association of alchemy with science in that alchemical texts often appear together with other scientific material, especially medical texts.[24] This connection between alchemy and medicine is particularly strong. Medico-alchemical tracts such as Rupescissa's *De consideratione quintae essentiae* blur a clear separation of the two disciplines, and later writings by Paracelsus and his followers underscore their continual intersection.[25] As we saw earlier, the Treatise's promise of curative elixirs may also have been the principal reason for the astrological physician Simon Forman to copy out manuscript A of the Treatise. Admittedly, alchemy was the object of much criticism as early as the Middle Ages, and aspects of it (especially the pursuit of transmutation) continued to be controversial in many quarters in the sixteenth and seventeenth centuries. Despite this criticism, alchemy obviously enjoyed widespread support, and even such in modern eyes unimpeachable scientists as Robert Boyle and Sir Isaac Newton read, copied, and even composed alchemical texts.[26]

The connection between alchemy and what later became the science of chemistry is one of the thorniest, yet potentially most interesting, issues for linguistic studies. There is no clear line of development from alchemy to chemistry. In fact, both the terms *alchemy* and *chemistry* were used interchangeably to denote a whole range of (al)chemical approaches during the early modern period. Not until the late seventeenth century and increasingly in the eighteenth century did the terms begin to go their separate ways. Although this was not a universal

---

[22] See Grund, "In Search of Gold," 265–66.

[23] Crisciani, "Alchemy," 10–12.

[24] Linda E. Voigts, "Scientific and Medical Books," in *Book Production and Publishing in Britain 1375–1475*, ed. Jeremy Griffiths and Derek Pearsall (Cambridge: Cambridge University Press, 1989), 345–402, here 345–49; eadem, "The 'Sloane Group': Related Scientific and Medical Manuscripts from the Fifteenth Century in the Sloane Collection," *British Library Journal* 16 (1990): 26–57; 'scientific/medical writings' in *eVK*.

[25] See Bruce T. Moran, *Distilling Knowledge: Alchemy, Chemistry, and the Scientific Revolution* (Cambridge, MA: Harvard University Press, 2005), 11–15.

[26] Webster, "Alchemical and Paracelsian Medicine"; Principe, *The Aspiring Adept*; Betty Jo T. Dobbs, *The Foundations of Newton's Alchemy or "The Hunting of the Greene Lyon"* (Cambridge: Cambridge University Press, 1975). See also Abraham, *A Dictionary*, xv, and the discussion of the sociohistorical context of the Treatise. For Newton's interest in alchemy, see http://webapp1.dlib.indiana.edu/newton/index.jsp (as accessed in 2006). For an eminently accessible discussion of alchemy and science, see Moran, *Distilling Knowledge*.

trend that everybody followed, *alchemy* came more and more to be applied to metallic transmutation, while *chemistry*, which rejected many of the central tenets of transmutational alchemy, was approaching its modern meaning.[27] Of particular linguistic interest here is the possible continuity and influences of textual and linguistic conventions. It remains to be explored whether texts from different disciplines covered under the term alchemy in the early modern period all exhibit similar linguistic characteristics and genre conventions, and to what extent texts on chemistry in subsequent centuries adopted and adapted these characteristics and conventions.

Alchemy's affiliation with both medicine and what later became known as chemistry and its status as a science in earlier periods thus make alchemical texts important objects of study for historical linguists interested in the vernacularization of scientific texts in the fifteenth and sixteenth centuries. Examining the dissemination, transmission, and conventions—linguistic as well as extra-linguistic—of early alchemical texts reveals illustrative patterns of vernacularization, as I have shown elsewhere.[28] A systematic study of the linguistic patterns and conventions of a large body of alchemical texts is unfortunately made difficult by the dearth of editions of vernacular writings. However, a micro-level study of the Treatise can begin to address that lack and point to some prominent characteristics of the language of early vernacular texts on alchemy.

Symbolic Language

Although it is probably the aspect of alchemical language that has received the most scholarly attention, the alchemists' use of symbolic language is still a major obstacle for the interpretation and understanding of alchemical writings.[29] One of the most powerful explanations for the use of symbolic language is concealment: the alchemists would reveal their art only to other adepts of their craft.[30] One reason behind this idea may be religious. Many alchemical writers express the fear of repercussions from God if the art of alchemy is revealed to unworthy

---

[27] Newman and Principe, "Alchemy vs. Chemistry," 63–64; Moran, *Distilling Knowledge*, 117–24.

[28] Peter Grund, "The Golden Formulas: Genre Conventions of Alchemical Recipes in the Middle English Period," *Neuphilologische Mitteilungen* 104 (2003): 455–75; idem, "'ffor to make Azure'."

[29] Within the concept of *symbolic language*, I include the use of symbols, metaphors, allegories, analogies, and similar modes of expression. I have preferred this term to "Decknamen," which is frequently used for at least some of these phenomena, mainly because some of the strategies that I touch upon go beyond simple words. See e.g., Principe, "Decknamen," 104–6.

[30] Taylor, *The Alchemists*, 56.

people.³¹ The Treatise is no exception. It stresses that the divine art of alchemy should not be bestowed upon people who do not enjoy God's sanction:³²

> For when by great laboure and moste painfull travaill the auncient phi*lo*-*sofe*rs had attayned to the pe*r*fecte mastery therof and then not willinge that their labours should be buried in the bottomlesse lake of oblyuion, haue co*m*piled great volums of the same worke, hydinge it vnder shadowe of misticall word*es* and names infinite, and haue sealed it in their writing from vnworthy people vnder the mistie mantle of phi*loso*fie, fearinge leste the greate wrath of God should com vpon them for openinge of soe greate a secreate to the people not worthy therof. For yf the seacreat working of this science were vniuersally knowen, ynnumerable people wold labour therin. And whoe can tell wher God wold giue the doing*es* therof to wicked men to plague the world therby. For ther can be noe greater plague to the wordle then wicked men to haue this glorious science, wherby ther might happen subuertion of kingdoms, rebellion of subiect*es*, dissobedience of seruant*es* and innumerable other scourges depending vpon the same beinge in the hand*es* of wicked men. Therfore, let noe man entend himselfe to labour in this moste highe scie*n*ce vnworthy leste his wicked harte be lifte vp to encrease his dampnation by wickednes, hauinge abundance, and destitute of the grace of God may be, besides his owne co*n*dempnation, the distruction of many other. But to such as entend to worke this worthie science to the glory of God, to the relife of those that haue need, and for the maintenance of the common welle of their country, to such I write; and those, ther is noe doubte, but the God of nature will open vnto them the gate of the pallace therof, wherby they may enter into the garden of the Ladie and Emprise of Phi*loso*fie, wherin ther is to be found infinit treasures [301v: 11–38]

Social and economic factors may have been as influential as religious factors. Since patents or intellectual property were unknown concepts in the late Middle Ages and the early modern period, one way to avoid revealing one's professional secrets and hence retain one's livelihood was to make one's writings unintelligible to ordinary people.³³ Although the Treatise hints at this kind of professional secrecy (e.g., 299r: 1–5), it stresses other similar incentives for intentional obscurity even more. Most importantly, as we saw earlier, Lock had a vested interest in making his Treatise opaque and in not revealing the whole secret of his alchemical practice: obscurity, Lock thought, would compel Cecil and Elizabeth to help him return to England.

---

³¹ Crosland, *Historical Studies*, 51–52.
³² See also 294r: 23–27, and 319r: 10–30.
³³ Crosland, *Historical Studies*, 52; Hannaway, *The Chemists*, 120; Hans-Werner Schütt, "Sprache der Alchemie," in *Alchemie: Lexikon einer hermetischen Wissenschaft*, ed. Priesner and Figala, 340–42, here 341.

The use of symbolic language does not always appear to reflect a wish for concealment. It may even have had the opposite purpose: to illustrate the procedures. Frank S. Taylor suggests that the use of symbolic language among alchemists was a way of explaining the chemical processes by analogies, drawing upon the alchemists' conception of the world:

> To give the combination of two substances to make a third the name or symbol of "a marriage and birth" was to fit the phenomenon into his [i.e., the alchemist's] world and so to make sense of it. He would then act on the principle that the phenomenon *was* a marriage and birth, and he would provide the sort of conditions which in his mind would be favorable to such a process.[34]

It is debatable whether such a strategy lies behind the language found in all alchemical texts, and whether all linguistic strategies can be accounted for in this way. But although the idea may not be universally applicable, it may still go a long way toward explaining the origin of some symbols. The Treatise provides at least partial support for this theory. Symbolic language occurs predominantly in the sections that deal with alchemical theory, but is conspicuously absent from the practical, procedural parts. (All alchemical texts do not follow this neat pattern, however.) The origin of much of the symbolic language may thus have been an attempt to illustrate in visual terms the complex theories that underpinned the practical procedures. For this purpose, the alchemists drew upon current conceptions of the world and different areas of human experience.[35] However, although the origin of some symbols and imagery may lie in a wish for clear exposition, it is doubtful whether this is how many later alchemists (to say nothing of the uninitated) perceived the tropes. Divorced from their original cultural or social context through textual transmission, such symbols and metaphors must have seemed opaque and part of the alchemists' propensity for secrecy.[36]

The Treatise contains a diverse, though limited, set of symbols, metaphors, and analogies. Some are obfuscatory, while others even seem to help the understanding of the procedures (as we will see later). The Treatise's symbolic language has a variety of sources.[37] The most frequent analogical imagery connects

---

[34] Taylor, *The Alchemists*, 157–58.
[35] For more esoteric, religious or experiential interpretations of alchemical language, see p. 72.
[36] See Schütt, "Sprache," 341.
[37] I will focus on major patterns in the Treatise, but some of the individual phenomena will only be discussed in detail in the actual explanatory note on the passage. By incorporating so many earlier texts (most of them relatively uncritically), the Treatise mixes several different alchemical traditions. As a result of this mixture, different passages of the text, even though they use the same symbolic language or imagery, may have very different significances; after being embedded in the Treatise, the symbols may even have

different aspects of the alchemical opus with the human being. Sometimes the Treatise makes explicit the connection between the two:

> for like as noe other thing engendreth man but man, nor a beaste but a beaste, soe all thes thi<n>g*es* [i.e., blood, hair, eggs, etc.], being far from the kind of mettalls, it is vnpossible that mettalls should be engendred of them, and forasmoch as mettalls ar not engendred but of their sperme, and quicksiluer is the sperme of all mettalls and their begininge [299v: 41–300r: 3]

In analogy with the creation of humans, the metals are seen as created out of a sperm, *mercury*. In other cases, the similes or metaphors relate more specifically to human behavior. The following passage underscores the harmonious state that ensues after the joining of opposing principles:[38]

> for all the p*a*rtes beinge made cleane, then nature by force of workinge, by his o<w>ne strength, ioyneth together co*n*trary thing*es*, as heat with cold, and moist with dry, and sulfur w*i*th argent viue, which being closed together in on vessell will not leave stryuinge vntill by their vehement wrastelinge ther appeare in the vessell drops of sweete descending vpon the mortifyed body black and stinckinge, [and] afterward the said drops, being changed into the form of p*e*rells or the eyes of fishes firmly standing on the earth, is the sign that, after that longe strife, they ar made frind*es* into such an vnity as cannot be sep*a*rated. [320v: 44–321r: 3]

Although this passage seems fairly illustrative and straightforward, there are more complicated cases where the interpretation of the symbols or analogies is more difficult. These instances seem to exemplify the alchemists' propensity for concealment rather than illustration.

> For in this manner is the sonn white in his appearinge and red by exp*eri*ence; and soe is the red man and the white wife maryed together, which ph*iloso*f*e*rs call by the name of gould [302r: 33–36]

The *red man* and the *white woman* ("the white wife") are alchemical commonplaces. In most alchemical descriptions, the red man symbolizes sulfur, which was seen as the male principle of the stone, and the white woman mercury, the female principle. The joining of these two principles would lead to the creation of the philosophers' stone.[39]

Another analogy that is common in the Treatise connects the alchemical opus and the tripartite division of the human being into *body*, *soul*, and *spirit*:

---

lost their original meanings and taken on new ones. This factor should be borne in mind in the subsequent discussion.

[38] Abraham, *A Dictionary*, s.vv. *peace* and *strife*.
[39] Abraham, *A Dictionary*, s.vv. *red man* and *white woman*, *chemical wedding*.

> Nor the sprite may not be ioyned wit*h* y*e* bodie but by a mean of the soulle. And thinke not that the bodie of whom the sprite shalbe ioyned is any straunge thinge that should be added therto, but such a thinge as was hid in the sprite in his first beinge [302r: 22–26]

During the Middle Ages and into the early modern period, the human body was believed to consist of three parts: the material body, the immaterial soul, and the spirit, which constituted a middle ground or a mediator between the other two.[40] The alchemists adopted this idea to represent one of their processes. They believed that when heat was applied to the raw material of the stone, the material (or *body*) was dissolved into its constituent parts, its prime matter. At this point, a volatile vapor, the *soul* of the raw material, proceeded from it. After the body had been purified in the bottom of the vessel, the soul was reunited with the purified body, which took place through cooling and condensation. This process was cryptically stated to be impossible without the mediation of the *spirit*. The end result was stated to be the philosophers' stone.[41] The Treatise frequently alludes to this relationship of *body*, *soul*, and *spirit*, although it often reverses the ordinary roles of the spirit and the soul (as in the earlier extract).[42]

Like many other alchemical texts, the Treatise also connects the constitution of metals with the human body. The philosophers' stone was often seen as a medicine that cured imperfect metals, such as copper, lead, iron, and tin, into perfection, that is, silver or gold. The Treatise sometimes refers to the imperfect state of a metal as a disease:

> But this elixar of the minerall stone [. . .] is mad elixar without putrifaction; wherfore he is of nature to mans bodie veri mortalle. But of the minerall bodies he healeth their leprouse diseas & maketh of an ympe*r*fecte bodie a pe*r*fecte bodie, abydinge every examination and tryall of the fier. [313v: 2–8]

This description illustrates the commonplace use of medical terms for alchemical processes. It also exhibits the use of the term *elixir* (derived via Arabic *al-iksir* from Greek *xerion* 'powder for wounds'), which is common in the Treatise as a synonym or near-synonym for the philosophers' stone. This use further indicates the blend between medical treatment and alchemical practice.[43]

---

[40] E. Ruth Harvey, *The Inward Wits: Psychological Theory in the Middle Ages and the Renaissance*, Warburg Institute Surveys 6 (London: The Warburg Institute, 1975), 7–8, 27–28.

[41] Abraham, *A Dictionary*, s.vv. *soul, spirit, distillation, chemical wedding, ferment, glue*.

[42] In fact, the reversal is already found in the source of the Treatise, the *Dicta*; see p. 41. See also Abraham, *A Dictionary*, esp. s.vv. *soul, spirit*.

[43] See Crisciani, "Alchemy and Medicine."

Besides analogies with the human body and human behavior, the Treatise also uses animal symbolism in its descriptions. These are mostly stock images such as (*the blood of*) *the green lion, the bird of humanity, the flying eagle, the dragon eating his own wings*, and *the raven's head*. They refer to both substances and processes. *The dragon eating his own wings* (see 309r: 5–6), for example, commonly refers to the process that is described above in the discussion of body, soul, and spirit, where the body is reduced to its primary constituents while a vapor ascends to the top of the vessel.[44] *The raven's head*, on the other hand, illustrates the initial stage of the alchemical opus where the base material is dissolved or putrefied into its prime matter, which was often thought of as a black substance:[45]

> And putrifaction is nothing ells but that heate, workinge in a moiste bodie, engendreth blacknes, w*hich* in the begini*n*ge of putrifaction is called the <u>ravens heed</u>. [308v: 28–30]

Although the symbols, metaphors, and similes in the Treatise in many cases impede a straightforward interpretation of the text, there are cases where symbolism seems to be clearly intended to help the reader understand the sequence or end of a process. Such helpful cues usually take the form of similes where processes or substances are likened to concrete phenomena or items. In contrast to most types of symbolism, these similes are more common in the technical procedures than in the theoretical discussions. Three objects of comparison are particularly common: *colors*, *wax*, and *pearls* or *fish eyes*. *Colors* play an integral part in the alchemical opus, signaling different stages of the alchemist's work. Black signals the initial stage, whereas white indicates, for example, that the alchemist has obtained the elixir that will turn a substance into silver (e.g. 302r: 18); red signals that the elixir which is used to produce gold has been reached (e.g. 302r: 32).[46] Colors also illustrate more specifically what is taking place in the alchemical opus:

> then being set in balneo, yt will be straightway <u>decked with a grene collour like vnto an emerauld</u>; and when the body is thus dissoued and moulten in y*ᵉ* sayd water, let the water be distilled from the oill in balneo, w*hich* oill wilbe <u>like oille oliue, enclyninge somwhat to blacknes</u>. [317r: 35–39]

Another important simile is *wax*, which illustrates the consistency of substances after they have been put through a certain process.[47] *Pearls* and *fish eyes* are usually used in combination, and they often occur as a further elaboration or description of *bubbles*: "ther appeare brubles vpon the earth in the forme of p*er*elles

---

[44] See Abraham, *A Dictionary*, s.vv. *dragon, bird, Bird of Hermes*.

[45] See Abraham, *A Dictionary*, s.vv. *raven, crow, black, nigredo*. For a discussion of the other terms, see EN 304r: 26–27, and 310v: 37–41.

[46] See also e.g. Roberts, *The Mirror*, 55–56.

[47] See e.g. 302r: 26, 311v: 37, 313r: 34, 317r: 6–7.

or the eyes of fishes" (316v: 22–23). The fish eyes or pearls often indicate that the alchemist has successfully reached the stage where the substance has been purified.[48] These similes highlight that some of the symbolic language can be used to help rather than confuse the reader. In other words, they provide an opportunity for the alchemist to verify empirically that the procedure has been successful.

Alchemical Sigils

One of the most common features associated with alchemical language is the use of sigils for different substances, such as '☉' for *gold* and '♀' for *copper*.[49] Sigils are found as early as Greek texts on alchemy, but their number increased and they grew more complex in the Middle Ages and early modern period.[50] The extent to which individual texts or indeed individual manuscripts of the same text employ sigils may vary greatly. The copies of the Treatise are an excellent case in point. While F,[51] S1, and S2 do not employ sigils at all, G frequently uses sigils for the seven metals: '☉' for *gold*, '☽'[52] for *silver*, '☿' for *mercury*, '♃' for *tin*, '♄' for *lead*, '♂' for *iron*, and '♀' for *copper*.[53] In addition to the sigils found in the G manuscript, S3 and S4 also contain, for example, the sigil '🜍' for *sulfur*, '🜁' for *air*, '🜂' for *fire*, '🜃' for *earth*, and '🜄' for *water*, which are not recorded in the other manuscripts. Manuscript A presents yet another pattern. Since this is the manuscript edited here, I will examine its use of sigils in more detail.

---

[48] See also Abraham, *A Dictionary*, s.vv. *pearls, fishes' eyes, ablution, albedo*.

[49] Fred Gettings argues convincingly for the use of the term *sigil* to refer to these forms: *Dictionary of Occult, Hermetic and Alchemical Sigils* (London: Routledge & Kegan Paul, 1981), 7–9. He argues that *symbol*, which is commonly used in this context, "carries a literary as well as an iconographic connotation," and is hence not specific enough. *Sigil*, on the other hand, "appears frequently in mediaeval magical contexts, and has even been used specifically for certain astrological symbols and devices which were supposed to be amuletic in power." From the eighteenth century, the term has been used with the specific meaning of 'a small image.' See also Voigts, "The Character."

[50] Crosland, *Historical Studies*, 228; Gettings, *Dictionary*.

[51] The copyist of F does sometimes use sigils in his marginal comments.

[52] In G, the sigil is lying on its back with the crescent upwards.

[53] In addition to the sigils for the seven metals, G also uses a star-shaped sigil for *sal ammoniac* once. At the beginning of the Treatise, it also seems to make a distinction between *mercury* 'the metal' and *mercury* 'one of the components of all metals,' and hence uses two variants of the sigil '☿.' The distinction breaks down, however, and hybrid forms of the two are found in the latter half of the text.

Table 4.1. Alchemical sigils in A[54]

Sigil	Referent	Occurrence	Full form	Occurrence
☿	mercury	120 (56%)	argent vive, mercury, quicksilver	94 (44%)
☉	gold	55 (29%)	gold, Sol, sun	136 (71%)
☽	silver[a]	40 (34%)	Lune, Luna, moon, silver	77 (66%)
♀	copper	7 (37%)	copper, Venus	12 (63%)
♄	lead	1 (11%)	lead, Saturn	8 (89%)
Total		223 (41%)		327 (59%)

[a] In the manuscript, the crescent is facing the other way, but technical limitations made it necessary to reproduce the sigil in this way.

Manuscript A only has a limited set of sigils, and only three occur with some frequency: '☿,' '☉,' and '☽.' The sigil for mercury, '☿,' is the only one that is more frequent than the full forms of the word; none of the remaining sigils occurs more than thirty-seven percent of the time.[55] Why A never uses the sigils for *iron* and *tin* is unclear, but part of the explanation may lie in the fact that the Treatise rarely mentions these substances at all.

The sigils in A may have several functions. Sigils have often been connected with the alchemists' desire to hide their art from uninitiated practitioners.[56] Although secrecy may of course have been a contributing factor, it seems unlikely that it would be the full explanation for their use in A. The sigils in A are few, and they are of the most common type, that is, sigils that derive from the signs denoting the seven planets. These sigils are commonplace in alchemical texts, and they are even found with some frequency in other types of scientific writing (most notably, but not surprisingly, in astrological/astronomical texts). Considering the widespread use, they would probably have been only a partially successful means

---

[54] From the figures in the table, I have excluded three instances of '☉' and two instances of '☽,' which refer to the sun and the moon as celestial bodies (see 303v: 17, 313r: 6, 319r: 32 (x2), 324r: 35). There are additional instances of the sigils in the marginal comments that are not found in the other manuscripts: ☿=13, ☉=9, ☽=7, ♀=4, ♄=1. I have also excluded the marginal comments made by Richard Napier, which contain some sigils; see p. 97.

[55] There are also other terms which can be used to refer to, for example, mercury, such as aqua permanent, and it is theoretically possible that the underlying form of the sigil is one of these symbolic terms. However, since this is difficult to establish, I have not included these terms in the tabulation of the full forms.

[56] Crosland, *Historical Studies*, 234; Roberts, *The Mirror*, 66–68.

of concealment.[57] Perhaps a more powerful explanation for the use is that the sigils were employed as convenient marks of abbreviation.[58] Support for this hypothesis comes from the use of sigils in A's marginal comments: sigils are much more frequent than the full forms of the words, probably because of space restrictions. More generally, the fact that the use of different types of sigils persists well into the eighteenth and nineteenth centuries in works of chemistry further supports such a claim. To some extent, sigils could be seen as the forerunners of subsequent systems of chemical abbreviation or annotation, such as the ones developed by John Dalton and J. J. Berzelius.[59]

Although the actual referent of the sigil may be easily recoverable, the specific lexical form that the sigil corresponds to may not be as easy to determine. Linda Voigts even argues that sigils are primarily extralinguistic, since they may correspond to a number of terms, in a number of different languages.[60] Since code-switching between two or more languages is quite common in early scientific texts, it would not be surprising for individual Latin or French terms to appear in a text that is otherwise composed in English, or vice versa.[61] The ambiguity is present in A as well. An illustrative example is the sigil '☿,' which could represent *mercury*, *quicksilver*, *argent vive*, or some other name for this substance. Although mostly ambiguous, instances do occur where the underlying word can be determined. A uses '☿ᵃˡˡ' twice, where the only interpretation possible seems to be *mercurial*, and there are four instances of '☽ᵃ,' which clearly shows that the intended word is *Luna*.[62] In addition to illustrating that sigils are not always extralinguistic, these instances seem to provide further support for the view that the sigils could simply be used as means of abbreviation.

There are two additional sigils of a more general or technical character in A: '℥' for *ounce* and '℞' for *take* or Latin *recipe*.[63] These sigils, which are found in other types of scientific and informative writing, were originally abbreviations. They are exclusively found in the Treatise's descriptions of technical procedures, formulated as recipes.[64] Like the planetary sigils, '℥' for *ounce* is probably used

---

[57] Principe, "Decknamen," 104, considers the planetary names synonyms of the actual names of the metals rather than symbolic cover terms, mainly because the connection between the two sets of names was generally known. Although he does not discuss sigils but only the written-out names, his claim is at least partially valid for sigils as well.

[58] Crosland, *Historical Studies*, 230, suggests that some of the sigils at least originate in abbreviations.

[59] Crosland, *Historical Studies*, 234–35, 256–81; Voigts, "The Character," 95.

[60] Voigts, "The Character," 91.

[61] Voigts, "What's the Word"; Pahta, "Code-Switching."

[62] There are also two instances of '☿ᵉˢ' (one occurring in the comments in the margin). But the interpretation may be *mercuries*, *quicksilvers*, or *argent vives*.

[63] These sigils also occur in all the other manuscripts except for G.

[64] See p. 90.

as a convenient mark of abbreviation. It occurs five times, whereas the full form *ounce* is never used.[65]

The function of 'R̞' may be more multilayered. I have argued elsewhere that the 'R̞' sigil functions as a conventionalized initial marker in a collection of alchemical recipes from the fifteenth century: the sigil appears to be a linguistic cue to the reader that a recipe is beginning.[66] This may be the function of 'R̞' in A as well, since it appears at the beginning of two chapters formulated as recipes, and at the beginning of a recipe within a chapter.[67] The underlying language is again unclear: it may represent *take*, but one of the examples occurs in a Latin introductory phrase: "R̞ ergo in nomine domine [sic] . . .," which suggests that the Latin *recipe* is also a possible interpretation.[68]

Technical Vocabulary

The fifteenth and sixteenth centuries witnessed a virtual explosion of new words related to science, as translators and writers struggled to come up with suitable terminology in English for scientific concepts, equipments, and procedures. Words were borrowed from foreign languages (primarily Latin, French, and Greek), new words were coined, and old words took on new meanings. The development of medical terminology in late Middle English and early Modern English has recently received a great deal of scholarly attention, and our knowledge of how writers of medical texts shaped the vocabulary of medicine has been enhanced substantially.[69] Unfortunately, the vocabulary of other sciences, including alchemy, has not been explored to a similar extent.[70] Again, the scarcity of editions seems to be to blame. In the meantime, the Treatise constitutes an excellent showcase for the developing lexical resources of alchemical writers and translators: it contains a number of words that have not been recorded in the

---

[65] 312r: 3, 4, 5; 324v: 35–36. An abbreviation "oz", with a loop extending from the 'z' over the 'o,' is also found once (312r: 3).

[66] Grund, "The Golden Formulas," 461; see also Voigts, "The Character," 101.

[67] 324r: 31, 324v: 35, 325r: 18.

[68] Another abbreviation that has characteristics similar to 'R̞' and '℥' is 'li' for *libra* or *pound*, which occurs fourteen times in A. Cf. Voigts, "The Character," 99.

[69] Norri, *Names of Sicknesses*; idem, *Names of Body Parts*; idem, "Entrances and Exits in English Medical Vocabulary, 1400–1550," in *Medical and Scientific Writing in Late Medieval English*, ed. Taavitsainen and Pahta, 100–43; McConchie, *Lexicography*.

[70] Juhani Norri has explored some aspects of (al)chemical terminology, but only as it is found in Middle English medical texts: "Notes on the Origin and Meaning of Chemical Terms in Middle English Medical Manuscripts," *Neuphilologische Mitteilungen* 92 (1991): 215–36.

English language before; and it demonstrates the adoption and adaption of already existing lexical items for alchemical uses.[71]

The Treatise uses a range of technical terms for inherently alchemical processes and procedures, including sublimation, calcination, multiplication, and projection. Appearing as early as Chaucer's *Canon's Yeoman's Tale*, many of these terms have been recorded in the *OED* since the first edition. Some of them are represented by multiple citations since they were taken over and adapted by later practitioners of chemistry. But like its coverage of medical terms, the *OED*'s coverage of alchemical terms is very far from exhaustive.[72] Words in the Treatise both antedate and postdate the earliest and latest *OED* recordings, and the text provides examples of previously unrecorded words. Most words are borrowings from Latin, an indication of the Treatise's dependence on its sources (see Table 4.2). They include nouns, verbs, and adjectives; and denote processes ("elixeration"), characteristics of substances ("aerouse"), and alchemical procedure ("encere").[73]

---

[71] Some symbols may perhaps also be claimed to be part of this technical vocabulary since they form lexical units with a specific meaning (e.g. *the raven's head* and *the flying eagle*), but since I have treated them elsewhere they will not be included here.

[72] For the *OED* and early Modern English medical terminology, see McConchie, *Lexicography*. McConchie's study was based on the first edition of the *OED*, however, and additions have undoubtedly been made in the subsequent revisions of the dictionary.

[73] *OED*'s primary sources of alchemical vocabulary in the fifteenth and sixteenth centuries seem to be Chaucer's *Canon's Yeoman's Tale*, Ripley's *Compound of Alchymy*, Norton's *Ordinal of Alchemy*, and a version of Rupescissa's *The Book of Quintessence*. The terms are variably and confusingly classified as *alchemical*, *old chemistry*, and *ancient chemistry*. The distinction between these labels is not always clear. I have also consulted the *MED*, which is largely based on the same alchemical material, although it does contain a few more texts (mainly verse tracts). Since the *MED* does not provide any additional information on the items in Table 4.2, I have not made a specific reference to this dictionary.

Table 4.2. Lexical items in the Treatise

Item	Meaning	OED
aerouse, ayerouse, ayrouse	'airy,' 'of the nature of air'	s.v. *airous*: first and only recorded in 1683
alkamistick, alkamistik	'alchemical'	s.v. *alchemistic*: first recorded in 1689
aptation	'constitution'	—
balination	'bathing,' 'exposing a substance to heat in a balneum'	s.v. *balneation*: first recorded in 1646, and only in the nontechnical meaning 'bathing'
confixat	'(firmly) fixed together'	—
elixerate	'turned into an elixir'	—
elixeration	'the process of turning [something] into an elixir'	—
encere	'turn [something] into the consistency of wax'	—
kymia	'an alchemical vessel [used primarily for distillation]'	—
mercuried, mercurized	'mixed with mercury'(?)	—
olefied	'turned into an oil'	—
phillet	'a raised rim on a table'	s.v. *fillet*: this specific meaning is not recorded. See, however, 11
precipitate	'settle as a precipitate or solid substance'	s.v. *precipitate*: first recorded in an intransitive meaning in 1626; see III.6.a.
tinturouse	'full of tincture'	—
volatiue	'volatile,' 'characterized by a natural tendency to dispersion in fumes or vapor'	s.v. *volative*: meaning not recorded

The Treatise also uses more everyday lexical items with an extended, technical meaning. *Earth* and *water* are illustrative examples of this. The Treatise helpfully introduces them as follows: "the ph*iloso*fers calle all fluxible thin*ges* water, and coaguled or thick thin*ges* & stife or dry, earth (302v: 6–7)."[74] *Earth*, in particular, occurs with some frequency in the Treatise with the meaning of 'a solid body,'

---

[74] This is the reading of A. The other manuscripts contain slightly different readings, leaving out the adjectives *thick*, *stiff*, and *dry*.

whereas *water* is only used occasionally in the general meaning of 'a liquid.'[75] Although this may seem to be fairly straightforward, the use does lead to ambiguity, especially because of a "semantic overload" of the terms. It is frequently unclear whether the Treatise has in mind *earth* or *water* in the general sense or in the sense of 'one of the four elements,' as the following extract demonstrates:

> Then distill the water from the earth by the heate of balneo, hauinge a verie softe fyer; which water then will haue noe taste of sharpnes or otherwise verie lyttle because all his hole fier will remain in the earth. Afterward let the earth be well dryed, and when it is congealed, suffer the vessell to be cold [321r: 40–44]

Alchemists can also assign a special meaning to a substance by adding the qualifying *our*, *philosophers'*, or *of the philosophers*.[76] In this way a non-technical lexical item can receive a new, highly technical significance. Again, like the semantic extension of commonplace lexical items, this strategy often leads to opacity. The Treatise frequently makes use of the modifiers. A prime example is of course *the philosophers' stone* itself. This substance is usually not considered a stone in the ordinary sense of the word; rather, it is commonly conceived of as a powder, or sometimes even a liquid: "For when our stone is liquid, flowinge, fleinge, and sufferinge, it is called ye woman."[77] At times, it is even simply referred to as "the stone."

Another substance that frequently occurs with a qualifying attribute is *mercury*. In connection with *our* or *of the philosophers*, mercury attains the meaning of one of the two primary components of metals. Even *mercury* on its own frequently signifies one of these two primary components rather than the metal *mercury* that was considered one of the seven main metals. The following extract appears to make a distinction between *crude* mercury and mercury (represented by a sigil),[78] the former being the metal, the latter the primary component.

> Other I haue knowen excellently well learned that haue taken ☿ crud and put him into a stronge glasse, and then haue set the glasse in a stronge furnace, and haue therto gyuen stronge fier vntill the crud ☿ became precipitate. Then haue they set it in balneo vntill it became crud again. Afterward

---

[75] Cf. Reidy, *Thomas Norton's* Ordinal, 37, 77–78; *MED* s.v. *erthe* 11. Principe, "Decknamen," 104.

[76] Crosland, *Historical* Studies, 25–27; Principe, "Decknamen," 105. See also Halversen, "*The Consideration*," 54; for *philosopher* in the sense of 'alchemist,' see *OED* s.v. *philosopher* 1a.

[77] 302v: 24–26. See also W. J. Wilson, "An Alchemical Manuscript by Arnaldus de Bruxella," *Osiris* 2 (1936): 220–405, here 298.

[78] Again, it is difficult to determine if the actual lexical item intended is *mercury*; see p. 85.

they removed it againe into the dry furnise to precipitate, and soe haue oftentymes made it precipitate and again crud, and at the laste with soe great a fier that the furnace rente in peaces. Yet found they nothing but precipitate. And the cause of the forsaid erroure is because <u>noe crud body is ☿</u>. Therfore, noe crud body can be changed excepte it can firste be brought into ☿, which is their center and begininge, but muste needes contynue and hould his vulgar nature &c. [301r: 18–31]

The Treatise also uses multiple terms to designate seemingly the same substance. This is perhaps most evident in the variable use of *mercury, quicksilver,* and *argent vive*, where the Treatise draws on both native resources and Latin borrowings. These terms are frequently employed interchangeably to denote either one of the seven main metals or one of the two components of all metals. *Quicksilver* seems most commonly to be used for the metal, and *quicksilver* and *mercury* are often juxtaposed, the latter being the basic component of all metals. Some patterns are also apparent as to where in the Treatise the terms appear. *Argent vive* is predominantly used in Chapter 3, but is also found occasionally in other chapters, especially Chapters 26 and 27. *Quicksilver* is predominant in Chapters 2, 4, 5, and 10, whereas *mercury* occurs predominantly in Chapters 1 and 2, and then intermittently in the other chapters. As we saw earlier, the sigil '☿' is very frequent throughout the Treatise, and it is mostly impossible to determine which linguistic form the sigil represents. The patterns of the written-out forms may point to the merging of different textual traditions: Chapters 2, 4, and 5, where *quicksilver* is common, derive from a single vernacular source (*Scoller and Master*). The use of *argent vive*, on the other hand, may have its origin in the texts drawn upon in Chapter 3.

Theory and Practice: Overall Discourse Strategies

As remarked earlier, when outlining the characteristics of alchemical language, previous studies have almost exclusively focused on the sigils, symbols, metaphors, and imagery commonly employed in alchemical writings. However, such descriptions highlight only some aspects of alchemical language, whereas others are ignored. Most importantly, they commonly disregard the linguistic features of the practical procedures that are found in most alchemical texts. The linguistic strategies of these practical sections are often strikingly different from those of the more theoretical discussions. Moreover, the language of the *practica* ties alchemical texts to other scientific texts of the period, which exhibit similar linguistic characteristics. It is the interplay between the language of the *theoria* and the *practica* that makes up alchemical language, not the language of either section alone.⁷⁹ The comparison between two chapters in the Treatise, one dealing with theory and the other with practice, underscores the two discourse strategies.

4.3 Comparison of Chapter 3 and Chapter 5

Expository Material: Chapter 3 (302r: 28–44)	Procedural section formulated as a recipe: Chapter 5 (307v: 4–24)
But in the begini*n*ge, the sprite of our mastery is coaguled into his body by the quality of heate, and then is it made ferme*n*te and tincture which ph*ilosofe*rs calleth their gould, which gould is white in deed and red in poware. And white is semip*erfe*cte, and is made p*erfe*cte w*i*th rednes without any other thinge then himselfe save fier. For in this manner is the sonn white in his appearinge and red by exp*erie*nce; and soe is the red man and the white wife maryed together, which ph*ilosofe*rs call by the name of gould. For after their totchinge, ther be too manners of gould: the on white and the other red. For without the ☉ and the ☽, this crafte of alkamie is not made because they be only the fermen*tes* of the ☉ and ☽, and ioineth the co*m*pound of elemen*tes* together into on fixed bodye, w*h*ich is noe other thinge after alkami to be vnderstand but moiste and dry, water & earth; and in the water is ayer, and in the earth fier. For after the sep*ar*ation of elemen*tes*, ther is none of them seene, but earth and water, and ayer & fier is in them as qualitie, and cannot be seen nor p*er*ceyued, but worketh inuisibly.	Take i^{li} of quicksilu*er* & grind it strongly with on pound of com*m*on salte. Then put all together into a pan of seathinge water, and ther let it take residence vntill the salte be all resolued. Afterward powre out the saulte water & salte, and put to other warme water, & soe washe it 2 or 3 tymes. Then dry it well by the heate of y*e* fier. Then take 2^{li} of vitriall finely ground and dryed against the fier. Then take i^{li} of salte alsoe well dryed and grind the vitriall and salte together strongely, and again dry them. Then take your quicksiluer and put it to the vitriall and salte by lyttle and lyttle, co*n*tynually grindinge them together soe longe as youe can p*er*ceiue any quicksiluer to appeare. And here is to be noted that to the compound of vitriall and salte beinge 3^{li}, youe muste put on pound of quicksiluer. And when they ar throughly encorporated together, put them in certain pot*es* of sublimation, set one their covers and lute them faste. The covers muste haue, every on of y^m, a lyttle hole in the top wherby the moistnes may pass out; and when the lute is dry, set your sublimatoryes vpon a furnace of sublimation, and vpon the hole of every cover lay a peace of glasse that youe may knowe when the moistnes is vapored away.

Chapter 3 of the Treatise is mainly concerned with the theoretical background of alchemical practice, whereas the goal of Chapter 5 is to instruct the reader on how to sublime quicksilver. Characteristic of Chapter 5 are the short clauses that mainly consist of imperative verb forms or modal constructions with *must*. The temporal sequence of the processes is stressed with the help of adverbs such as

---

⁷⁹ Naturally, there are texts that are exclusively theoretical or practical in nature, and hence only exhibit features of either strategy; see Halleux, *Les textes*, 73–86.

then and *afterward*. The syntax is fairly uncomplicated, although subordination does occur, primarily in the form of temporal subordinate clauses that highlight the chronological sequence. The vocabulary is seemingly straightforward, without obvious symbolic language (however, see further below). This section is best characterized as an alchemical *recipe*, since it follows the structural and linguistic conventions of recipe writing.[80]

The language in the extract from Chapter 3, on the other hand, is conspicuously different. It contains symbolic language in the form of the "red man and white wife," and "is it made fermente and tincture which ph*ilosofe*rs calleth their gould" (which presumably differs from ordinary gold). The syntax is frequently more complex than in Chapter 5; there are sometimes several layers of subordination, as in e.g. "then is it made fermente and tincture which ph*ilosofe*rs calleth their gould, which gould is white in deed and red in poware." The verb phrases are in the present tense, frequently in the passive.

The Treatise contains thirteen chapters (5, 10–14, 16–19, 23–24, 33) that are largely structured like the extract from Chapter 5, showing the same type of linguistic characteristics. An additional six chapters (21–22, 25–28) embed practical sections in more expository material that deals with alchemical theory; some chapters even include two or more separate recipes. The transition between theory and practice is usually introduced by formulations such as "Therfore, in the name of Christe, let vs goe to the practice vpon the calce of the bodie" (313v: 25–27), or "Nowe to com to the practice as we entend vpon the said oille" (314v: 29; see also 319v: 7, 320r: 25, 321r: 24). The Treatise is not in any way peculiar in employing these discourse strategies. In fact, the mix of expository and instructive material is found in many types of scientific texts from the Middle Ages and early modern period. Taavitsainen, for example, has pointed out that texts or passages of text formulated as recipes in medicine do not always appear in collections or as single items in manuscripts; they can be found incorporated in longer tracts.[81]

---

[80] For the genre conventions of recipes, see Grund, "The Golden Formulas"; Manfred Görlach, "Text-Types and Language History: The Cookery Recipe," in *History of Englishes: New Methods and Interpretations in Historical Linguistics*, ed. Matti Rissanen et al. (Berlin: Mouton de Gruyter, 1992), 736–61; Ruth Carroll, "The Middle English Recipe as a Text-Type," *Neuphilologische Mitteilungen* 100 (1999): 27–42; Irma Taavitsainen, "Middle English Recipes: Genre Characteristics, Text Type Features and Underlying Traditions of Writing," *Journal of Historical Pragmatics* 2 (2001): 85–113; Ruth Carroll, "Middle English Recipes: Vernacularisation of a Text-Type," in *Medical and Scientific Writing in Late Medieval English*, ed. Taavitsainen and Pahta, 174–91; Martti Mäkinen, "On Interaction in Herbals from Middle English to Early Modern English," *Journal of Historical Pragmatics* 3 (2002): 229–51.

[81] Taavitsainen, "Middle English Recipes," 94–98.

Although the theoretical and practical sections of the Treatise exhibit marked differences, it is also important to note that the two strategies are not discrete in all cases. The instructive passages, for example, are not always completely straightforward or free of symbolic language, as is shown in the following extract.

> And when the earth is well dryed into powder, gyue it drinke 7 tymes by imbybition, as it is showed in the other chapters; and then again putrify the wholle matter. But here thou muste vnderstand this drinke to be noe other thinge but only <u>the blod of the lyon</u>, from the w*hi*ch <u>blod</u> the watrines muste be taken away. [318r: 23–28]

A straightforward interpretation of the recipes also presupposes that terms such as *quicksilver* and *vitriol* (as in the extract given from Chapter 5 above) are the actual substances and not used as cover terms for something else.[82]

Despite the occasional overlap of these discourse strategies, they clearly represent two different modes of expression. But although most alchemical texts contain both types of discourse, the language described as alchemical in previous research is commonly only the type of symbolic language that is found in descriptions of alchemical theory. It is to some extent true that symbolic language sets alchemical texts apart from many other types of (scientific) writing, although medical texts sometimes employ similar strategies of symbolic language and analogic imagery as well.[83] But the identification of alchemical language with symbolic language is misleading. In late medieval and early Modern English codices, recipes were one of the most common types (if not *the* most common type) of alchemical writing, and recipes are frequently found embedded in treatises. To disregard this aspect of alchemical language would thus be to neglect and misrepresent the body of extant alchemical writings and their linguistic characteristics.

---

[82] See also Principe, *The Aspiring Adept*, 57.
[83] Taavitsainen, "Transferring Classical Discourse," 53–54.

Manuscripts

In the description of the physical characteristics of the manuscripts, I pay special attention to manuscript A, since this is the version of the Treatise presented in the edition. Similarly, as the the main collation manuscript, S2 is described in detail. The remaining manuscripts are discussed more briefly, though features of prime interest are still covered, such as general appearance, date of copying, and early history of the manuscript. The descriptions are all based on the original manuscripts, which I have consulted in person.

Oxford, Bodleian Library MS. Ashmole 1490 (A)

Manuscript A contains a large number of booklets in different sizes.[1] The booklet which incorporates the Treatise and some additional poems and recipes constitutes fols. 291r–331v in the present binding.[2] The paper leaves of the booklet measure approximately 290 × 200 mm, and are foliated in Arabic numerals in the bottom right-hand corner; they are in pencil in a modern (nineteenth–twentieth-century) hand. An earlier pagination from five to seventy-seven (fols. 293r–331r) appears in the hand of the one-time owner of the manuscript, Elias Ashmole.[3] Ashmole is most probably also responsible for marking the first four leaves of the

---

[1] For a comprehensive description of the contents of the manuscript, see William H. Black, *A Descriptive, Analytical, and Critical Catalogue of the Manuscripts Bequeathed unto the University of Oxford by Elias Ashmole* (Oxford: Oxford University Press, 1845), 1354–68; Traister, *The Notorious Astrological Physician*, 200–1; Kassell, *Medicine*, 236–39.

[2] Traister, *The Notorious Astrological Physician*, 201, suggests that the Treatise, including the dedication, starts at fol. 291r but ends in 306r. She labels the rest of the Treatise and the items following the Treatise "[a]ppendices to the previous item." Traister seems to have been misled by the fact that Forman has added a colophon on fol. 306r (see below).

[3] Not all of the pages in 293r–331r are marked. The following pages are marked (the current folio number is given first): 293r=5; 294r=7; 295r=9; 295v=10; 296r=11; 298r=15; 300r=19; 301r=21; 302r=23; 303r=25; 304r=27; 305r=29 [earlier "21" rubbed out]; 306r=31; 307r=33; 308r=35; 309r=37; 310r=39; 311r=41; 312r=43; 313r=45; 314r=47; 315r=49; 316r=51; 317r=53; 318r=55; 319r=57; 322r=59; 323r=61; 324r=63; 325r=65; 326r=67; 327r: 69; 328r: 71; 329r: 73; 330r: 75; 331r: 77. It is unclear why some folios are not paginated. Folios 320r and 321r seem to have been skipped completely in the pagination.

booklet (fols. 291r–94r) with "XVIII" in the upper right-hand corner, indicating the number of the booklet in the volume (see Plate 1).

We are very fortunate in that the main copyist of the manuscript, Simon Forman, has dated the booklet, and the Treatise in particular, on several occasions. In fol. 306r, Forman has added a colophon: "1590. Noviter scriptum per Simonem Forman. Iuni i9" '1590 written again by Simon Forman June 19.'[4] The date "1590" also appears in fol. 320r and fol. 331v; the latter dating indicates that the whole booklet was finished in 1590. The watermark of the paper does not provide any additional information: it is a sphere, which is close but not identical to Briquet 14055, dated 1598.[5] This watermark is not found elsewhere in the manuscript. All bifolia except those of fols. 308–13 (totaling nineteen leaves) contain a leaf with a watermark (see also below for the quires of the manuscript).[6] Of the leaves that seem to be singletons, fol. 329 contains a watermark, whereas fols. 330 and 331 do not.

The tightness of the binding makes it impossible to determine the gatherings with certainty. The booklet seems to consist of one main quire, comprising eighteen bifolia (fols. 293–328). Fol. 329 appears to be a singleton, but it may have been a bifolium at one point. The first leaf of the bifolium would have appeared between fols. 292 and 293, which contain the dedication of the Treatise. The loss of this leaf would conveniently explain why sixteen stanzas of the dedication are now missing in A. Fols. 291 and 292 may be two singletons or one bifolium. The two final leaves of the booklet, fols. 330 and 331, appear to be singletons. The counter-leaf of fol. 330 is probably a torn leaf, a piece of which is still remaining in the manuscript. This leaf was most probably without writing, since the poem "The way to calcen perfectly," which begins in fol. 330, continues uninterrupted in fol. 331r. Finally, although now the last leaf in the booklet, fol. 331 may not have been so originally: the poem "The way to calcen perfectly" appears to end imperfectly, lacking four stanzas that appear in S2's and F's versions of the poem. Importantly, such a missing leaf may also have contained some of the items that are found in F and S2 but are now missing in A.[7] On the other hand, since Forman has signed

---

[4] This note appears after Chapter 3 of the Treatise. Since it does not appear at the very end of the booklet, the note should probably not be taken as an indication that the whole of the booklet was finished on exactly 19 June. Another slightly ambiguous aspect of the note is the meaning of "noviter." I have interpreted it as meaning 'again.' The meaning of "noviter scriptum" would then be 'copied.' However, "noviter" may also have the meaning 'in a new way,' though this seems unlikely, especially since A is not radically different from any of the other copies.

[5] Charles-Moïse Briquet, *Les filigranes*, The New Briquet, Jubilee Ed., 4 vols. (Amsterdam: Paper Publications Society, 1968).

[6] The following folios contain watermarks: 291, 295, 296, 297, 303, 306, 309, 310, 314, 316, 317, 319, 320, 321, 322, 323, 327, 328, 329.

[7] See p. 15.

"Τελως 1590 S. Forman" 'end 1590 S. Forman' in red ink in fol. 331v, it is perhaps more likely either that the poem in fol. 331 was not copied in its entirety, or that the exemplar of A did not contain the four final stanzas of the poem.

The booklet is written in four main hands (see Plates 1–3).[8] The majority of the booklet, and especially the Treatise, is written in Forman's hand, a fairly regular secretary hand with some italic features. There are also some additions in his hand that seem to have been supplied at a later date, a common phenomenon in texts copied by Forman.[9] These additions are in brown-black ink (as opposed to black in Forman's original writing); the letter forms are slightly more rounded, written perhaps with a quill with a broader nib. They seem to include the chapter headings of Chapters 21–23 and 27, which are not found in any of the other copies, and some marginal notes.[10] A number of corrections and glosses may also belong to these later additions.[11]

Besides Forman's hand, the hands of three unidentified copyists are found in the booklet (see Table 5.1 below). Two of the hands occur only in the Treatise. Copyist 2 writes in a small rounded secretary hand with many idiosyncratic letter forms (see Plate 2); this hand is also characterized by many peculiar words (most of them seemingly errors of copying) and spellings. Copyist 3 writes in a small angular secretary hand, exhibiting many flourished letter forms, and seemingly otiose tildes (see Plate 3). The hand of Copyist 4 is attested only in five lines of the manuscript (see Plate 2), and is similar to that of Copyist 3, but the two are clearly distinguished by some letter forms, the use of abbreviations, and orthography.

It is obvious that the four copyists of A must have been working on the manuscript at roughly the same time. On several occasions, the hands of two or more copyists are attested on one and the same page in the Treatise, the dedication, and the additional texts in the booklet: in fols. 292r, 294v, 329v, 330r, Forman and Copyist 2 have both copied out text; in 295r Forman and Copyist 3; and in 298r Forman, Copyist 2, and Copyist 4. Forman sometimes corrects the work of Copyists 2 and 3, which illustrates that he was supervising their copying to ensure accuracy.[12] As has been argued earlier, the multiple copyists and their tag-team cooperation makes manuscript A unique or at least very unusual among the extant manuscripts that once belonged to Forman.

In addition to Forman, another identifiable copyist has made contributions to the Treatise, though not to the text per se. Dr Richard Napier, Forman's one-time disciple and longtime friend, has added a recipe in fol. 324r between the

---

[8] See also p. 23, for a discussion.
[9] See Cook, *Dr Simon Forman*, xii.
[10] See 299v: 34, 300r: 9, 316r: 15, 34, 318r: 15, 320r: 22, 321v: 3, 15, 323r: 39.
[11] See 307v: 31, 310v: 6, 17, 18, 26, 311r: 11.
[12] See e.g. 296v: 31.

heading of Chapter 28 and the main text,[13] and marginal comments, underlining, and an interlinear gloss appear in his hand.[14] Attesting to Napier's reading of the Treatise, these additions were presumably made when A was in his ownership (see below).[15]

---

[13] Versions of the same recipe also appear in Napier's hand later in the manuscript (fol. 352v).

[14] See 310v: 16, 19, 20, 21, 23, 311r: 35, 324r: 32, 324v: 33, 35, 325r: 2, 5, 6, 18.

[15] I have compared the handwriting in A with autograph letters by Napier in MS. Ashmole 1730 to verify that the notes in A are by Napier.

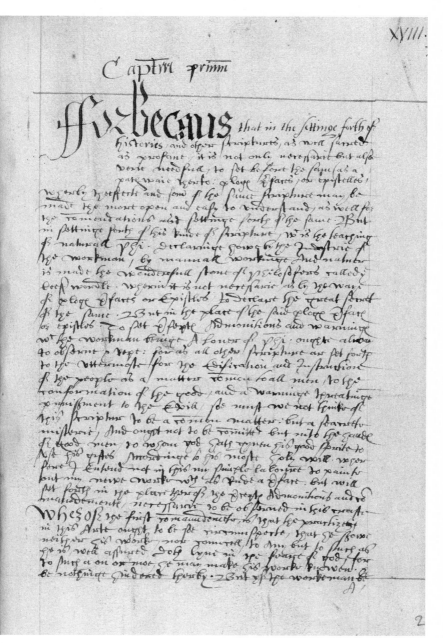

PLATE 1.
Bodleian Library MS. Ashmole 1490, fol. 294r (Simon Forman).
Reproduced with the permission of The Bodleian Library, University of Oxford.

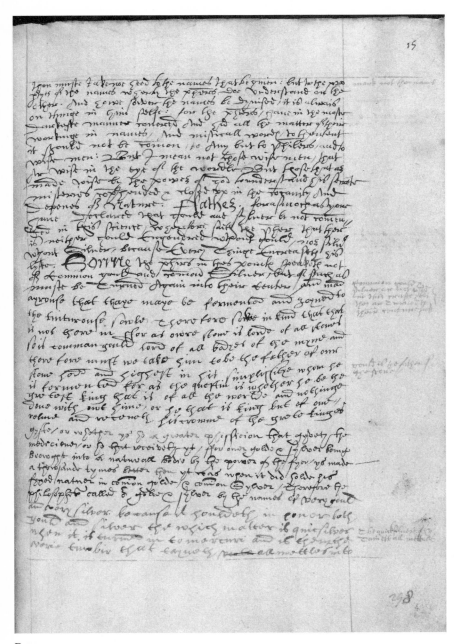

Plate 2.
Bodleian Library MS. Ashmole 1490, fol. 298r (Simon Forman, Copyist 2, Copyist 4).
Reproduced with the permission of The Bodleian Library, University of Oxford.

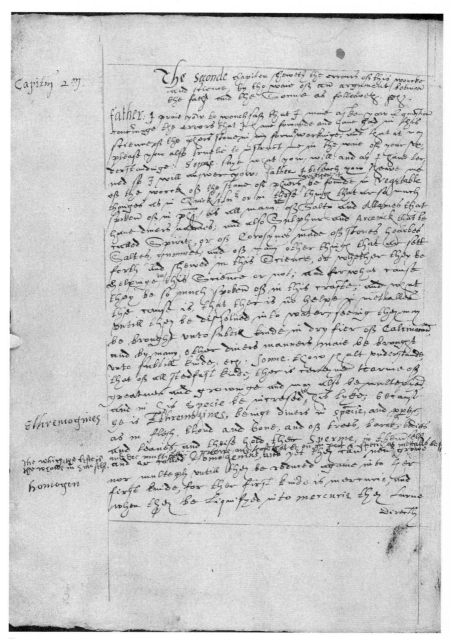

PLATE 3.
Bodleian Library MS. Ashmole 1490, fol. 296v (Copyist 3 with additions by Forman).
Reproduced with the permission of The Bodleian Library, University of Oxford.

The contents of the booklet and the hand in which the particular items are copied are shown in Table 5.1.

TABLE 5.1. CONTENTS OF A (FOLS. 291R–331V)

Folios	Item	Copyist
291r–293r	Dedication (stanzas 35–50 missing)	Simon Forman: Address, Stanzas 1–18 Copyist 2: Stanzas 19–34, 51
294r–325r	The Treatise (untitled)	A. Simon Forman — 294r–v: 1 (ends with "whom") — 295r: 1–20 (ends with "other") — 298r: 1–19 (ends with "their center") — 298v–325r (except 306v; see E) B. Copyist 2 — 294v: 1(begins with "he")–36 — 297r–297v — 298r: 19 (begins with "and mad")–29 — 298r: 35 (begins with "of very gould")–39 C. Copyist 3 — 295r: 20 (begins with "is")–296v D. Copyist 4 — 298r: 30–35 (ends with "by the names") E. Blank page — 306v
325v	"The standing of the glass in the fier" "Finis q^d H L Et rescriptu*m* *per* S Forman"	Simon Forman
326r–327v	"Here followeth the sain*ges* of ph*ilosofe*rs"	Simon Forman
328r	"To calcyne truly"	Simon Forman
328r	"The calcynation of the salte"	Simon Forman
328v	"To drawe the quintescentia of antimonie"	Simon Forman
328v	"To make sal armoniak"	Simon Forman
329r	"I haue also hearevnto added sertayne abreuiaciones in myter"	Copyist 2

Folios	Item	Copyist
329r	"When the ferment thow wilt p*r*epare"	Copyist 2
329r–329v	"Of the stone and ferment"	Copyist 2 (Stanzas 1–2, 3 [ll. 1–3]) Simon Forman (Stanzas 3 [ll. 4–6], 4)
329v–330r	"De auro potabile"	Simon Forman (Stanzas 1–5, 6 [l. 1]) Copyist 2 (Stanzas 6 [ll. 2–7], 7)
330r–330v	"Replies castele"	Simon Forman (Title, Stanza 1 [ll. 1–2]) Copyist 2 (Stanza 1 [ll. 3–7], 2–7)
330v–331v	"The way to calcen perfectly"[A] "Τελως 1590 S. Forman"	Copyist 2 (Title, Stanzas 1–5, 6 [ll. 3–7], 7–10) Simon Forman (6 [ll. 1–2], signature)

[A] In Black, *A Descriptive*, 1367, this poem is included under "Replies Castle." However, it is clearly a separate item, as is indicated by the title in the manuscript and its appearance as a separate item in other manuscripts.

A's writing space is framed, the length varying between 195 and 220 mm, and the width between 135 and 165 mm. The number of lines on full pages goes from thirty-three to fifty-one, and writing is frequently found below the lower frame. Comments abound in the outer margins throughout the Treatise, mostly in Forman's hand and often written in bright red ink. The comments are rarely evaluative or analytic; they commonly repeat phrases or clauses in the text, probably for the purpose of drawing attention to the passage. Only on rare occasions do they provide explanations and clarifications of some readings. In Chapters 1–3 of the Treatise, Forman uses an elaborate scheme to highlight passages: a line in red ink sometimes appears between two lines in the manuscript, extending from a section in the manuscript to a corresponding note in the margin. The marginal comment here seems to serve almost as a heading to what follows. This scheme seems to have been more or less abandoned after Chapter 3, unless Forman found less of note in the later chapters. Forman also employs red ink and/or a larger type of script (often italic script instead of the normal secretary) in the text proper to emphasize words or phrases, or to structure the text. The whole manuscript, especially the parts copied by Forman, thus gives an impression of careful copying, with special attention paid to layout features and the physical appearance of the manuscript. The neat appearance of A is not unusual for books copied by Forman. As Barbara Traister remarks, "[m]any of his manuscript books seem ready for a printer, neatly copied, ruled, and sometimes even marginally referenced [. . .]."[16]

The exact history of A is unfortunately unknown, but it is likely that it followed the course of the majority of extant manuscripts once belonging to Forman. After Forman's death in 1611, most of his books and manuscripts seem to have come into the hands of Richard Napier, after whose death the books passed on to Napier's nephew Sir Richard Napier (1607–1676).[17] When Sir Richard died in 1676, his son Thomas Napier (*b.* 1646) gave the books to Elias Ashmole, who after acquiring the books and manuscripts had the booklets bound up in the bindings most of them appear in today. Ashmole's books were subsequently bequeathed to Oxford University, deposited in the Ashmolean Museum after Ashmole's death in 1692, and later transferred to the Bodleian Library.[18]

London, British Library MS. Sloane 299 (S2)

S2 contains fifty-two paper leaves, measuring approximately 340 × 225 mm, foliated in the upper right-hand corner in Arabic numerals written in pencil. An earlier pagination is also visible in the upper right-hand corner. In fols. 20r–52r, this earlier pagination replaces an even earlier foliation in Arabic numerals starting from one. This foliation is of particular significance, since these are the folios occupied by the Treatise, the dedication, and some of the recipes and poems most probably produced by Lock. In addition to this earlier foliation, there are other indications that the Treatise and some of the concomitant items originally constituted a separate item or that they appeared elsewhere. Written in a seventeenth-century hand, probably that of the one-time owner Gabriel Gostwyk (see below), a note at the beginning of the manuscript (fol. 1r) reads "Humfry Locks Collections of Alchime who were the first p. 19."[19]

---

[16] Traister, *The Notorious Astrological Physician*, 123.

[17] Traister, *The Notorious Astrological Physician*, xiv–xv.

[18] Josten, *Elias Ashmole*, 1: 301.

[19] Although it seems clear that the note points to a different placement of the Treatise and the concomitant items at some point, there are problems in interpreting the note. It is unclear what "p. 19" refers to. There is a possible way of accounting for this reading if "p. 19" is taken as meaning or as a mistake for *fol. 19*. On fol. 1r, Gabriel Gostwyk, who most probably added the note in question, also provides a table of contents and a statement of how many pages the manuscript contains. According to the statement of the number of pages, S2 comprises "102 pages" (fol. 1r). However, in its present foliation S2 contains 104 pages (52 folios). This discrepancy can be explained if Gostwyk started his count on fol. 2r, which is the first page containing text. If this is the case, fol. 19 would be fol. 20 in the foliation that S2 now has. This is the folio where the dedication of the Treatise begins. However, it must also be recognized that the note may not refer to S2 but to a phenomenon in another manuscript.

The question is when and where the Treatise and the additional material that follows it appeared as the initial items in a manuscript codex. The quiring of the manuscript offers no clear clues to its history, especially as regards the placement of the items in question. In the present binding, the first three quires contain eight folios each; quires four and five contain six each; quire six consists of fourteen folios; and finally, quire seven has two folios. In this binding, the dedication and the beginning of the Treatise appear in the same quire as another text: "The moste excellente and trewe booke of the Reuerent Docter Almante & Lorde Barnarde Earle of Marche and Travison of the philosophers stone" (quire three). But it is not certain that this quiring is original. In 1972, S2 was rebound and given a new cover. The Manuscript Superintendent of the Manuscript Reading Room at the British Library assures me that the quires were not changed in the rebinding of the book.[20] Nevertheless, since the leaves are mounted on stubs, it is impossible to ascertain whether any of the leaves now fitted together were once singletons. Perhaps, like A's version, the S2 Treatise and its accompanying poems and recipes originally constituted an independent booklet, which was later incorporated into what is now S2. If the note by the seventeenth-century hand refers to the version of the Treatise and the concomitant texts in S2, this must have happened as early as the seventeenth century.

The foliation and the state of the quires are also of relevance for the dating of the manuscript. One of the texts in S2 is a translation of Heinrich Khunrath's *Confession*, stating "giuen forthe at Magdeburghe 1596," which was the year of the first Latin edition.[21] Although this text is written in a different hand from the one found in the Treatise, it is the same hand as the Bernard of Trevisa text, with which the beginning of the Treatise shares a quire. The existence of the *Confession* in the codex may thus indicate that the Treatise was not copied until after 1596, but this is not necessarily so if the Treatise did not earlier occupy the place that it now occupies (as suggested above). Important in this context are the watermarks of the paper: many of the leaves contain a watermark in the shape of a letter 'B' inside a shield, with the name "Nicholas Lebe" written beneath the shield.[22] The watermark is very close to Briquet 8080, dated 1578, but it does not have the same back-to-front writing as the Briquet watermark. Although the date 1578 thus seems to be a fairly safe *terminus post quem* for S2, it may have been written as late as after 1596.

The manuscript is written in three main hands, all probably late sixteenth-century secretary hands: 2r–2v (Hand 1); 2v–19r (Hand 2); 20r–52r (Hand 3). The copyists are all unknown. The initials "W. G." appear after two of the texts in Hand 2 (fols. 2v, 18v), but it is unclear what the initials stand for. Hand 3,

---

[20] Personal communication 2002.

[21] James B. Craven, ed., *Doctor Heinrich Khunrath: A Study of Mystical Alchemy*, limited ed. (Glasgow: Printed and Bound by Adam McLean, 1997), 28.

[22] For this watermark, see also S1.

which copied out the Treatise, writes in a small professional-looking engrossing Elizabethan secretary hand.[23] In addition to the three main hands, there are two to three supplementary hands. A seventeenth-century hand, mostly like that of the one-time owner Gabriel Gostwyk, adds an index and comment on fol. 1r and single-word comments in the margin on a large number of folios. There is also a later hand that has added a number of titles in pencil to the alchemical texts in the volume, including the Treatise.

The contents of the manuscript are given in Table 5.2.

TABLE 5.2. CONTENTS OF S2

Folios	Items
1r	List of contents, signatures, notes
1v	Blank
2r–2v	Alchemical recipes
3r–7v	"The practise of the greate worke by the mouncke of Berrye"
8r–10r	"The confession of Henry Khunrathe of Lippeswicke Docter of ether phisicke"
10v–19r	"The moste excellente and trewe booke of the Reuerent Docter Almante & Lorde Barnarde Earle of Marche and Travison of the philosophers stone" [also includes illustrations from the *Rosarium* cycle on fols. 16v–17v; two of them have been crossed out.]
19v	Blank
20r–22v	Dedication
23r–42v	The Treatise
42v	"I haue also hearvnto added certayne abreuiacions in miter"
42v	"Of fermente	When thy ferment thow wilt prepare"
42v	"Of the stone and fermente &c"
42v	"Of aurum potabile &c"
43r	"Ripplies castell"
43r–v	"The way to calcine perfectly"
43v–50r	George Ripley's *Compound of Alchymy*
50v	"The standinge of the glas..."
50v–51r	"Heer followeth the sawes of philosophy"
51r–51v	"To calcine lune &c"
51v	"The calcynacion of {the} salte"

---

[23] Anthony G. Petti, *English Literary Hands from Chaucer to Dryden* (London: Edward Arnold, 1977), 17.

51v	"Hereafter follow*e*th a bal for alkamist*es* to play withall &c"
51v	"To make flos eris"
51v	[Title cut away] *inc.*: "Take the flower of Venus..."
51v	"The sublimac*i*on of sall armo*ni*ac for the redd tincture followinge &c"
51v	"Heer follow*e*th the 5intaecencie of antimony &c"
52r	[Title cut away] *inc.*: "Take the oyle of crocumferre..."
52r	"To make sall armoniac"
52r	"After the putriffac{c}*i*on and congelac*i*on{s} of the ferment..."

There are faint traces of a frame ruling in S2. The measures of the writing space vary: the length is between 305 and 313 mm, and the width is between 155 and 168 mm. The number of lines per page varies between forty-five and fifty-eight. There are very few marginal comments in the Treatise; most of them explain abbreviations in the text.

Little is known about the early history of the manuscript. The index to the Sloane collection records S2 being in the early seventeenth century in the ownership of Gabriel Gostwyk, whose name appears on fol. 1r.[24] He may be the same as the Gabriel Gostwick that entered Sidney College, Cambridge, on 4 June 1627, at the age of fourteen, received his BA and MA in 1630/1 and 1634 respectively, and became minister of North Tawton, Devon.[25] However, I have found no evidence in standard reference works, nor in genealogical surveys to corroborate this connection.[26] In addition to S2, another manuscript of the Treatise seems to have been in Gostwyk's ownership, S1, together with eight other manuscripts: MSS. Sloane 249, 393, 398, 403, 404, 405, 521, and 634. These manuscripts contain medical writings in Latin, Middle English, and early Modern English; S2 and S1 are the only alchemical volumes. In most of the manuscripts, the same hand, presumably the hand of Gabriel Gostwyk himself, writes the name Gabriel Gostwyk, adds a note on the number of pages in the codex, and supplies a list of contents. After being in the ownership of Gostwyk, S2 in due time found its way into the collections of Sir Hans Sloane (1660–1753), and was part of the collection of manuscripts that came to make up the original collections of the British Museum (now the British Library).[27]

---

[24] Edward John L. Scott, *Index to the Sloane Manuscripts in the British Museum* (London: The British Museum, 1904), 222.

[25] J. A. Venn, *Alumni Cantabrigienses*, 4 vols. (Cambridge: Cambridge University Press, 1922–1927), 2: 243.

[26] Frederic T. Colby, ed., *The Visitation of the County of Devon in the Year 1620*, Publications of the Harleian Society 6 (London: The Harleian Society, 1872); Stuart A. Raymond, *Devon: A Genealogical Bibliography*, 2nd ed. (Exeter: Raymond Genealogical Bibliographers, 1994).

[27] *ODNB*, s.n. Sloane, Hans.

Glasgow, Glasgow University Library MS. Ferguson 216 (F)

F comprises fifty-four paper leaves, paginated from one to one hundred and seven (one hundred and eight is unpaginated) in Arabic numerals. The leaves measure approximately 335 × 206 mm. The volume is written in the same very regular seventeenth-century hand throughout, which gives the impression that the manuscript is a fair copy or at least a very polished copy. The appearance of a marginal note written in this hand on p. 90 suggests that the manuscript was not completed before April 1614 (although the exact meaning of the note is unclear): "in cangar[?] of S. f. at Ih[?] ye 1^s day of Aprill 1614." The watermarks of the paper regrettably provide very inconclusive evidence for the dating of the manuscript, even as watermarks go. The main watermark is a foolscap accompanied by a countermark in the shape of the letters "PHO" in an interlaced oval form. Although there is no exact match, the closest fit in watermark collections is a foolscap found from the late sixteenth century to the eighteenth century, though most commonly in the seventeenth century. Neither have I been able to find the exact countermark. Used as a countermark, the letters "PHO" are listed together with other types of main watermarks by Edward Heawood, although never enclosed in an oval form as in F.[28] The countermarks given by Heawood stem from London, in 1682, 1684, and 1670 respectively. The watermark evidence is thus inconclusive, but the countermark might point to a date as late as the 1670s or 1680s, although the handwriting seems to indicate an earlier date in the seventeenth century.

Several names appear in the manuscript, but most of them seem to be names of the originators or authors of individual recipes or other items. The only name that appears in the Treatise is "John Atherton" (p. 36), which occurs after the third chapter together with a quote from the Bible (Philippians 4: 11–13). This John Atherton does not appear in standard reference works.

The Treatise takes up pages seven to eighty in the manuscript, and is followed by a large number of other items, including some of the poems and recipes presumably written by Lock. The writing space in the Treatise is framed, and the Treatise is heavily annotated in the same hand as the manuscript proper. The annotations comprise clarifications, highlighting, and cross-references. Hands with pointing index fingers frequently appear in the margins with different-sized and different-shaped cuffs. The annotation reveals a meticulous reading and analysis of the Treatise.

The early history of F is unknown. It was formerly in the possession of John Ferguson (1838–1916), Professor of Chemistry at the University of Glasgow, and was acquired by Glasgow University Library in 1921.[29]

---

[28] Edward Heawood, *Watermarks, Mainly of the 17th and 18th Centuries*, Monumenta Chartae Papyraceae Historiam Illustrantia 1 (Hilversum: Paper Publications Society, 1950), nos. 374, 637, 2268.

[29] Glasgow University Library at http://special.lib.gla.ac.uk/manuscripts (as accessed in 2006).

Copenhagen, Royal Library of Copenhagen MS. Gamle Konglige Samling (Gl. Kgl. Saml.) 1784 (G)

G contains forty-four paper leaves, measuring approximately 193 × 139 mm. Several features of the manuscript contribute to an idiosyncratic appearance, starting with the cover of the volume, which consists of a piece of parchment covered with scribbles in a number of hands. The parchment contains an earlier text starting from the left-hand side of the back of the cover. Unfortunately, the text is largely illegible apart from scattered words. The end of the document, where a stretch of text is legible, provides some very interesting information, suggesting that the text was of a legal nature: "the viith day of Marche in the xxxith ye<ar> of the regne of our so<v>erayne Lady Elizabeth by the grace of God Queene of England France & Ireland &c in p*re*sens of those whose names are subscrybed." This statement is followed by a number of more or less illegible signatures claimed to be "atturneys."

G contains a very limited set of texts, with the Treatise as a kind of centerpiece. Preceding it are twelve alchemical recipes on fols. 1r–3r (3v is blank). Following the Treatise, on fol. 44v, are two more alchemical recipes, preceded by a blank page (fol. 44r). The Treatise appears in between, on fols. 4r–43v. What is noteworthy is that the Treatise contains an earlier foliation from one to forty-one (thirty-three has been omitted) in the upper right-hand corner in Arabic numerals; G is otherwise unfoliated. This foliation suggests that the Treatise previously appeared elsewhere, or that the alchemical recipes preceding and following the Treatise are later additions to the codex. The date given on the binding indicates that the items as they now appear were not bound together until after 1589 (the thirty-first year of Elizabeth's reign), but there are no clear signals when or where the Treatise appeared as the first item in a codex. Previous dating of the manuscript is problematic. Irma Taavitsainen describes G as "[a]n early 16th-century alchemical book."[30] She informs me that the dating was made on the basis of the general impression of the manuscript.[31] Since there is very strong evidence that the Treatise was compiled in the 1570s, it is very unlikely that G is from the early sixteenth century. I have found no codicological or paleographical indications of such a date. In fact, the handwriting is consistent with handwriting in the second half of the century.[32] But in the absence of firm evidence, the date of G must remain uncertain.

The copyist of G is unknown. The name "Ja. Standysh" appears in an interlaced pattern upside down on fol. 1r, and "I" (for James?) occurs in a small

---

[30] Taavitsainen, *The Index*, 13.
[31] Personal communication 1999. It was not possible for me to check the watermarks, if any, of the paper used in G when I inspected the manuscript.
[32] See also Grund, "In Search of Gold," 274–76.

secretary hand on the inside of the back of the cover. It is unclear if this is the name of the copyist or possibly an owner.[33]

The history of G is largely unknown. The Royal Library may have purchased it in the seventeenth century at the behest of the Danish king Frederik III, who was very interested in alchemy.[34]

London, British Library MS. Sloane 288 (S1)

Comprising two hundred and thirteen paper leaves, S1 is a voluminous manuscript book, measuring approximately 355 × 225 mm. The Treatise appears as the first item in the codex (fols. 1r–32v), foliated in the upper right-hand corner from one to thirty-two. There is also an earlier foliation from fourteen to forty-five, possibly in the hand of the copyist of the Treatise. This earlier foliation is important in that it illustrates that the first chapter and most of the second chapter, which are now missing, were originally present in the version. It is likely that these leaves were missing at an early stage, since Gabriel Gostwyk, who seems to have owned the manuscript in the seventeenth century, adds his name and a note on the number of pages on fol. 1r (14r in the old foliation).

One of the most interesting aspects of S1 is that its paper contains exactly the same watermark as S2's paper, the shield described earlier. Like S2's version of the Treatise, S1 is also written in a small adorned late sixteenth-/early seventeenth-century secretary hand, though not the same as S2's. These features may point to a common origin of the two, but what exactly that origin was is difficult to pinpoint. The common paper stock may indicate that they were professionally copied or at least commissioned.[35]

Although the watermark suggests the *terminus post quem* of 1578, other factors need to be taken into consideration. On fol. 1r, there is a note stating "Johannes Pauper anno 1600 p. 53." In the old handwritten catalogue of the Sloane collection, kept in the Manuscript Reading room at the British Library, this date has been taken as an ownership note. However, it is more likely that it refers to an English version of the *Breviloquium* ascribed to Johannes Pauper that follows

---

[33] There is an interesting connection here to another manuscript. The name *James Standysh* also appears in British Library, MS. Additional 32621, a version of the famous Ripley Scroll. The signature is very similar to the signature found in G. The handwriting of the manuscript is, however, very different from that of G. Perhaps Standysh was the owner of both manuscripts (if it is indeed the same person), or perhaps he is even the copyist of one of them. The connection deserves more attention than the scope of this study allows.

[34] Taavitsainen, *The Index*, xiv–xv.

[35] See p. 105, for further discussion.

the Treatise.³⁶ Although it is written in a hand different from that of the Treatise, some of the leaves in the *Breviloquium* contain the same watermark as in the Treatise.³⁷ Exactly what 1600 refers to is unclear, but it may indicate that the Treatise was copied out sometime between 1578 and 1600.

The copyist of S1 is unknown. He appears to have returned to his text sometime after it was finished, and corrected the text on a number of occasions in red-brown ink.³⁸ Also, after stints by other hands, the Treatise copyist reappears in a number of other places in S1, most notably on fols. 199r–205r, where several of the items that are shared by other manuscripts of the Treatise are found.³⁹ Several other hands have made contributions to the Treatise. In addition to the seventeenth-century hand that seems to belong to Gabriel Gostwyk (as noted above), there is a note in pencil(?) in the bottom left-hand corner on fol. 32v: "Her Note E lib[?] Johns[?]." This note may simply refer to the text that follows the Treatise, written by John Poore (or Johannes Pauper).⁴⁰ This hand seems to be the same as the second supplementary hand that provides some notes in S2.

The known history of S1 is basically the same as that of S2. After being in the ownership of Gabriel Gostwyk, it ended up in the collections of Hans Sloane, and later on in the holdings of the British Library.⁴¹

---

³⁶ Singer, *Catalogue*, 191–92. The *Breviloquium* does not appear until page 66 in the present binding (or on page 92 according to the old, original foliation). It is possible that the reference to p. 53 is simply a mistake, but it might suggest that the note refers to a text or phenomenon outside S1.

³⁷ The collation of the manuscript is not of any help in this case. The manuscript seems to have been rebound at some point, most likely before the manuscript came into the ownership of the British Library, as the Manuscript Superintendent of the Manuscript Reading Room at the British Library informs me: personal communication 2002. The quires now mostly consist of singletons or single bifolia. However, in fols. 8v, 16v, 24v, and 32v there is numbering from two to five (possibly in Gabriel Gostwyk's hand). Perhaps this is an indication of an earlier binding. If that is the case, there are four quires of four bifolia each. However, in this sequence, the number one is missing. Possibly, this number appeared on an earlier page, indicating that an entire quire is missing. If the original foliation, which begins with 14, is correct, then this quire must have contained six bifolia and a singleton leaf, or some other constellation of folia that is not recoverable.

³⁸ See p. 120.

³⁹ As mentioned earlier, the quires of S1 might have been changed. However, the poems and recipes do not seem to have immediately followed the Treatise in the codex at any point. The poems and recipes begin in the lower part of fol. 199r. The top of 199r is occupied by the end of unrelated material that begins in fol. 198v.

⁴⁰ There are several other instances in S1 where this hand adds references at the end of one text indicating the beginning of the next.

⁴¹ For details, see the description of S2.

London, British Library MS. Sloane 3180 (S3), and MS. Sloane 3684 (S4)

Since S3 and S4 display quite a number of features in common, I have merged the descriptions of the two manuscripts. Densely written in one unidentified seventeenth-century hand throughout, S3 is a fairly small manuscript codex whose paper leaves measure approximately 226 × 175 mm. A foliation in the upper right-hand corner in red ink replaces an earlier pagination. The binding is very tight, and several words are illegible in the inner margins of the pages. Similarly, owing to the tight binding, the watermark of the paper is difficult to discern, as it appears in the inner margin and is cut in half by the binding. It appears to be a shield incorporating a fleur-de-lis, with a crown on top of the shield. The letters 'WR' appear at the end of a line at the bottom of the shield. The watermark resembles Briquet 7210, dated 1585. However, variations on this particular watermark are found as late as the second half of the seventeenth century.[42] A seventeenth-century date (perhaps even second half) is more in line with the general appearance of the handwriting.

S4 contains fifty-two paper leaves measuring approximately 213 × 145–150 mm. Originally paginated from one to ninety-two, the manuscript contains a later foliation in Arabic numerals in pencil from one to fifty-two. The remaining leaves (fols. 47r–52r) are paginated from one to ten. There are three or four different seventeenth-century hands in the manuscript: Hand 1 fols. 1r–46v (including the Treatise) and fol. 52r; Hand 2 fols. 47r–49v (which may possibly be two hands: 47r and 47v–49v); and Hand 3 fols. 50r–51r. Encrypted writing in a darker ink than that of the main hand appears in the margin of some of the texts, predominantly in the Treatise. It is unclear whether these encrypted notes have been added by an additional hand or by the same hand as the Treatise, but possibly at a later time.

Like S3's watermark, S4's watermark also appears in the inner margin and is often difficult to make out owing to the tight script. It is a crescent form with a cross on the top. I have not been able to find this exact watermark in standard collections. The shape of the cross and the crescent resembles Briquet 5249–51 and 5253–54, suggesting a date between 1549 and 1591. However, S4's crescent does not contain letters, which the Briquet watermarks do. In addition to this inconclusive evidence, the general appearance of the handwriting clearly indicates a seventeenth-century origin.

---

[42] William A. Churchill, *Watermarks in Paper in Holland, England, France etc. in the XVII and XVIII Centuries and their Interconnection* (Amsterdam: Menno Hertzberger, 1935), nos. 427–28.

S3 and S4 are intimately connected: they have the same repertoire of alchemical symbols, and they also correspond in a number of layout aspects. More importantly, they exhibit almost exactly the same content and writer colophons.[43]

Table 5.3. Contents of S3 and S4

Text	S3	S4
The Treatise	fols. 1r–11v	fols. 1r–36r [36v blank]
Albertus Magnus's *Compositum de compositis* [In English]	fols. 12r–13r	fols. 37r–41r [41v blank]
Arnold of Villanova's *Perfectum magisterium* (?) [Incomplete; in English]	fols. 13v–14v	fols. 42r–46v
Arnold of Villanova's *Semita semitae* (?) [In English]	fols. 15r–15v	fols. 47r–51r

S4 contains a final text in fol. 52r "A Tincture or Elixir of an unknowne ~~Author~~ Philosopher," which is not found in S3, whereas the texts in S3 are followed by some hundred blank leaves. In addition to the textual similarities, the two manuscripts share the following notes:

> The which said litle booke or work I, Edward Dekingston, copied out verbatum as above written with the hand of Mr Morryse; which said work was bound in with the litle booke of mr Guyllyam de Cenes that he sent to the Reverend Father the Archbishop of Raynes, & was brought vnto me to coppie by Mr Anthony Brighame. [S4, fol. 41r; cf. S3, fol. 13r]

> The which essencial & accidental questions was copied out by me, Edward Dekyngstone, forth of an olde booke, writen with the hand of one John Norryse, a practicsioner, that dwelt by St Joneses the 8th of March 1582, & was brought vnto me by one Anthony Brygeham, from whome I had this same aforesaid coppie, & also of his aforesaid essential & accidental questions, which I also in like manner coppied out verbatum, so near as I could. [S4, fol. 46v; cf. S3, fol. 14v]

These colophons and especially the mention of Edward Dekyngstone are particularly important since they connect the manuscripts with several other alchemical manuscripts. The index to the Sloane collection also records the name Edward Dekyngston in MS. Sloane 2170 (fol. 66r), where a note is followed by "E. Dekyng<s>ton," which is now barely legible.[44] Whether this is actually the same

---

[43] A collation of these additional texts would undoubtedly provide further interesting information on the relation of these manuscripts. However, the scope of the present study does not allow such a investigation.

[44] Scott, *Index*, 138.

man remains unclear. A stronger connection, not noted by the Sloane index, is found between S3/S4 and MS. Sloane 3630. Parts of MS. Sloane 3630 (fols. 70r–86r) are written in Hand 1 of S4, including Albertus Magnus's *Compositum de compositis* in the same hand and with exactly the same layout as S4 followed by the same colophon as in S3 and S4. Sloane 3630 also presents an additional text entitled "A tretice of Count Treversan" in Hand 1 of S4. Perhaps the most striking link, however, is to Bibliotheca Philosophica Hermetica MS. M199 (BPH M199). This manuscript has been described as the commonplace book of Edward Dekyngstone, and his name appears at least twice in the codex. Although it remains unclear whether Dekyngstone has actually copied out some of the texts in BPH M199 or whether he simply owned it, it is clear that Dekyngstone was a late sixteenth-/early seventeenth-century alchemist.[45] The Sloane index suggests that he was based in London and gives the date 1582–83. This information seems to derive from the colophons cited above, and must be treated with some skepticism. The date given in the colophon seems to refer to when the book was copied by John Norryse or Morryse, the practitioner whose book Dekyngstone refers to as old, rather than to the date when Dekyngstone made his copy, and the connection to London is uncertain.[46]

Extracts: Oxford, Bodleian Library MSS. Ashmole 1407 (E1), Ashmole 1424 (E2), and Ashmole 1494 (E3)

Extracts of the Treatise appear in three manuscripts: E1, E2, and E3. E3 (fols. 160r–160v), which contains extracts from chapters 2, 3, and 27 of the Treatise, was produced by Simon Forman, probably in the early seventeenth century.[47] E1

---

[45] For more information on this manuscript, see Peter Grund, "A Previously Unrecorded Fragment of the Middle English *Short Metrical Chronicle* in Bibliotheca Philosophica Hermetica M199," *English Studies* 87 (2006): 277–93, here 278–80.

[46] Glasgow University Library, MS. Ferguson 205 may also have been owned by the same Dekyngstone; see Halversen, "The Consideration," 73. In the provisional catalogue of the Ferguson collection available at http://special.lib.gla.ac.uk/ferguson/fergms201–250.html, it is suggested that the Dekingstone of MS. Ferguson 205 may be Edmund Dickinson, whose name appears in a large number of manuscripts in the Sloane collection. Among the manuscripts referred to in the entry is Sloane 3630 (but not S3 and S4). This Edmund Dickinson is most likely the alchemist and physician of Charles II who lived between 1624 and 1707; see *ODNB*, s.n. Dickinson, Edmund. However, I have not been able to verify a connection between Edmund Dickinson and the Edward Dekyngstone whose name appears in the three Sloane manuscripts. I have inspected an autograph manuscript by Edmund Dickinson in British Library MS. Stowe 748 (fol. 209r). The handwriting is not that of any of the manuscripts discussed here.

[47] The item that follows the extracts in the manuscript is dated 1607. However, this item appears on a different leaf, and it may not have been written at the same time. The leaf (fol. 160) does not appear to have a watermark.

and E2 contain two extracts each. E1 (Part 2, pp. 17–21) presents a version of the first chapter of the Treatise under the title "A declaration of certaine preceptes or comandements to be observed in the arte of alchimye by all practitioners herin and also rules necessary to be deligently noted of all young philosophers or practitioners in naturall philosophy"; and pp. 39–50 contain Chapters 4–5[48] and 10–17 under headings very similar to the ones found in the Treatise.[49] Similarly, E2 comprises Chapters 15–17 (fols. 31v–34r), and chapters 4–5 and 10–14 (fols. 38r–41v), which also appear with headings similar to those of the Treatise, though less close than in E1. E1 and E2 are both written in the hand of Thomas Robson (*fl*. beginning of the seventeenth c.[?]).[50] The extracts were probably copied at the beginning of the seventeenth century: several of the items in E1 bear dates between 1605 and 1612; and in E2 several items are dated 1614 or 1615.

---

[48] The extract begins with a few lines from the beginning of Chapter 4 (307r: 6–9), but after these lines the extract derives exclusively from Chapter 5.

[49] The description of Part 2 of E1 is slightly confusing in Black, *A Catalogue*, 1109–10. Black suggests that the first extract appears on fols. 6r–8r and the second on 17r–22v, although Black does not recognise that 17r–22v is all one item. This description rests on the first eight pages of Part 2 being classified as fol. 1a. What seems to have happened in Ashmole 1407 is that several disparate items containing separate foliation or pagination have been bound together. In the binding as it is now found, the two extracts appear on pp. 17–21 and 39–50, although not paginated as such. I have opted to follow the present binding in my description.

[50] I have found very little information about Robson. Black, *The Catalogue*, records many manuscripts in the Ashmole collection that are in Robson's hand, and Kassell, *Medicine*, mentions him several times, but neither provides any detailed information about him. Referring to Aubrey's *Brief Lives*, Webster seems to confuse Thomas Robson with Charles Robson, a fellow of Queen's College, Oxford, at the beginning of the seventeenth century; Webster, "Alchemical and Paracelsian Medicine," 311; see *Aubrey's Brief Lives*, ed. Oliver L. Dick (London: Secker and Warburg, 1949), 263. Michael Macdonald mentions a Thomas Robson as an alchemical adept who was in contact with Dr Richard Napier, but provides no more information: *Mystical Bedlam: Madness, Anxiety, and Healing in Seventeenth-Century England* (Cambridge: Cambridge University Press, 1981), 189.

Collation

Textual Transmission and Editorial Rationale

There are several problems inherent in the collation of manuscripts. Editors often claim that it is difficult, if not impossible, to reliably map the interrelationships of the manuscripts of a text. A number of well-known factors may obscure the interrelationships, such as missing branches in a particular manuscript family, incomplete versions of the text, conflation, multiple authorship, or the use to which the text was put. In particular, in the case of alchemical texts, we must recognize that many of the versions were most likely used for practical purposes: the alchemical adept would follow the instructions in the text in the hope of producing the philosophers' stone. In the process of carrying out the procedures, the alchemist may have changed the text to comply with his or her own experiences or with other sources, changes that were later incorporated into the text as it continued to be copied.[1] Such practice has the potential to complicate the collation and the interpretation of the interrelationships of the manuscripts.

The copies of the Treatise provide an illustrative example of the complex relationships that sometimes exist between alchemical manuscripts. The seven main manuscripts agree with each other in a number of overlapping groupings.[2] The "messiness" of the picture defies a straightforward stemma, and instead it is more appropriate to see the relationships in terms of tendencies or persistent groupings. As I will outline in more detail later, the manuscripts fall into two major groupings, AFG and S1S2S3S4, and further subgroupings such as AF, S1S2, and S3S4. But neither of the two major branches is clearly "superior" to the other, and although some manuscripts seem to be closer than others to the hypothesized original text or the last common ancestor, none is clearly the "best" copy. Some copies (especially G and S1) also show signs of changes based on reconsultation of the original sources or conflation, which, needless to say, complicates the picture. The complex and uncertain affiliations between the manuscripts make it difficult to present a critical edition that tries to recover or approximate the author's original or a common ancestor. Such an edition would have to emend a base text frequently, and many of the emendations would be

---

[1] Grund, "'ffor to make Azure'"; idem, "Manuscripts as Sources," 108–10.
[2] For a quantitative overview, see Appendix 2. I will discuss the extracts and *The Picklock* separately below.

based on very uncertain evidence. For this edition, I have therefore chosen to present one of the manuscripts, A (MS Ashmole 1490), and to print it without attempting to reconstruct an earlier stage of the text.

In presenting one manuscript, my edition joins the effort, invigorated over the last decade or so, to make available individual manuscript witnesses of texts rather than solely one reconstructed, canonical text. For example, projects are underway to edit all extant manuscripts of famous texts such as Chaucer's *Canterbury Tales* and Langland's *Piers Plowman*, and editors have also turned their attention to individual manuscripts of lesser known texts, such as the *Short English Metrical Chronicle*.[3] One of the most important goals of such editions is to allow us to explore how a text was received and used by a particular copyist, reader, or community; in other words, these editions make available a version of the text that was *the* text to one or more readers. More recently, scholars of English historical linguistics have made an urgent call for "linguistic editions." These editions would be produced without the modifications of language inherent and inevitable in a critical, eclectic edition. These "linguistic" texts would be used for a variety of purposes, not least for research into the development of the English language.[4] By printing one manuscript, I provide an early modern text on alchemy as it appeared at a specific point in that period. I preserve the original linguistic form of the manuscript as far as possible, which facilitates the use of the text for linguistic research.[5]

Several features commend the A manuscript as the best choice for this edition. Some of the other manuscripts present less complete versions of the Treatise, such as S1, in which Chapter 1 and most of Chapter 2 are missing, and G, which lacks Chapter 33. Other versions contain many unique readings, most of which clearly represent later revisions and/or are obvious errors, such as G and S3 or the group S3S4. Although my aim is not to approximate the author's original, editing one of the less complete versions or one that is clearly idiosyncratic creates many problems in terms of the accessibility and understanding of the Treatise as

---

[3] Una O'Farrell-Tate, ed., *The Abridged English Metrical Brut: Edited from London British Library MS Royal 12 C. XII* (Heidelberg: C. Winter, 2002).

[4] Roger Lass, "*Ut Custodiant Litteras*: Editions, Corpora and Witnesshood," in *Methods and Data in English Historical Dialectology*, ed. Marina Dossena and idem (Bern: Peter Lang, 2004), 21–48; Richard W. Bailey, "The Need for Good Texts: The Case of Henry Machyn's Day Book, 1550–1563," in *Studies in the History of the English Language II: Unfolding Conversations*, ed. Anne Curzan and Kimberly Emmons (Berlin: Mouton de Gruyter, 2004), 217–28; Grund, "Manuscripts as Sources"; Merja Kytö, Terry Walker, and Peter Grund, "English Witness Depositions 1560–1760: An Electronic Edition," *ICAME Journal* 31 (2007): 65–85.

[5] For the recent interest in scientific texts among historical linguists, see Irma Taavitsainen and Päivi Pahta, eds., *Medical and Scientific Writing in Late Medieval English* (Cambridge: Cambridge University Press, 2004); Grund, "The Golden Formulas"; idem, "Albertus Magnus"; idem, "'ffor to make Azure'." See also pp. 74–77.

a text. Of the more complete manuscripts, A is the only manuscript that is explicitly and exactly dated, and the date, 1590, is only some twenty years after the original text must have been compiled. We also know the identity of the copyist who made A: Simon Forman. Because of Forman's prolific copying and his meticulous case books and diary, detailed information is available on his life and interests. Although manuscripts copied by Forman survive in great numbers, there are few editions and in-depth studies of individual manuscripts copied or used by him.[6] As I have shown above, manuscript A contains numerous marginal comments by Forman,[7] and there are additional scribbles indicating that Forman's disciple and friend Richard Napier read the text and possibly carried out some procedures described in it. We thus have first-hand evidence that A was consulted by two contemporary physicians who were interested in alchemy and had a wide knowledge of the art. Printing A will thus supply material for further studies of Forman's copying activities and his interest in alchemical texts. It also provides an opportunity to investigate Forman's linguistic behavior and relate his linguistic preferences to various extra-linguistic criteria.

Overview of Manuscript Relationships

Although the general impression of a collation of the Treatise manuscripts is one of chaos, there are some persistent groupings of manuscripts.[8] The clearest pattern is the bipartite division into AFG and S1S2S3S4. The two groupings exhibit striking lexical differences, restructured material, and passages that must have been omitted in one group or added in the other. Comparisons with the sources of the Treatise show that both at times agree with the source text against the other branch, which makes clear that neither of the branches is markedly "superior" in terms of representing the original or the last common ancestor. A few examples may suffice to demonstrate this point:

AF: wyth these bodies is the spiritt of mercurie <u>forced</u> fixed andd fermented [295r: 35–36]

---

[6] See, however, Mary Edmond, "Simon Forman's Vade-Mecum," *The Book Collector* 26 (1977): 44–60; Schuler, *Alchemical Poetry*, 49–70; Kassell, *Medicine*, chap. 8.

[7] One of the most intriguing notes appears in 304r: 20, where Forman underlines the word "vaine" in the text and points out that he has changed the reading found in his exemplar: "Wine alias vain as I tak it but ye word was <u>wine</u>." This comment perhaps indicates that Forman was a conservative copyist, preferring to keep the original reading, and if he decided to change it, he was careful to note the change. This seems to run counter to Barbara Traister's claim that "Forman had no reservations about altering a text he was copying or translating to make it 'more perfect'": Traister, *The Notorious Astrological Physician*, 128.

[8] See Appendix 2.

S2S3S4: wi*th* these bodies is the sperit of mercurye <u>foresaide</u> fixed & fermented [S2 23v: 43–44][9]
Mirror of Lights: with them are spe*rity*s <u>enforced</u> & fermented [Cambridge, Trinity College MS. R. 14. 45, fol. 67r]

AFG: The <u>3</u> spirit is arcenack [296r: 13]
S2S3S4: The <u>white</u> spirit is arcenick [S2 24r: 37][10]
Mirror of Lights: The <u>thridde</u> spirit is orpement [Cambridge, Trinity College MS. R. 14. 45, fol. 68r][11]

AFG: of every stedfaste kinde [308r: 19]
S1S2S3S4: <u>for th'inspira*ci*on of th'ayre is the life</u> of euery stedfast kinde [S2 31v: 31–32]
Scoller and Master: <u>for þe life</u> of all stidfast kyndes co*m*myth <u>of i*n*spiracion of þe eyre</u> [Oxford, Corpus Christi College MS. 226 fol. 32v]

AFG: the menstrue resolutiue springeth of wyne or the <u>water</u> therof [320v: 9–10]
S1S2S3S4: the menstrew resolatiue springeth of wine or the <u>tarter</u> therof [S2 39v: 17–18]
Medulla: menstruum resolutivum oritur ex vino, ut dicit Raymundus, sive ex <u>tartaro</u> ejusdem [the resolutive menstruum originates in wine, according to Raymund, or in its <u>tartar</u>] [Ripley, *Opera omnia*, 164–65]

The subgroupings within the two major categories are more difficult to delineate with certainty. The strongest grouping seems to be S3S4.[12] These two manuscripts share a number of unique readings, including absent single words and instances of modernization of morphology and lexis. This is just a small selection: **S3S4** *absent*] water, rightly, in ballneo, take, is made, naturall **AFGS1S2** (310v: 31; 310v: 37; 314v: 19; 316v: 28; 324r: 28; 324v: 4); **S3S4** additi*on*] additamente **AFGS1S2** (302r: 13); **S3S4** woman] wife **AFGS1S2** (302r: 35); **S3S4** fitt] apte **AFGS1S2** (313v: 29); **S3S4** apply'd] administred **AFGS1S2** (313v: 40).

S1 and S2 are another strong grouping, sharing a number of similarities in lexis, such as **S1S2** saltinge] setlinge **AG**, setting **FS3S4** (311r: 18); **S1S2** ferma*ci*on] ferme*n*tation **AFGS3S4** (311r: 36); **S1S2** subtrahe, subtragh] substracte, substract **AFGS3S4** (313v: 42, 322r: 12).[13]

---

[9] This instance appears in Chapter 1, where S1 is missing. In G, the word is absent.
[10] This instance appears in Chapter 1, where S1 is missing.
[11] See EN 296r: 13 for the difference between the source and the Treatise in referring to *arsenic* and *orpiment* respectively.
[12] See p. 113 for extralinguistic connections between these two manuscripts.
[13] S1 shows some traces of correctional conflation or possibly reconsultation of the original sources. For example, in 304r: 1, the copyist adds "worke" (found in AFGS3S4) above "worlde" (also found in S2), suggesting that the copyist may have had access to A,

Despite these tendencies, it is difficult to establish more precisely the relationships among the S1S2S3S4 manuscripts. S3 cannot alone have served as the exemplar of S4, owing to the many clear omissions in S3 where S4 agrees with the other manuscripts. The reverse is probably also true: both S3 and S4 group with other manuscripts in such a way as to undermine the assumption that S3 derives from S4. The two may share the same exemplar or at least a fairly close common ancestor.

Both variant readings and the date of S3 and S4 make clear that neither of them can have served as the exemplar of S1 or S2. It is also unlikely that S1 or S2 is the basis of S3 or S4, since S3 and S4 sometimes agree with AFG against S1S2. Whether S1 and S2 have an exemplar-copy relationship (S1 being copied from S2 or vice versa) is difficult to determine. Again, there are several variational groupings that undercut such an assumption, and if there were such a relationship, quite a few readings would have to be accounted for as coincidental or the result of conflation. It is nonetheless highly probable that S1 and S2 share a very close common ancestor.

The relationships within the AFG branch are even less clear than those of S1S2S3S4. All three possible subgroupings of these manuscripts (AF, AG, and FG) occur relatively frequently. The strongest grouping numerically is AF, which exhibits some striking lexical similarities, including e.g. **AF** place] plaite **GS1S2S3S4** (311v: 10); **AF** coulbuste] combust **GS1S2**, combustible **S3S4** (314r: 30); **AF** inclinable] enclined **GS1S2S3S4** (322r: 24). A and G also exhibit some agreements, but these seem to be less substantial than the agreements between A and F. Most of the variation concerns morphological features such as **AG** dryveth] drive **FS2S3S4**, M **S1** (295v: 16); **AG** ded thing*es*] a dead thing **FS2S3S4**, M **S1** (295r: 9). With FG, as in the case of AF, the agreements are primarily lexical: **FG** wealth] welle **A**, weale **S1S2S3S4** (301v: 35); **FG** her] his **AS1S2S3S4** (310v: 39); **FG** yow] we **AS1S2S3S4** (316r: 18); **FG** matter] water **AS1S2S3S4** (316v: 20); **FG** water] matter **AS1S2S3S4** (316v: 29).

It is quite clear that none of the AFG manuscripts can have served as the exemplar for the other, as they all contain a number of unique readings where the other manuscripts agree with S1S2S3S4. What is not clear, however, is exactly how they relate. The likeliest scenario is perhaps that A and F share a close ancestor, while G represents a branch of its own, G having by coincidence ended up with the same readings as A or F in a few cases. This conclusion remains tentative, however, considering the criss-cross agreements among the manuscripts.

F, G, S3, S4, or another manuscript related to one of them. In 322r: 32–33, the copyist adds "naturall" above "latall" (found in the other Treatise manuscripts). *Natural* conforms to the source reading in Ripley's *Medulla*.

The Extracts

The nature of the extracts varies. E3 comprises several short extracts from Chapters 2, 3, and 27 of the Treatise,[14] whereas E1 contains two clearly separate extracts: Chapter 1, and Chapters 4–5,[15] 10–17. E2 contains the same chapters of the Treatise as E1 (except for Chapter 1), although the chapters are separated by unrelated material in E2.

While E3's extract from Chapter 2 appears to be abbreviated and to some extent restructured, the extracts from Chapters 3 and 27 are fairly close to the text of the Treatise, although a section is absent in the extract from Chapter 27.[16] There is no clear evidence of a close relationship with any one manuscript. There is an extralinguistic connection between E3 and A as both were copied out by Simon Forman, and E3 does share a unique, although possibly coincidental, reading with A: **AE3** forth] *absent* **FGS1S2S3S4** (304r: 9). E3 furthermore contains a number of unique readings, which suggests either that E3 is based on A, but that Forman made some changes to the exemplar, or that E3 is based on another manuscript related to A (possibly, though improbably, the exemplar of A).

E1's copy of Chapter 1 is a much abbreviated and revised version of the text. There are many obvious omissions and several major revisions. A few readings suggest that E1 is more closely related to the AFG manuscripts than to S2S3S4[17], as e.g. **AFE1** forced] foresaide **S2**, afores*ai*d **S3S4**, *absent* **G** (295r: 35); **AFGE1** liquefaction] lyonyffacc*i*on **S2**, lyonificat*i*on **S3S4** (295r: 4); **AFGE1** 3] white **S2S3S4** (296r: 13). But a close relationship between E1 and any one manuscript cannot be established. In the two readings that can be verified in a source (295r: 35 and 296r: 13), the AFE1 or AFGE1 reading is supported. This pattern might indicate that E1 is not necessarily part of the hypothesized AFG branch, but that it might go back to a manuscript preceding the split into AFG and S1S2S3S4.

Like E1's Chapter 1, Chapters 4–5 and 10–17 in E1 and E2 present substantially reworked versions. The two versions are very close to each other, and E2 is possibly based on E1, or the two share a close ancestor. When AFG and S1S2S3S4 differ, E1 and E2 overwhelmingly agree with AFG. In the one reading that can be verified in a source, E1 and E2 agree with AFG. Again, this agreement may indicate that E1 and E2 are not necessarily part of the hypothesized AFG branch, but may instead derive from a manuscript that predates the split into AFG and S1S2S3S4.

---

[14] 297r: 10–25; 304r: 3–12; 323r: 7–28.

[15] The extract begins with text from the beginning of Chapter 4 (307r: 6–9), but continues with material exclusively from Chapter 5.

[16] Cf. 323r: 9–14.

[17] Since Chapter 1 is missing in S1, this manuscript is not included in the collation.

The Picklock

The two manuscripts of *The Picklock*, Ba and Bw, share an exemplar-copy relationship, Ba being copied from Bw.[18] Ba is a very faithful copy of Bw: they disagree only some forty times, and the disagreements are never of a substantial kind (e.g. *unto* vs. *to*, and *burble* vs. *bubble*). As expected, the Bw copy always agrees with the manuscripts of the Treatise in these cases. Since there is very little variation between Bw and Ba, I will concentrate on Bw here.

There is no clear relationship between Bw and any one manuscript of the Treatise: Bw agrees with all of the manuscripts in different constellations. Most importantly, however, Bw exhibits readings peculiar to both AFG and S1S2S3S4, as in **AFG** and let ther be made balination as before and when the water is tayned with the oill of gould (314v: 6–9), **Bw** and set yt in bal*neo* as before and when the water is teyned with yᵉ oyle of golde] *abs* **S1S2S3S4**; **S1S2S3S4** with a softe fire let it be calcined with a stronge fire which thinge don increace it oftentimes & dry it, **Bw** with a softe fyer calcine yt with a stronge fyer which don incere yt often and dry yt] *abs*. **AFG** (315r: 6); **AFGBw** fyer] force **S1S2S3S4** (320v: 4).

The agreements of Bw with both AFG and S1S2S3S4 readings are significant in that they demonstrate that none of the extant manuscripts of the Treatise can have served as the sole exemplar of Bw, and consequently *The Picklock*. The originator of *The Picklock* may of course have had more than one manuscript at hand, one representing each of the two groupings, and conflated the two branches. However, this seems unlikely. The source of Bw was probably a single manuscript of the Treatise that contained features of both AFG and S1S2S3S4. Collations with the sources of the Treatise supports this hypothesis: they show that Bw always contains the reading that is supported by a source where AFG and S1S2S3S4 present different variants.

---

[18] See p. 27.

Editorial Principles

The aim of this edition is to present the version of Humfrey Lock's alchemical Treatise that is preserved in A (MS Ashmole 1490). I have not attempted to reconstruct the author's original text or the text of the common ancestor of all the extant manuscripts.[1] The main guideline for this edition has been to present A as faithfully as possible. At the same time, some features have been changed to facilitate the reading of the text. These are the specific principles followed:

I) Lineation. The linear arrangement of the manuscript has been retained, while the folio and line numbering is editorial. If the manuscript line extends beyond one line in the edited text, the continuation has been marked by indentation. One or more asterisks after a line signal that one or more comments appear in the EN (see also IX and X).

II) Punctuation and Capitalization. I have replaced the manuscript punctuation and capitalization with modern punctuation and capitalization. Applying modern punctuation and capitalization to a sixteenth-century text is of course not without difficulties, and readers should be aware that by modernizing I have to some extent imposed my own interpretation on the text.

III) Letters. The manuscript use of 'u' and 'v' has been kept. A long 'i' sometimes occurs in numerals, and once in the word "coagulijd" (295v: 14), which has been transcribed as 'i,' as in the Roman numeral *three*, 'iii.'

IV) Word Separation. In general, word separations have been modernized, mainly because it is difficult to determine whether two graphic units are intended to be written together or separately. For instance, *him selfe* has been given as *himselfe*. There are a few exceptions to this principle. Forms with *will* or *shall* + a form of *be*, such as *shalbe*, have been retained since these forms are common in the period. Another exception is the treatment of the word *quintessence*. Forman often spells it "quintaessentia," which seems to be a Latinate or even a Latin spelling. Although the Latinate spelling may have warranted "quinta essentia,"

---

[1] See p. 117.

I have retained A's spelling since it is so consistent. I have kept it as two words when it appears as such in the hand of Copyist 3.

V) Abbreviations. Abbreviations have been expanded and the letters added have been italicized. The expanded form of the word represents the most common spelling of the word when it is found in its full form in the manuscript. For example, "phi." with a tilde above has been rendered as "phi*losofie*."

Instances of "y^e," "y^t," and "y^m" (once in 307v: 19), representing *the*, *that*, and *them* have been retained. Superscript letters have been kept in ordinals, such "4th." In the few instances that superscript letters are used without an indication of abbreviation, such as "goue^{r}nance" (303r: 40) for *governance*, and "e^{r}go" (324r: 31) for *ergo*, I have simply reproduced them without superscripts, i.e., as "ergo" and "gouernance."

The letters 'n' and 'm' in final position often end with a flourish. Since there is no clear evidence that the flourish represents an 'e,' I have transcribed them without a final 'e.'[2] Similarly, seemingly otiose tildes, which are particularly common in text produced by Copyist 3, have been left out without comment.

Sigils, such as ☉ for gold, have been retained, both because they are an inherent part of the text as it stands, and because it is difficult to determine what underlying form is intended.

VI) Editorial Brackets. Curly brackets, '{},' enclose material that has been added above the line in the manuscript. If a caret, '^,' is used in the manuscript, it has been reproduced. Angular brackets, '<>,' signal that letters are unclear in the manuscript, owing to damage or ink blotting. Square brackets, '[ ],' have been used for emendations (see VII) and sometimes for editorial comments in the critical apparatus (see IX). Canceled words are given in the apparatus.

VII) Emendations. In accordance with the general principle of the edition, no emendations have been made on the basis of the readings in the other manuscripts or conjecture, even if A contains a clear error. Differences between A and the main collation text, S2, have been noted in the critical apparatus (see IX), and important differences are further discussed in the EN (see also X). In four instances (294r: 37, 296v: 34, 313r: 45, 320v: 50), words that seem to have been intended to be in the text are found only as catchwords. I have inserted these catchwords in the text within square brackets '[ ]'. Minim confusion in the form of loss or addition of a minim is fairly common in the manuscript, and I have silently added or subtracted a minim in these cases, guided by the spelling(s) the word appears in when fully expanded.

---

[2] See Petti, *English Literary Hands*, 22–23.

VIII) Marginalia. Marginal comments in A have been recorded in the critical apparatus. Since they are usually phrases or abbreviated clauses, I have not provided any punctuation. Unless otherwise stated, the marginal note is written in the same hand as that of the main text, and it is written in ordinary ink; use of red ink and the annotation of the passage in a hand other than the hand of the text proper are recorded in the apparatus. A bar '|' indicates a line break in the marginal note.

IX) Critical Apparatus. The critical apparatus focuses on variation between A and one other manuscript, S2. Noting all the variation between all the manuscripts would have made the apparatus unwieldy. (As shown in Appendix 2, G alone contains over 500 unique readings in selected chapters!) Instead, since the aim is to edit A, I have concentrated on showing A's place in the manuscript tradition. However, when A and S2 differ, I have supplied the readings of all the other manuscripts to show whether A's reading is unique or whether it belongs to a grouping of manuscripts. (The fragments and *The Picklock* have not been cited in the apparatus.) In the dedication, on the other hand, I have recorded all substantive variants in both S2 and F, since, apart from A, only these two manuscripts contain the dedication.

In the apparatus, the variant reading that is closest to A is cited first. S2 is always cited first if a number of manuscripts contain the same reading, followed in order by F, G, S1, S4, and S3. S4 is cited before S3 since the exact reading of S3 is not possible to recover in a limited number of instances owing to the tight binding. The abbreviation '*m.*' (for *missing*) is used in the apparatus before S1 in Chapter 1 and most of Chapter 2, and before G in Chapter 33, since S1 and G do not contain these chapters. The abbreviation '*abs.*' (for *absent*) means that the manuscript or manuscripts do not contain the word or phrase. In cases where one or more of the manuscripts F, G, S1, S3, and S4 have a different passage of text that stretches outside the lemma discussed, I have simply noted that a longer passage is absent or different (*longer pass. abs./diff.*), since readings in these manuscripts are of only secondary importance in this edition.

I do not comment on accidental differences between the manuscripts, such as orthography, capitalization, punctuation, abbreviation, etc. In two cases of substantive readings, I have not recorded the differences between the manuscripts. A consistently uses a form of *sprite* where the other manuscripts use *spirit*; and A frequently uses *vegital* where the other manuscripts employ *vegetable*. Since these two words are very common, recording every instance would make the apparatus unwieldy.

The apparatus also includes comments on the physical appearance of A, marginalia (see VIII), and other related issues. Editorial comments in the apparatus are given in italics. In some cases, I have given editorial comments in italics within square brackets to separate the comment more clearly from the text and to show that the comment is not an emendation. These comments mainly appear when there is a change of ink in the middle of a marginal note (for examples, see e.g. the critical apparatus on the marginal notes in 297r: 1 and 10). The comment pertains to the text that follows the bracketed description. Readings that are further discussed in the EN are followed by an asterisk (see also I). Observe that a note may not occur exactly for the line number of the lemma but may be included in a note on a longer passage.

X) Explanatory Notes (EN). The focus of the EN is on differences in readings between the Treatise manuscripts, and the relationship between the Treatise and its sources, especially if the source supports a manuscript other than A. The transcription of the source texts follows the principles that have been outlined above. The translation of text passages in Latin is my own, unless otherwise stated.

XI) Glossary. Following the edited text and EN is a selective glossary, which covers both alchemical terminology and obsolete or archaic words. It is concluded by a list of proper names.

Dedication

To the right honorable and his singular
good Lorde, S*ir* Willi*a*m Cycill, Lord of Burglay, Knight *
of the Order of the Garter, and and Highe Tresurer
of England, Humfrey Lock wisheth longe lif
with encrease of moch godly honour &c.   5

To youe, my good Lord, I doe directe
my labour thus compild,
in simple sorte, vnclarly don,
in terms not finly fild.

But since the matter of itselfe   10
is worthy to be louid,
more worth then gould it may be made
yf practise mighte be prouid

of him that nowe is far awaie
from his owne natiue lande   15
that wold righte gladly he were ther
to labour with his hand,

and showe himselfe to his country   *
a member worthy place,
which froward fortune in tyme paste   20
out of that land did chase,

by wante, the cruell enimie
of every p*er*fecte minde
that vnto arte and sciences
moste aptly ar inclinde.   25

---

    1 singular] *abs.* **F**
    3 of the Garter] *abs.* **F**
    3 and and] and **S2F**
    7 compild] complide **F**
    8 vnclarly] vnclarckly **S2F**
  10 since] yet **S2**, *abs.* **F**
  20 tyme] times **F**
  24 arte] artes **S2F**

Yet repent I not; my longe absence
in vain yet hath not byn:
such knowledg here to me hath come
as at home I haue not seen,

wherby my seruice mended is   *   30
and apployed for my parte   *
in such affaires of my prince
as longe vnto my arte.

Soe should I showe my dewty bound
vnto my prince and land,   35
w*hich* I desier to prossequte
with the seruice of my handes.   *

With longinge harte I expecte the tyme
of my retorne again
into the place I moste desier   40
in this life to remaine.

For lothsumnes hath weryed me
in place wher I nowe lyue,
for thing*es* that ther ar dayly done
my verie soulle doth griue.   45

Therfore, my ghoste, in pacience
contynewe yet a whille,
for the tyme shall com that thou shalt be
deliuered from exylle

and broughte again into the place   50
wher I had life and breath,
and ther when I haue run my race
to end my life by death.

---

27   yet] it **S2F**
31   and] yf **S2F**
31   apployed] applyed **F**
33   my] myne **S2**
34   should] shall **S2**
37   handes] hand **S2F**
45   griue] ~~quicke~~ griue **A**
48   the] *abs.* **S2**
48   shall] will **S2F**
52   when] where **F**

Dedication

This hope shall still remain with me,
thoughe tyme drawe out in length, 55
that once I shall deliuered be
by God, my only strength.

For other hope I see is none
wherin I haue to truste.
Therfore, my God and Lord alone, 60
as thou wilte, soe I muste.

Yet help, good Lord, for Iesus sake,
yf God haue seen it good,
that home I may be broughte again
in grace and louinge moode, 65

and wrath retorned from the chase,
and displeasure laid asyde,
and frindshipe vsed in their place,
me homward for to guide.

To perfecte ther my natiue land * 70
I longe therfore and locke,
and to put in practize thes same things
that ar writen in this bock,

which ar as trewe as lif doth leade
the body heare on earth 75
from place to place by natures strength
even from his verie breath.

My good Lord, therfore, my humble suete:
my laboure thus directe
may by your Honor in your cheste 80
be from evill men protecte

that rune about deseyuinge suche *
as in them put theare trust,
and ar a slaunder to the good
whose meaninge is true and iust. 85

---

55 in] at **F**
62 Yet] But **F**
70 perfecte] proffitt **S2F**
72 thes] the **S2F**
77 breath] berth **S2**
84 slaunder] scandall **F**

Seinge natewres wai<g>h<s> such neuer knowne
how showlde the bringe to passe   *
the sekret workinge of the stone
hid in a globuse masse.

And yet strange thinges thaye take in hande   90
and worke thearon theare will.
Of nothinge do thay make good end
but all there matter spill.

One sault*es* and alomes do thay worke,   *
one heare and eake on blood,   95
gooths hornes also and allam plu*m*be,
that neuer com to good.

In iron some do thinke to finde
the philozofors stone,
and worke theareon w*i*th greate expence,   100
yet better let alone.

In vinniger and other thinges,
yn tartur burned whight,
thay wen to finde ph‸{il}ozofie,
and thear thay losse thear light.   105

A hundred thinges I coulde reherse
that thay do labor in,
but this suffise for to declame
how bye theare worke thaye wine.   *

Allthoughe I speake of veneger   * 110
that is right sharpe in taste,
I know it now as it was known
to philozophors past.

---

    86 wai<g>h<s>] *<g> written over earlier 'y'; <s> written over earlier 'e'* **A**
    86 knowne] knewe **S2F**
    87 the] they **S2F**
    89 globuse] globeous **F**
    93 matter] matt*er*s **F**
    96 allam] *written over* allm **A**
    99 philozofors] *'i' written over earlier 'e'* **A**
    103 tartur] *'a' written over earlier 'e' (?)* **A**
    104 thay] the **F**
    108 suffise for to] sufficeth to **F**
    108 declame] declare **S2F**

Of argalle I wright, for that it is
ofte in this crafte namid, 115
and put in place for to disseaue
such as ar not ordayned

the secretes to knowe of natewrs workes *
enclosed in one masse,
but to wander still and go astraye 120
and bring nothinge to passe. *

Yf all men knew this holie workes,
then were the worlde forlorne,
for none wolde then for other doe,
but on another starue. * 125

Like as the water of the see
is now whitt, n\<ow\> reed, now gren,
as tempest doth it change eftsonne,
so is owr water clen,

ofe colleres dyvers by his kinde 130
as natewer doth him change
in times and plasses of his worke
apperinge wonderes strange,

vntill he come to stablenes
that fier cannot fraye, 135
becomming then most p*er*fecte whit
when the blakenes is awaye.

Then is he silfer most of price;
no yerth is then his like,
not gould himselfe in all his glorye. 140
The tr\textsubscript{∧}{e}w{th} whou lyste to seke

---

115 namid] Ƚnamid **A**, inamed **S2**, I named **F**
118 workes] worke **S2F**
122 this] their **F**
124 other] others **F**
125 starue] skorne **S2F**
127 n\<ow\>] *corrected from an earlier word (?)* **A**
133 wonderes] wonderous **S2F**
138 silfer] sulpher **S2F**
141 tr∧{e}w{th}] tr∧{e}w̶g̶t̶h̶{th} **A**
141 lyste] E̶ lyste **A**

[*The following stanzas, which are not found in A, have been taken from F*]

 shall find him wondrous in his worke   *
 when he is fully fedd,
 and brought from whitenesse into white
 and after into red,   145

 then two the golden drinke, I say,
 a medicine full of might.
 Noe phisicke needeth where he is,
 for he is king by right

 aboue all phisicke druggs and drames   150
 and all that therto belong.
 He breedeth health and pleasant life:
 his workeing is soe stronge.

 He refraineth age, that crooked carle,
 when he should many assaile,   155
 and youth reneweth in his place:
 his might doth soe prevaile,

 in repulseing that that is in man
 whereon sickenesse oft doth grow
 that bringeth age in youthfull yeares,   160
 if phisick bee to slow.

 His worthinesse cannot, I write,
 with penn bee praisd at length,
 nor in short verse of many laies
 his working and his strength.   165

 Therefore, I will returne againe,
 and mine humble suite renewe
 vnto your Honoer for to helpe
 a subject good and true.

 Although evill will and scandall both   * 170
 haue sought all that they may

---

 155 should] woolde **S2**
 162 I write] inoughe **S2**
 164 laies] lines **S2**
 167 mine] my **S2**
 170 scandall] slaunder **S2**

to exile me farr from home to dwell
as a member cast away,

yet your Lordshipps helpe may worke my will
and bring mee home againe,   175
and make those men to be my friends
that brought me in disdaine,

and such to giue mee loueing lookes
that were my friends before,
which vntrue reports haue made to frowne   180
vpon me verie sore;

whose friendshipps once recovered
I meane noe more to loose,
but to keepe it with such diligence
that I may therein repose,   185

a sure state and staffe most strong
to vphold mee in my right,
and to defend mee in my trueth
against my enemies spight.

Then shall I with a joyfull heart   190
assay to bring to passe   *
the pleasant things by nature hidd
within the globeous masse,

which is more worth, when he is wrought,
then the king in all his wealth,   195
and noething in the world more apt
to keepe a man in health.

Therefore, regard that is to come
and not thats gone and past.
Soe may you reape assuredly   200
good fruite now at yᵉ last

that was not sowne in every heart
that husbandrie hath sought,
but in such ground as God would not
cast in his seed for nought.   205

---

174 will] weale **S2**
194 worth] ₍{rich} ~~worth~~ **S2**
199 thats] that **S2**

[*The final stanza is from A*]

    My good Lord, therfore, your Lordshippes helpe
    most humblie I ^{doe} craue
    and thuse allwayes I praye the Lorde
    your Honnor euer saue.

    Finis H Lock.   210

---

    207 ^{doe}] *insertion in Forman's hand (?)* **A**, *abs.* **S2F**
    208 praye] praise **F**
    210 Finis H. Lock] Finis Humphrey Lock **F**, Finis **S2**

Cap*itulu*m primu*m*   *   294r

For becaus that in the settinge forth of
histories and other scriptures, as well sacred
as profane, it is not only necessarie but alsoe
verie needfull to set before the sam, as a
pathwaie therto, p*r*olog*es*, p*r*efaces or epistelles,
wherby the effecte and som of the same scripture may be
made the more open and easy to vnderstand as well for
the co*m*mendations as settinge forth of the same. But
in settinge forth of this kinde of scripture, w*hich* is the teaching
of naturall phi*losofie* declaringe howe, by the industrie of
the workman, by manuall workinge and natuer,
is made the wonderfull stone of philosofers called y*e*
leess wordle, wherin it is not necessarie as by the waye
of p*r*olog*es*, p*r*efaces or epistles to declare the great secret
of the same, but in the place of the said p*r*olog*es*, p*r*efaces
or epistles to set p*r*esept*es*, admonitions, and warning*es*,
w*hich* the workman, beinge a louer of phi*losofie*, oughte alway
to obserue & kepe. For as all other scripture ar set forth
to the vttermoste for the edification and instruction
of the people as a matter co*m*mon to all men, to the
confermation of the good, and a warninge threatni*n*ge
punishment to the evill, soe must we not thinke of
this scriptur to be a co*m*mon matter but a seacrete

294r:
1   Cap*itulu*m primu*m*] Capit 1 **F**, Ex Semita Alberti Magni et Medulla Ripleyij Ca-
     pitula 1 **S2**, Ex Semita Alberti Magni et Medulla Ripleyij Chap 1 **S4S3**, The hole
     Compound of Allchimya or Elixzeres of the philosophres wrigtten by H Lock **G**,
     Philosophicall Manusc˄{r}ips Large ffol° **S1** *
7   same] saide **S2S4S3**, *m*. **S1**
8   to vnderstand] to the vnderstandinge of the readers **S2FGS4S3**, *m*. **S1**
     as well. . .as] as well as. . .and **S2F**, as well as. . .of **S4S3**, as for. . .and **G**, *m*. **S1**
9   co*m*mendations] comendation **S2FS4S3**, *m*. **S1**
     as] *written on top of earlier and (?)* **A**
15 secret] secret*es* **S2FG**, *longer pass. abs.* **S4S3**, *m*. **S1**
18 alway] alwas **S2FGS4S3**, *m*. **S1**
19 scripture] scripture and histories **S2F**, scriptueres & historyes **GS4S3**, *m*. **S1**

misterie, and ought not to be committed but into the hand*es* 25
of good men to whom God hath gyuen his good sprite to
vse his giftes accordinge to his moste holy will; wherfore I entend not in this my simple laboure to painte
out my newe worke w*i*th as rude a p*re*face, but will
set forth in the place therof the p*re*cept*es*, admonitions, and co*m*- 30
manedement*es* necessarie to be obserued in this crafte;
wherof the first co*m*maundeme*n*te is that the practize<r> *
in this arte ought to be soe circumspecte that he showe
neither his worke nor councell to any but to such as
he is well assured doth lyue in the feare of God. For 35
to such a on or moe, he may make his worke knowen &
be nothing hindered therby. But yf the workeman be [a] **
flebergeber, hauinge noe regard to whom he openithe his secretes, * 294v
suche a man shall not be without troble and late or never
bringe anythinge to good ende. The second precepte or
commandiment is that the master beware he worke not
but in preuye and stronge places far from the sight of men 5
within 3 or 4 chamberes that in one he maye make fier for *
sublimac*i*on and calsinac*i*on, the second for his instrumentes, and
the 3 most secret for the putrffacon of the stone. The third precept *
is that all manner of vessell that the medicin most be soluid in
ought to be of glasse. For if it showlde be don in any other vessell * 10
of copper or lead, it hurtith and wasteth the medcin. But this
knowe that you maye sublime in vesselles made of earth being
wel glased. The iiii precepte is that no manner of workeman *
{in} this arte entermedle himselfe w*i*th princis or lordes. For if
he doo, thay will be allwayes comming or sending to hime {to see or} 15
knowe what he dooth, and when his worke will be at an ende,
and what he doth so longe time bringinge nothing to posse, and *

---

29   newe] rude **S2FGS4S3**, *m.* **S1**
32   practize<r>] 'r' *smudged, written on top of earlier letter(s)* **A**
37   be nothing hindered] nothing hindered **S2FGS4S3**, *m.* **S1** *
      [a]] *catchword* **A**
294v:
8   most secret] most most fecret **S2**, *abs.* **G**, *m.* **S1**
12   earth] 'ea' *written on top of earlier letters* **A**
13   manner of workeman] manor workman **S2**, worckman **G**, *m.* **S1**
14   {in}] ynes {in} **A**
15   {to see or}] seer {to see or} *correction in Forman's hand (?)* **A**
17   posse] passe **S2FGS4S3**, *m.* **S1** *

thus the workeman ˄{shal} be disturbed and in great hassarde to
be vndon. And this not<e>: if thow bringe thy worke to good ende,
there is nothinge gotten thereby but bondage. For ether thow   20
most remayne with him as his th<ria>ll or eles be put in prisson
and so be in danger of thy life. The 5 precepte is that the workman enter not into this crafte except he haue spending
at the l<e>st ˄{for too yers or more, yet nevertheles} he may doo it in
    lesse time if he worke well.
The 6th precepte is that the workeman obserue dueli the   *   25
times of soluc{i}on, distillacion, sublimacon, and calcynacon,
and that he continewe his worke vntyll it be ended.
For if he worke one month and leaue another, in so dooinge
he shall not onely lese his time and expencis
but allso his worke; wherefore this worke, being once   *   30
begone, may not be discontinued vntell the whight
sulp<hu>r be mad, which is the ende of the first parte
of the worke and the begininge of the second. For after
the su˄{lpher} is once created, he maye be norished at the will   *
of the workeman. The 7 precepte is that {in}   *   35
this crafte the worker must deuly keepe the orderes of
this crafte; wherof the firste is in the preparations, the second   295r
in sublimations, the 3ᵈ in calcination, the 4ᵗʰ in coniunction,
the 5ᵗʰ in putrifaction, the 6 in fixation, the 7ᵗʰ in
imbibition, the 8ᵗʰ in fermentation, the 9 in liquefaction
of the medisone, and the 10ᵗʰ in proiection. For yf proiection   5
be made with pouders sublimed and not fixed, they tarry

---

18   the workeman ˄{shal}] shall the workman **S2FGS4S3**, *m.* **S1**
21   th<ria>ll] 'ria' *written on top of earlier letters* **A**
24   l<e>st] 'e' *written on top of earlier* 'a' **A**
       {for too yers or more yet nevertheles}] {for too yers or more yet nevertheles he may doe it in less} [*In the left margin in Forman's hand*] tyme yf he work well **A**
27   vntyll] 'y' *written on top of earlier* 'e' **A**
29   lese] lose **S2FGS4S3**, *m.* **S1**
       expencis] epxences expencis **A**
32   *In the left margin in the hand of Copyist 2(?)*] Sulphur **A**
       sulp<hu>r] *changed from an earlier word, in Forman's hand (?)* **A**
34   su˄{lpher}] super˄{lpher} **A**
35   {in}] you youse {in} **A**
36   worker must] *changed from* worke men *in Forman's hand(?)* **A**
295r:
2   sublimations] sublymacion **S2S4S3**, *m.* **S1**
       in coniunction] is coniunction **S4S3**, conionccion **S2**, *m.* **S1**
4   liquefaction] lyonyffaccion **S2**, lyonyfication **S4S3**, *m.* **S1**

not but fly awaye; and yf youe make proiection with
pouders fixed and not {solued} & firmented, they enter   *
not into the bodies but remaine as ded thinges because   *
they ar not made meltinge and lack their percinge vertue   *   10
wherby they should enter. Alsoe ther ar 4 manner of con-   *
ditions, som naturall, ₍materiall₎, some formall, and som preceptuall,
that ar to be noted in this crafte of alkamy; wherof we nede
to make relation but of too, that is to saye, of the condicions naturall and materiall. The naturall conditions   15
totcheth all thinges wherof this stone is made and holp{e}ne *
withall in his workinge, as tyne, quicksiluer sublimatum,
water, oylle, gould, siluer, and alsoe in the beginninge vitriall
and salte. Therfore, knowe that with thes thinges and none
other is the stone both begon and ended. Materealls noteth as well *   20
the thinges aforesayde as all other vsed in this crafte, as bodies,
spirites, saltes, stones, powlders, glasses, furnisses, sublima-   *
tores, water, oylle, mannes vryne, maidens vryne, andd all
manner of thinges necessary vsed in this crafte, bodies   *
or the 7 mettalls, that is to saie, golde, silver, copper,   25
tynne, leadd, yron, and quicksilver, which philosophers hauee
called after the names of the 7 planetes; as golde they
haue called Sol, silver Luna, copper Venus, tynne
Iupiter, lead Saturne, yron Mars, quicksilver   *

---

8   {solued}] sv<olo>ved {solued} **A** *
9   ded thinges] a dead thinge **S2FS4S3**, *m.* **S1**
11   *In the right margin in red ink; a line in red ink between lines 10 and 11 points to the note*]
      4 conditions **A**
      ther ar 4 manner of conditions. . .that ar to be noted in this crafte of alkamy] there
      are foure manner of condicons some naturall some formall and some preceptuall
      that are to bee noted in this craft of alcumy **F**, thear be 4 maner of condishones
      som naturall som formall and some preceptuall which ar to be noted in this crafte
      of alkamy **G**, there are iiii manner of condicions that are to bee notyd in this craft
      of alkamye wherof som bee naturall condicions som materiall som formall and som
      preceptuall **S2S4**, there are 4 manner of conditions yᵗ are to be noted in this craft of
      alchymie whereof some be naturall some materiall some formall some preceptuall
      S3, *m.* **S1** *
16   this stone] the stone **S2FGS4S3**, *m.* **S1**
      holp{e}ne] holpi{e}nge **A** *
      In the right margin in red ink; a line in red ink between lines 16 and 17 points to the note]
      Thinges wherwith yᵉ | stone is holpen ar | tynn quicksiluer &c **A**
24   necessary] necessarily **S2S4S3**, *abs.* **G**, *m.* **S1**
25   or] are **S2FGS4S3**, *m.* **S1** *
28   haue called] called **S2S4**, *longer pass. abs.* **S3**, *m.* **S1**

Mercurie. Of these bodies some be p*er*fecte and some bee 30
vnp*er*fect. P*er*fect bọdies be golde and silver; wherfore they *
onlie app*er*tayne to the worcke of alkamye in the great
worcke, but in the lesse worcke Venus is vsed as a
meane betweene the sonne and the moone. W*y*th these *
bodies is the spiritt of mercurie forced, fixed, andd * 35
fermented, and so is made elixar, whitt or redd; for **
Sol and Luna be turned into sowles. W*y*th the volatylle
spiritt is made a stone that reu*y*veth deed bodies of his kinde 295v
and chaungeth them from ther owne nature vnto p*er*fect
Sol and Luna. The elixar, being yoined w*y*th vegetable *
quinta essentia, by circulac*i*on is made the great elixar of
lif, w*hi*ch gladeth man*n*es hart, conserveth youth, healeth the 5
leprocy and all mann*er* of diseases of the body, as well
cankared woundes as inwarde sycknes, w*hi*ch no golde alkamy
can do, altho*u*gth he be as p*er*fecte in coulor and examinac*i*on
as gold naturall and lasteth for ever. And heare intendinge
somwhat to speake of the fouer spirit*es*, I will therfor begin 10
w*y*th mercury because he is the spirit that onlie belongethe
to the stone of ph*ilosoph*ers, and is a wilde water, runni*n*ge in **
the veynes of the earth, a subtill substaunce of a *

---

30   these bodies] the bodies S2, *m.* S1
31   *In the right margin in red ink in Forman's hand; a line in red ink between lines 31 and 32 points to the note*] ☉ & ☽ ar p*er*fect bodies A
34   *In the right margin in red ink in Forman's hand; a line in red ink between lines 33 and 34 points to the note*] ♀ is vsed in the lesse | work as a meane | betwen ☉ & ☽ <S> A
35   forced] foresaide S2, aforesaid S4S3, *m.* S1
37   be turned] beinge turned S2S4S3, *m.* S1 *
     In the right margin in red ink in Forman's hand] ☉ & ☽ ar turned into | souls w*i*th the volatill | sprite A
295v:
2   vnto] into S2FGS4S3, *m.* S1
3   *In the left margin in red ink in Forman's hand; a line in red ink between lines 2 and 3 points to the note*] Elixar ioined w*i*th the | vegetall qui*n*tessens A
6   leprocy] lepory S2, *m.* S1
     mann*er* of] manor S2, *abs.* S3, *m.* S1
9   lasteth] lastly S2, *m.* S1
     ever] *an 'r' is written above the second 'e'(?)* A
10   *In the left margin in red ink in Forman's hand; a line in red ink between lines 9 and 10 points to the note*] 4 sprit*es* A
11   *In the left margin in red ink in Forman's hand*] Of mercurie what | it is and how it is | the first sprite A
     that] which S2S4S3, *m.* S1
13   a subtill substaunce] a̶s̶ a subtill substaunce A

whit coulor, coaguliid wyth his sulphur so imperceptably that
he resteth not but flyeth from plaine places and cleaveth * 15
not therto. For althougth his waternes dryveth him vpp
and downe, yet the drines of his sulphur will not suffer him
to cleave to any thinges, neyther to rest in any place playne,
and by this his swifte course in the veines of the earthe, *
he engendreth all metalles. For whersoeuer he fyndeth an earth 20
accordinge to his nature he yoineth wyth yt and so groweth
into a mettall perfect or vnperfect accordinge to his nature
and the purity of the earth and the comforte of thee
ellementt then aboundinge. For as the eleament habound-
deth in the comixion of the earth, so is the mettall en- 25
gendred perfect or vnperfect, according to the quality
of the saide earth; and althought his coulor doe shewee *
him to be of complection colde and moyste, yet by the
drynes of his sulphur he holdeth the nature of euery
element, and for that cause he is called homogaine and * 30
the mother of all mettalls. For wythout him may noe
mettall be engendred, for he only is ther begininge &
well. Therfor, because of his subtill nature, he hath his
name of the planet mercury, and is the fountayne from the which 296r
the stone of philosophers ought to be taken. The 2 spiritt is
sulphur, and it is a fatnes of the earth by temperatt
heate and decoction made thicke, and is of a harde sub- *
staunce, and by the reason of his hardnes he is called sul- 5
phur, and is of a stronge workinge full of drye fewme;
for the which cause oylle cannot be made of him as of other

---

14   imperceptably] ymperceptable S2S4S3, impercepttablely G, *m.* S1
16   his waternes dryveth] his watreines driueth G, the watrines dri{v}e S2S4S3, his
        waterings drive F, *m.* S1
18   any thinges] anythinge S2FGS4S3, *m.* S1
        place playne] plaine place S2GS4S3, *m.* S1
20   *In the left margin in red ink in Forman's hand*] The generation of | mettalls A
30   *In the left margin in red ink in Forman's hand; a line in red ink between lines 29 and 30
        points to the note*] ☧ is homogen A
        homogaine] homogenie S2GS4S3, homogene F, *m.* S1 *
32   begininge & well] begine & weell G, well and begininge S2S4S3, *m.* S1
296r:
1   *In the right margin in red ink in Forman's hand*] Mercuri is the fountain | from which
       the philosofers | stone is taken A
2   the stone] the~~e~~ stone A
3   *In the right margin in red ink in Forman's hand; a line in red ink between lines 2 and 3
       points to the note*] The secound sprit is | sulfur & of his | generation A

spirites but the helpe of stronge waters. And of sulphur
ther ar diuers manners and of dyvers coulours, as whit,
redd, grene, setrine, and black, and also of one effect being
well sublimed; the which spiritt we neede not in this worck
of the stone, for our mercurie holdeth in himself his sulphur whit and redd. The 3 spirit is arcenack and is of
such a percinge nature that any red body, beinge molten
in the fier, he will enter into him and coullor him
like vnto silver; but in the seconde or thirde examinacion
he will flye away. The same spirit, beinge ioined wyth
mercury and so fixed with him and fermented, may be
made a good elixar. Yet we neede him nott
in our worck of the stone, for our matter is
homogene and requireth no strange thinge to be yoined wyth him. The ˄{iiii} spirit is sal armonack, which spirit
we nede not in our worcke, for we neyther sublyme
nor dissolue wyth him; wherfore he is little needfull
in the worck of philosophy. But for makinge of verdegreece
he is very profitable, for wyth him may bee much
made in a smalle tyme etc.

The seconde chapiter sheweth the errours of this worcke
and science by the waie of an argument between
the father and the sonne as followeth etc.

"Father, I praie yow to wouchsaf that I maie aske yow a question
concerninge the errors that I haue fowunde and haue had in this
science of the philosophers stone in my former workinge, and that it may
please yow also truelie to instruct me in the waie of your vn-

---

8   the helpe] by the helpe **S2GS4S3**, with the helpe **F**, *m.* **S1** *
10   also] all **S2FGS4S3**, *m.* **S1**
13   The 3 spirit] The white spirit **S2S4S3**, *m.* **S1**
      In the right margin in red ink in Forman's hand; a line in red ink between lines 12 and 13 points to the note] The 3ᵈ sprit is | arsenike **A**
17   ioined] 'i' *changed from* 'y' *in red ink by Forman(?)* **A**
21   homogene] homogeny **S2GS4S3**, *m.* **S1**
22   *In the right margin in red ink in Forman's hand; a line in red ink between lines 21 and 22 points to the note]* The 4 sprit is | sal armoniak **A**
27   *In the right margin in red ink in Forman's hand]* And thes 3 last sprites | ned not in this | worke **A**
296v:
1   *In the left margin]* Capitulum 2ᵐ **A**
4   to wouchsaf] voutsafe **S2FGS4S3**, *m.* **S1**

derstandinge." "Sonne, saye what yow will and as I haue lerned, I will answer yow." "Father, I beseach yow, shewe me
of the worck of the stone of phi*losoph*ers ˄{whether it} be founde   *
    in vegetable   10
thinges as in quicksilu*er*, or in those thin*ges* that ar so much
spoken of in phi*losophy*, as all mann*er* of salt*es* and allames that
haue diuers names, and also sulphur and arcenick that be
called spirit*es*, or of corosyues made of stones, hearbes,
saltes, gummes, and of many other thin*ges* that ar sett   15
forth and shewed in this science, or whether they be
helpinge {to} this science or not; and for what cause
they be so much spoken of in this crafte; and what
the cause is that ther is no helpe in mettalles
vntill they be dissolued into water, seeing they may   20
be brought vnto subtill kinde in dry fier of calcinac*i*on,
and by many other diuers manners maie be brought
vnto subtill kinde etc." "Sonne, thow shalt vnderstande   **
that of all stedfast kinde ther is certeyne tearme of
greatnes and growinge and may also be multeplyed   25
and in his specie be incresed ˄{to} his lyke, because   *
he is ethremogines, beinge diuers in specie and p*ro*pertye,   *
as in flesh, bloud, and bone, and of trees barck, bodies,
and leaues; and theise hold their sperme in themself

---

9   I will] ~~so~~ I will **A**
      shewe] to shewe **S2FS4**, *m.* **S1**
10 of the worck of the stone of phi*losof*ers ˄{whether it} be founde] of y*ᵉ* worcke of y*ᵉ*
      stone of y*ᵉ* phios*oph*e*res* if it be fownd **G**, if the worke of the stone of phi*losoph*ers bee
      founde **S2FS4S3**, *m.* **S1** *
11 in those thin*ges*] of those thin*ges* **S2S4S3**, *m.* **S1**
      ar] bee **S2FS4S3**, *m.* **S1**
17 {to}] in **S2FGS4S3**, *m.* **S1**
21 vnto] into **S2FGS4S3**, *m.* **S1**
23 vnto] into **S2S4S3**, *longer pass. diff.* **G**, *m.* **S1**
26 ˄{to} his lyke] his licke **G**, in his like **S4S3**, the like **F**, his life **S2**, *m.* **S1** *
27 *In the left margin in Forman's hand(?)*] Ethremogines **A**
      ethremogines] ~~ethrogenius~~ {etherogenius} ˄{hetrogenius} **S2**, heterogenious **S4S3**,
      heterogenie **F**, *m.* **S1** *
      p*ro*pertye] party **S4S3**, in partie **S2F**, in parte **G**, *m.* **S1**
28 in] is **S2S4S3**, *abs.* **G**, *m.* **S1**
      bodies] bodie **S2FGS4S3**, *m.* **S1**
29 hold] houldeth **S2GS4**, *m.* **S1**

{and doe multiplie & growe. And those that be on in part &
   specie, as mettalls be, of   *   30
the which the leste is the wholle in himself,} and ar called homogenius;   31
   and yet they can neu*er* growe
nor multeply vntill they be reduced againe into ther   32
first kinde; for ther first kinde is mercurie, and
when they be liquifyed into mercurie, they turne [directly]   *
to there sentur wher thay offer themselfes vnto thee   297r
in the forme of watur, wherof maye be mad a naturall
bodi of the fier; wherefore the ph*ilozofe*rs sayed: Blessed
be the highe God of kind that gaue his vnderstanding   *
to make in few dayes aboue the yerth that kinde   5
maye not doo wi*t*hin the erth in a thowsand years.
And as towching all thes salltes and allames and
all maner of strange thinges that be not of the kinde
of mettelles, thay be not helping to this crafte, but
in the beginninge of preperatione, so*m*me be nedfull,   10
as vitryall and commane sault, or sault armoniack; and   *
these be all that we ned in this worke. For if thaye
be put in any manner of degre or otherwise
then in sublimac*i*on of quicsilver, all is lost without

---

30   doe] doth **S2GS4**, does **S3**, *m*. **S1**
        those] these **S2S4**, *m*. **S1**
31   the which the leste is the wholle in himself] *added in the left margin in Forman's hand* **A**
        homogenius] homogenies **S2G**, homogenie **F**, *m*. **S1**
32   into] to **S2S3**, vnto **FS4**, *m*. **S1**
        In the left margin in Forman's hand(?)] Homogen **A**
34   [directly]] *catchword* **A**
297r:
1   *In the right margin in red ink in Forman's hand*] Mettalls help not till they | be turned into water | or ☿ [*in black ink*] vz into salte **A**
2   watur] 'u' *written on top of earlier* 'e'*(?)* **A**
4   his] this **G**, vs **S2S4S3**, *m*. **S1** *
6   doo] *second* 'o' *changed from an earlier* 'e' **A**
9   to this crafte] in this crafte **S2GS4S3**, *m*. **S1**
10   *In the right margin in Forman's hand; lines in red ink between lines 10 and 11, and 13 and 14 point to the note*] Vitriall | common salte | sal armoniak [*In red ink*] thes 3 ar | vsed in | the begini*n*ging | in p*r*eparatio*n* ells not | and non other and | that is in the sublima*t*ion of quicksiluer in y*e* | animall & vegetall | stone **A**
11   sault armoniack] salt of almoniacke **F**, sal armo*ni*ac **S2GS4S3**, *m*. **S1**
13   put] put to **S2FGS4S3**, *m*. **S1**
        or otherwise] otherwise **S2FS4S3**, *abs*. **G**, *m*. **S1** *

remedye. But here thow must vnderstand that I   *   15
speake now of the a{n}i‸{m}all and vigitable stone and
not of the mynerall. For in the minerall stone,
the ardent water, being ioyned with the spirit of
vit{riall} and vermelon, is ingrossed and made
thicke and more cleuinge vnto the bodye, with a   20
swifter fixac*i*on then before, by the ress{o}n whereof
his fixac*i*on is compleate in a shorter time then in
the other stones. And for our matter in the a*ni*mall   *
and vigitable stones, after thay are once dissolvid,
willith no stranges thinges, for etheromogines be   **   25
enymies to thinges dissolvid. And furthermore, sault
vnkindlie doth rott in putrificac*i*on and letteth   *
congelatione and gendreth corruptione and turneth
matteres vnto ruste. And allso arsnicke, sulpher,
and orpiment, the which be sperites, and tuthie and   30
calami{n}a{r}yes, which tayne and geue coullour, but because thay be not of the kinde of mettelles, there
colleres tarry not, but in the examinacion it
flyeth away.
And as for es vst and crocfer, allthoughe thaye be made   297v
of copper and iron, thaye be but as ded thinges no
better then yerth, because thay be vnkindlie rotted
and with strange thinges to moch burnte, where-
by there radicaulle humydities is distroyed. There-   *   5
fore, none of these thinges are helpinge to this

---

16   vigitable] *second* 'i' *written on top of earlier* 'e' **A**
      In the right margin in Forman's hand] Animall | vegitall **A**
19   vit{riall}] vitriall *changed from* viterall; *word smudged and correction/clarification added above the line in Forman's hand(?)* **A**
25   *In the right margin in red ink in Forman's hand*] After disolution vse | not strange thing*es* **A**
27   putrificac*i*on] putrifaccion **S2FS4S3**, *abbreviated* putr. **G**, *m.* **S1**
29   vnto] into **S2FS4S3**, *m.* **S1**
      In the right margin in Forman's hand] Sulphur | arsnek | orpiment sprit*es* [*In red ink*] & ar | not vsed **A**
31   tayne] staine **S2S4S3**, *m.* **S1**
      In the right margin in Forman's hand] Lapis calaminaris **A**
33   colleres] coloure **S2**, *m.* **S1**
297v:
3   vnkindlie] 'k' *written on top of earlier* 'd' **A**
4   burnte] brunte **S2G**, burned **F**, *m.* **S1**
5   humydities] humiditye **S2FGS4S3**, *m.* **S1**
6   are] is **S2**, *m.* **S1**

science, for in the preuitye of phi*lozo*fie, thay be all
condemned. For silver and gould is not commended
in the preuitie of this craft, as thay be silver and
gould, but by the {waye} of a mene dispocic*io*n   10
{be} brought to it by the rules of phi*lozo*fie, and otherwise
are named to the hyding of this sience. For the
ph*ilozo*fers in times past did knowe the grete danger
and mischife that might com by this science, did
hid it depely that no vngodly or vnressononable   15
peapele should com thereto. For otherwise it
should haue bene commane to all peapel, and then
had the ph*ilozo*fers bene the cause of that distructione.
For licke as a good man worketh it to the profit
off life and soule, even so workemen would worke   20
it to there one damnacione and the destruc-
tione of mayny other. Therefore, to euery
degre of working thay haue geuen a diuers
name, and the effecte thereof is fownd by that
which is next vnto it in compleccione, and the   25
qualitie can nether be felt nor sene but in
his working. The diuersitie of names are geuen
bycause of the worthynes of his kinde in
his meruelus nature knit together be equallitie
of elementes. There the ph*ilozo*fers called there ston   30
by all names of names, hauing one only oper-
atione and houlding in his belly all that
he nedeth to his *p*erfectione. Therefore,
thou muste take noe heed to the names that be giuen, but to the pro-   298r
p*er*tyes of the names wherby the phi*loso*fers doe vnderstand on the

---

10 {waye}] weighe {waye}; *changed in Forman's hand(?)* A
11 {be}] are {be} A, are **S2FGS4S3**, *m.* **S1** *
13 times] time **S2S4S3**, *m.* **S1**
19 good] godly **S2S4S3**, *m.* **S1**
20 even] 'v' *written on top of earlier* 'u' A
    workemen] wicked men **S2FGS4S3**, *m.* **S1** *
30 There] Therfore **S2GS4S3**, *m.* **S1**
31 names of names] manner of names **S2FGS4S3**, *m.* **S1** *
298r:
1 giuen] giuen to him **S2FS4S3**, geuen him **G**, *m.* **S1**
    In the right margin in red ink] Mark not the names **A**
2 doe vnderstand on the other] vnderstoode one another **S2FG**, vnderstand one another **S4S3**, *m.* **S1**

other; and howesoever the names be dyuised, it is alwais *
on thinge in himselfe. For the phi*losof*ers haue in the moste *
darkeste manner couered and hid all the matter of their 5
workinge in names and misticall word*es* to thentent
it should not be co*m*mon to any but to phil*osof*ers and to
wise men. But I mean not those wise men that
ar wise in the eye of the wordle, but those that ar
made wise by the power of God to understand his secrete 10
misteries co*m*pre*h*ended & closed vp in the co*n*cauity and
depenes of nature." "Father, forasmoch as youe
haue declared that gould and syluer be not com*m*ended in this science, wherfore saith the ph*ilosof*er that ther
is neither gould engendred w*i*thout gould, nor silu*er* * 15
w*i*thout siluer, because every thinge encreaseth his
lyke?" "Sonne, the phi*losof*ers in thes point*es* speaketh not *
of common gould and co*m*mon siluer, but of such as
muste be turned again into their center and mad
ayrouse that thaye maye be fermented and ioyned to 20
the tinturouse sowle. Therefore, so<w>e in kind that that ***
is not therein. For as owre stone is lorde of all stones, *
so is comman gould lord of all bodyes of the myne, and
therefore must we take him to be the father of our
stone, hedd, and highest in his simplissitye whom he * 25
is fermented. For as the questin is whether he be the *
gretest king that is of all the worlde and nothinge *
done withowt hime, or he that is king but of one

---

3   howesoeuer] however **S3**, howsoeuer ˰{that} **S2S4**, *m*. **S1**
    dyuised] diuersed **S2FGS4S3**, *m*. **S1** *
    alwais] alwaie **S2**, *m*. **S1**
7   to wise men] wise men **S2FGS4S3**, *m*. **S1**
8   those] 't' *written on top of earlier* 'c'(?) **A**
14 ph*ilosof*er] philosophors **S2FG**, *abbreviated* phy **S4S3**, *m*. **S1**
15 neither] neuer **S2S4S3**, *longer pass. diff.* **G**, *m*. **S1** *
17 speaketh] speake **S2FS4S3**, *m*. **S1**
19 into] to **S2FS4S3**, *m*. **S1**
    and mad ayrouse] ayreouse **S2S4S3**, *m*. **S1**
    In the right margin in red ink in Forman's hand] Common gould & | siluer ar not vsed | in this crafte but | they ar turned to | their centure first **A**
21 so<w>e] sowe not **S2FGS4S3**, *m*. **S1** *
23 and therefore] therfore **S2S4S3**, *m*. **S1**
25 whom] when **S2FGS4S3**, *m*. **S1** *
    In the right margin in red ink in Forman's hand] Gould is the father of | the stone **A**
27 that is] that is kinge **S2FGS4S3**, *m*. **S1**

relme and receueth his crowne of the grete kinges
gyfte, or whether ys he a greater phissicion that gyveth the * 30
mediciene, or he that receiveth yt. For ouer golde & sylver, being *
browght into a naturall bodie by the power of the fyer, ys made
a thowsande tymes better then yt was when it did holde his
fyxed natuer in common golde & common sylver. Therefore, the
philosophere⟨s⟩ called ouer golde & sylver by the names of very gould 35
and very silver because it houldeth in pouer both *
gould and silver; the which matter is quicsilver
when it is turned into mercuri and is the⟨n⟩ the
verie tin⟨c⟩bir that tayneth all mettles into *
gould and siluer after the makinge of the elixar white or red. And 298v
therfore is ☿ called the mother of all mettalls. For the wanting
of perfection in vnclean bodies is the lyttlenes of mercury and
good inspissation therof with permanent fixation in cleannes
of earth. Therfore, the philosofers saye that he is animall to * 5
mettalls, and the mean that ioyneth and encreaseth tyncture.
Therfore, common gould, before his alteration, is noe more to
the entente of philosofers then is a kinge of on land to him that
gouerneth all the wordle. For gould in himselfe, before **
he hath receyued tincture, hath nothinge to moch nor 10
to lyttle, for he is even of elementes and fulfilled
with hote and moiste, could and dry, evenly
temperate, and nothinge is in him but that he neadeth

---

30 a greater phissicion] ₍{yᵉ}₎ greter phision **G**, greater phisition **S2**, greatest phisitian **S4S3**, *m.* **S1**
35 called] call **S2GS4S3**, *m.* **S1**
37 *In the right margin in red ink in Forman's hand*] The quicksiluer of ☉ | taineth all mettalls **A**
38 the⟨n⟩] 'n' *changed from earlier* 'm' *by cancellation of the last minim (?)* **A**
39 tin⟨c⟩bir] tincture **S2FGS4S3**, *m.* **S1** *
   tayneth] tayneth ~~vnto~~ **A**
298v:
3 lyttlenes] lightnes **S2S4S3**, *m.* **S1**
5 saye] saith **S2GS4**, said **F**, *m.* **S1**
   In the left margin in red ink] ☿ is animal to all mettalls | & doth encrese tincture **A**
6 mettalls] mettall **S2FG**, *m.* **S1**
9 all the wordle] all the whole worlde **S2S4**, *m.* **S1**
12 could and dry] could and ~~heate~~ dry **A**
13 temperate] attempered **S2F**, tempered **GS4S3**, *m.* **S1**

to himselfe. Therfore, gould cannot giue that which is
not in him. For everie mettall that is corrupte is
more helpinge to this science then is common gould
and common siluer because they be nere their centure, and their sulphure and ☿ is not soe soer fixed
in them as it is in gould and siluer, which ar the perfecte
bodies."
"Farther, it seameth to me that stinckinge mettalls, as
tyne, iron, copper, and leed, should not be profitable to
anythinge in the perfection of this science seinge they be
corrupte in their kinde. For it is againste reason that a
measell should bringe forth children of clean complection."
"Sonn, nowe thou speakeste of those mettalls that be corrupte in their generation; of those I mean not. Yet shalt
thou never haue perfection but in those that be corruptible.
For ther is noe lyuinge thinge that may be made newe
excepte it first die, for every thinge beinge dead corrupteth.
And our bodies muste soe corrupte that noe elemente
may be with other, for otherwise he shall never be made
perfecte to yeald that ˄{bi art that} he may not of kinde. For corruption
of the one is the generation of the other. And in this matter
shall the mettall corruptid be made perfecte medison to yeld
his vertue without number."
"Father, forasmoch as common gould and siluer is not profitable
to this science, what is the cause that tyn is put to this
worke in the begininge rather then lead or other mettalls?
For I haue proued by practice that ther be 4 principalle
mettalls, vz., ☉, ☽, ♀, & ♄. Yf thes bodies haue noe cause of solution

---

14 gould cannot giue that which is not in him] nothinge maie goulde giue that is not in him **S2S3**, {nothing} may ~~gold~~ give {gold} that is not in him **S4**, may gold nothing giue that is not himselfe **F**, can ☉ geue that which is not in him to geue **G**, *m.* **S1***

16 *In the left margin*] All other mettalls y^t ar | coruptible be more vnto | this work then common | gould or siluer **A**

19 as it is in gould and siluer which ar the perfecte bodies] as in the perfect bodies **S2F-GS4S3**, *m.* **S1***

25 complection] complection {&c} **S2S4S3**, *m.* **S1**

28 that be corruptible] which are corruptable **S2S4S3**, *m.* **S1**

33 {bi art that}] *abs.* **S2FGS4S3**, *m.* **S1**
  In the left margin in red ink] Coruptio vnius est generacio | alterius **A**

34 matter] maner **S2GS4S3**, *m.* **S1**

36 number] number &c **S2FGS4S3**, *m.* **S1**

38 *In the left margin*] Tyn put to y^e | worke in principio **A**

41 *In the left margin in red ink*] 4 principall mettalls | ♄ ♀ ☽ ☉ **A**

into ☿, what is the cause that tyn is put to this worke rather then lead or other mettalls?"

"Sonne, this thou must vnderstand that yf this science might haue bin found in common gould & siluer, in leed, copper or yron, it had not nowe byn to seke among the people; for that is on of the principall hydinges of this science of philosofie. And the cause that tyne is soe helpinge in this crafte is for that tyn is clean and nigh vnto ☿ in complection; for by his porosity and hollownes, he houldeth abundance of ayer within him; wherfore he receyueth of ☿ a qantiti of cold and dry, and putteth from him a quantitie of heete and moisture. And this is the cause of our solution."

"Father, wherfore said youe in the begininge 'Make ☿ of quicksiluer,' seinge that ☿ and quicksiluer be both one in knowledge to all men that worke in this science?"

"Sonne, althoughe ☿ and quicksiluer be both one in kinde, yet ar they diuers in specie. For thou shalte vnderstand that ther be too ☿^{es} and on quicksiluer. For ther is firste the ☿ of kinde, engendred in the earth, and is called sperma metallorum and aqua viscosa, of the which mercuri is engendred. Another ☿ ther is that with mettell doth kindly dissolue himself without anythinge of other gender cominnge betwene; and when they be loosed in this manner, then be they verie ☿ themselfe and aqua permanent, hauinge in himselfe all that he neadeth to

299r

*

5

*

**

10

*

15

*
*
*

20

299r:
1   science] scecret **S2S4S3**, *m.* **S1**
2   siluer] comon siluer **S2FS4**, *m.* **S1**
5   And the cause] *in red ink* **A**
7   porosity] porosyties **S2S4S3**, perosities **F**, poriousnis **G**, *m.* **S1**
8   *In the right margin in red ink*] The cause whi tyne | is helpinge herin **A**
9   ☿] the mercurie **S2S4S3**, *m.* **S1**
10 heete and moisture] hote & moyst **S2FGS4S3**, *m.* **S1**
11 our] this **S2**, his **S4S3**, *m.* **S1**
12 Father] *in red ink* **A**
    said] saye **S2FGS4S3**, *m.* **S1**
15 Sonne] *in red ink* **A**
16 *In the right margin*] Too mercuries [*In red ink*] and | on quicksiluer **A**
17 For ther is firste] *in red ink* **A**
19 sperma metallorum] sparma **S2FGS4S3**, *m.* **S1**
20 Another ☿] *in red ink* **A**
22 loosed] dissolued **S2S4S3**, *m.* **S1**

his perfection. For this is the mercury of crafte that the   25
philosofers vsed and the same that he was before he was congealed into mettalle, for then be they turned downe
againe into the begininge of their center. And the third is
quicksiluer."
"Father, me thinke it a great wonder of the coniunction of   30
the stone because ther is soe moch more water then earth,
and that soe lyttle earth should congealle soe moch water
into a dry bodie."
"Sonne, thou shalte vnderstand that of all things that   *
groweth, things that may be felte ar beste, as may be pro-   35
ved by fier. For take a tree, that is, a carte load, and
when he is burned into ashes, all that is feeling of the tree   *
youe may put into a lyttle place. Therfore, in growinge
things the departing of elementes is death. When vegitable
things ar dissolued by death, they turn kindly eyther   40
into earth or water as to their center. For as all ve-   *
gitable things, being burned in the fier, ar turned to earth,
soe all minerall bodies, by the poware of the fier, in their
sparme ar turned into water as to their center and firste   299v
matter. Sonne, thou shalte alsoe vnderstand that the   *
philosofers hath lefte in their bockes too manners of workinge:
the one true and the other false. The true waie y^e

---

26   that he was before] that hee was before he was before that **S2S3**, he was before ~~he was before that~~ **S4**, *m.* **S1**
27   into] vnto **S2**, *m.* **S1**
       In red ink] for then **A**
28   into] to **S2FS4S3**, *m.* **S1**
30   Father] *in red ink* **A**
       me thinke] me thincketh **S2S4**, me thinks **S3**, I thincke **G**, *m.* **S1**
       In the right margin] Coniunction **A**
34   Sonne] *in red ink* **A**
37   he] she **S2**, it **GS4S3**, *m.* **S1**
40   *In the right margin*] The conuertion of | mineralls | and vegitalls **A**
42   to] {in}to **S2GS4S3**, *m.* **S1**
43   soe all] *in red ink* **A**
299v:
1   firste] primer **S2FGS4S3**, *m.* **S1**
2   matter] matter &c **S2FGS4S3**, *m.* **S1**
       Sonne thou] *in red ink* **A**
3   workinge] workenges **S2FGS4S3**, *m.* **S1**
       In the left margin] Too manners of | workinge **A**
4   The true] *in red ink* **A**

haue set forth in darke wordes, as by similitudes of thinges
knowen to the philosofers, and the false waie in lighte
wordes and easey, because none should vnderstand their
crafte but by the poware of Godes grace and gifte from aboue.
The false way is lightly wroughte with sulphur, arseneke, argente viue, and sal armonake sublimed, alsoe with
saltes, alloms, blode, heare, bodies, and such other, by
menglinge, seathinge, boylinge, dissoluinge, and congealing.
With many other manners of workinge haue they set
forth this false waye to defend the true from naughty
people that they might falle into errours and be deceyued.
And the cause of their errours I shall in order showe
hearafter that thou maiste therby the better knowe them;
wherfore, my sonne, thou muste vnderstand that our
crafte and science is the science of the elementes and of the
tymes and quallities of them by reuolution turned
everie one into other. And knowe for certainty that in
every thinge vnder heaven ar the 4 elementes, not in
sight but in vertue and poware of workinge; wherfore the
philosofers haue giuen forth this science darkly, as it were
vnder a cloude, and soe men haue wroughte after the
letter with blod, heare, egges, yron, and many other thinges,
and haue drawen of them the 4 elementes, that is to say, they
distilled of them first cleare water and afterward oille
yellowe aboue, and that, they sai, conteyneth fier and ayer.
And then remayneth in the bottom of the vessell a blacke
earth, vpon the which they poware the cleare water by
lyttle and lyttle, mengling them together, seathing, and

---

9   *In the left margin*] The false way **A**
12 menglinge] menginge **S2F**, *longer pass. abs.* **G**, *m.* **S1**
    boylinge] *abs.* **S2FS4S3**, *longer pass. abs.* **G**, *m.* **S1**
13 set] shot **S2**, *m.* **S1**
15 errours] error **S2**, *m.* **S1**
16 in order showe hearafter] hereafter shewe in ordre **S2FS4**, shewe herafter in order **G**, hereafter shew **S3**, *m.* **S1**
18 sonne] *in red ink* **A**
21 *In the left margin in red ink*] The 4 elementes ar in | all thinges not in sight | but in vertue **A**
25 soe] so soone **S2**, so some **FGS4S3**, *m.* **S1**
28 *In the left margin*] Errours **A**
29 fier] the fire **S2S4S3**, *m.* **S1**
32 them] them them **S2**, *m.* **S1**

congelinge soe longe vntill the earth be made white;
and then they poware vpon the earth his oille and soe
distill, seath, and congealle vntill it hath dronke vp all his fier;   35
and then they haue calcioned the earth by a longe tyme, and haue
caste the same medison vpon ☿ moulten and vpon other
bodies, and haue found nothinge. Yet haue they wroughte
as the philo*sofe*rs haue taughte in their bock*es*, and haue fallen
into great errours. And the cause of their erroure is   40
thus knowen: for like as noe other thing engendreth man
but man, nor a beaste but a beaste, soe all thes thi<n>g*es*,
being far from the kind of mettalls, it is vnpossible that mettalls
should be engendred of them, and forasmoch as mettalls ar   300r
not engendred but of their sperme, and quicksiluer is the
sperme of all mettalls and their begininge after the doctrine
of all ph*iloso*fers. Therfore, it followeth that neither blude,
heare, eggs, vrin, sulfur, arcenike, sault*es* or any such-   5
like thinge is quicksiluer; and therfore none of thes is our
stone.
Ther ar others that tak the 4 sprit*es*, that is to saie, ☿, sulfur,
arcenike, and sal armoniak, w*hich* ar called sprit*es* because
they fly from the fier & turn into smocke. Thes they toke   10
to be the 4 element*es*, forsomoch, say they, as the ph*iloso*fers
haue writen that their science is in sprit*es*, and thes ar sprit*es*.

---

34 *In the left margin in Forman's hand(?)*] Manner of | workinge **A**
35 hath] haue **S2FGS4S3**, *m.* **S1**
36 and haue caste] and then they haue cast **S2S4S3**, *m.* **S1**
37 medison] *abs.* **S2S4S3**, *m.* **S1**
42 man] a man **S2**, *m.* **S1**
  a beaste] his like **S2FGS4S3**, *m.* **S1**
  soe all thes thi<n>g*es*] *in red ink* **A**
300r:
1 *In the right margin in red ink*] Nota bene | mettells ar not engend<re>d | but of their sperme **A**
4 Therfore it followeth that] *in red ink* **A**
5 suchlike] such **S2FS4S3**, *m.* **S1**
8 Ther ar others] *in red ink* **A**
  others] other **S2FS3**, *m.* **S1**
  sulfur arcenike] arcenick sulphur **S2FG**, *m.* **S1**
9 *In the right margin*] 4 sprites **A**

Those they sublimed vntill they became of the kind of the fier,
and they resolued them and brought them to be of the kind of
water and ayer, and calsioned them and fastened them
vntil they came to be of the kind of the earth; and then they
mengled them together and calcioned them strongly &
caste their medison vpon Venus moulton and found
nothinge but erroure as in the firste worke. And the
cause of their errours were many, but thes ar sufficient
at this tyme: for as I said before that mettalls ar
not engendred out of their kinde, soe ar none of thes y^e
seed of mettalls outtaken of mercurie. Therfore, it is imposible
that of them should be made the generation of mettalls; and
because sulfur and arcenike ar sone burnte, howe should
any good thing com of them that ar sone absente and turned
into a colle? Therfore, it behoueth every man craftely
to vnderstand that neyther by themselfe, nor ioyned together with
    bodies, nor mengled
with oille drawen from eggs, vrin, heare or any such matter, beinge
put together or otherwise, is our stone or medisone. For, as in the
bringinge forth of the man ther is nothing added to the sparme or
seed but a receptakell and norishmente, soe lykewise in bringinge forth of any other perfecte thinge, nothinge is mingled that
is contrary to kinde; wherfore it is plain that our stone needed
not the minglinge of any strang thinge out of his kinde.
Some argue and alsoe thinke that our science should be in saultes;
wherfore thei resolue them, calcine them, melte them, and also

---

13 the fier] fire **S2GS4S3**, *m.* **S1**
20 cause] causes **S2F**, *m.* **S1**
      thes ar] this is **S2FGS4S3**, *m.* **S1**
22 engendred] gendred **S2FS4S3**, *m.* **S1**
23 imposible] vnpossible **S2GS4S3**, *m.* **S1**
24 and because] *in red ink* **A**
25 burnte] brunt **FG**, beaten **S2S4S3**, *m.* **S1**
26 sone] so soone **S2S4S3**, *m.* **S1**
27 Therfore it behoueth] *in red ink* **A**
29 oille] oyles **S2FGS4S3**, *m.* **S1**
31 the man] man **S2GS4S3**, *m.* **S1**
34 needed] nedeth **S2GS4S3**, *m.* **S1**
36 Some] *in red ink; changed from* Sonne *by cancellation of one of the minims in black ink* **A**
      In the left margin] Some **A**
37 also] *abs.* **S2FGS4S3**, *m.* **S1**

dresse them in diuers manners, and then caste them vpon
imperfecte bodies. Other ther be that mingle them with bodies   *
by sublyminge, calcioninge, and congealing of the said thinges. And   40
som ioine bodies and sprites together, and all in vain. The
cause of their errours is the same aforesaid. Som also
ther be that think our medison to be made only of bodies; and they calcin
them, dissolue them with sal armoniak, and then congeall them,
and cast them vpon bodies, and ar deceyued. And the cause is y^t 45
at the begininge they take not the sperm of mettalls. For noe bodie
   of himself   300v
is the sperme of mettalls but, beinge dissolued in his owne kinde,   *
   is made y^e
☿ of philosofers. Therfore, it showeth that our medison is not in blode, heare,
eggs or vrine, or any such thinge, neither in sprites ioyned toge-   *
ther, or by himselfe or by themselues, with the forsaid elementes, nor   5
yet in saultes or any such matter. Som ther weere that more
subtilly vnderstode the matter & sawe that ☿ was the sperm of mettells
and their begininge throughe the heat of sulphur, seathinge it in the
bowells of the earth, haue taken the said ☿ and haue mortified

---

38 then] they **S2**, *abs.* **FS4S3**, *m.* **S1**
39 imperfecte] vnperfect **S2G**, *m.* **S1**
  Other ther be] *in red ink* **A**
40 by] *abs.* **S2S4S3**, *m.* **S1**
42 aforesaid] aforesaide &c **S2FS4S3**, *m.* **S1**
  Som also] *in red ink* **A**
  Som] Sonn **S2**, *m.* **S1**
  also] *abs.* **S2FGS4S3**, *m.* **S1**
43 medison] medicines **S2F**, *m.* **S1**
300v:
1 *In the left margin in red ink*] Mettalls dissolued | in their owne | kinde ar made | the ☿ of philosofers **A**
4 or vrine] vrine **S2FS4S3**, *longer pass. abs.* **G**, *m.* **S1**
5 or by himselfe or by themselues] by himselfe **F**, by themselfe **S2S4S3**, *longer pass. diff.* **G**, *m.* **S1** *
  the forsaid] th'aforesaide **S2FS4S3**, *longer pass. diff.* **G**, *m.* **S1**
6 matter] thinge or matter **S2S4S3**, *longer pass. diff.* **G**, *m.* **S1**
  Som ther weere] *in red ink* **A**
  Som] sonn **S2**, *m.* **S1**
7 vnderstode] vnderstande **S2**, *m.* **S1**
  & sawe] sawe **S2FS4S3**, *m.* **S1**

it by itselfe, haue fastened it, & caste it vpon mettalls, and found 10
nothinge. And the cause of their erroure is that lik as the sperme
of man profiteth not nor bringeth forth noe fruite excepte it be
caste into the matrix of the woman that it may be norished ther
& broughte to perfection in the wombe, which is heare to be taken for y^e
glase. 15
Som ther be that mingell ☿ with the perfecte bodies, and washe
  them with
freshe water, and distill them vntill they thinke the vnclean bodye
be turned vnto ☿ and into his kinde; and then they seath it, hoping
that the ☿ will dwell with the bodie, and they find nothinge but the
bodie fowler then he was before and the ☿ vanished awaye. 20
And the cause of this errour is that yf no crud body can be made
the generation of a newe kinde vntill it be returned again into
his first matter mortified, reuegitatid, and elyuated into sulfur.
And soe shall our medison be procreated.
Some haue mingled vnperfect bodies, and haue put them in as- 25
say, thinkinge that the clean parte of the vnperfecte bodie wold dwell
with the perfecte bodie, and the other parte should vanishe awaie; and
they alsoe found nothinge. And the cause of their errour is the
vncleannes of ☿ and the combustibilyti of sulfure in the vnclean
bodies &c. 30
Some I haue kowen that haue drawen a stronge water of
vitriall, salpeter, and allom, and in the said water they dissolued
siluer, clensed it, and distilled it from his flegma; and afterward they drewe another strong water of vitriall, sal armoniak, & allom, and in the same water they dissolued gould, clensed 35

---

10 & caste] cast **S2F**, *m.* **S1**
12 *In the left margin*] The glase is the wombe **A**
15 glase] glasse &c **S2FGS4S3**, *m.* **S1**
16 Som ther be] *in red ink* **A**
  Som] Sonn **S2**, *m.* **S1**
18 vnto] into **S2FGS4S3**, *m.* **S1**
21 yf] of **S2FGS4S3**, *m.* **S1** *
23 elyuated] relyuatid **S2**, relynatid **S4S3**, *m.* **S1**
24 procreated] procreated &c **G**, procreate &c **S2FS4S3**, *m.* **S1**
25 Some haue] *in red ink* **A**
  mingled] mengyd **S2F**, mixt **S3**, *m.* **S1**
  vnperfect bodies] vnperfecte bodies with perfect bodies **S2S4S3**, *m.* **S1** *
26 parte] 't' *written on top of earlier* 'f' **A**
31 Some I haue] *in red ink* **A**
  kowen] knoune **S2FGS4S3**, *m.* **S1** *

it, and distillud it from his flegma as in the other water, and
then they ioyned the too waters together in on glasse, and closed
vp the glasse, and set him in fier of the first degre; and ther standing
a longe season, they loked for putrifaction and when they perceyved they were fallen into erroure, they toke their medison and
in disspite caste it out vpon a dunghill, and the hole matter
therof became good stufe to caste vpon landes, medowes, and
gardens, to encrese and multipli the quantity of herbes, grass, &
corn, according to the quantitie of the medison proiecte vpon
the said places. And the cause of their errour was that, as ther is nothing
engendred out of the kind of his owne sparme, soe likewise, forsoemoch
as thes thinges be not the sparm of mettalls, it is vnpossible that mettalls should be generated out of them howe wel soever they
be prepared and conioyned with perfecte bodies &c.
Other I haue knowen that haue amalgamed ☽ & ☿ together, &
then haue put them in a glasse, stopped the glasse with cotten wole,
and set him in his furnace. Then haue they made another malgam of gould and ☿, put him into his glasse, stoped the glasse
lykewise with cotton wolle, and set him in the fornace with the
firste glasse. Then made they a fermente of rawe ☿ and
gould, and haue fermented their medison without any manner
of alteration at all, eyther of the medison or fermente, and soe
haue contynued their medisone in fyer vnto thend of too yers;
and the matter remayned still cruddy nature. And when
they sawe their wholl worke turned into erroure, they
gaue over their labour with blered eyen, shakinge hedes, &
quaking handes, pore, impotent, ignorante, and as far from

---

37 and closed] closid **S2FGS1S4S3**
39 perceyved] perceued that **S1**, preceuid that **S2**, se **G**
41 caste] they cast **S2FGS1S4S3**
44 quantitie] qualitety **S2**, qualitye **S1**
45 as] *abs.* **S2GS1S4S3**
46 forsoemoch] forasmuch **S2S1S4S3**, *abs.* **G**
301r:
4   Other I haue knowen] *in red ink* **A**
7   put] and put **S2GS1S4S3**
9   ☿ and gould] mercury and siluer a<nd> rawe mercurie and goulde **S2S1S4S3**
11 or] or or **S2**
    fermente] of the ferment **S2S1S4S3**
12 fyer] firing **S2FS1S4S3**, *abs.* **G**
13 cruddy nature] a crudye natuer **G**, in his crude nature **S2S1S4S3** *
15 blered] 'l' *written on top of earlier* 'r' **A**

the trewth of the vnderstandinge of this science as they were
at the begininge. Other I haue knowen excellently
well learned that haue taken ☿ crud and put him
into a stronge glasse, and then haue set the glasse in a 20
stronge furnace, and haue therto gyuen stronge fier vntill
the crud ☿ became precipitate. Then haue they set it in
balneo vntill it became crud again. Afterward they removed
it againe into the dry furnise to precipitate, and soe haue
oftentymes made it precipitate and again crud, and at the laste 25
with soe great a fier that the furnace rente in peaces. Yet found
they nothinge but precipitate. And the cause of the forsaid
erroure is because noe crud body is ☿. Therfore, noe
crud body can be changed excepte it can firste be brought
into ☿, which is their center and begininge, but muste 30
needes contynue and hould his vulgar nature &c.
Many other errours haue alkamistes fallen into but
thes ar sufficiente for this tyme and place. Therfore,
my sonn, yf thou canste escape all thes former errours,
thou shalte be excused of many others that ar not 35
heare rehersed." Finis.

    The 3 chapter showeth forth the   301v
    theorik of this bock as heareafter followeth.

Forasmoche as at this presente entendinge to entreate of
the wonderfull & secreat science of naturall philosofie,
I haue thoughte it good in a small volume somwhat to make 5

---

18   Other I haue knowen] *in red ink* **A**
20   stronge] *abs.* **S2S1S4S3**
21   haue therto gyuen] haue there given **F**, there haue giuen **S2GS1S4**, haue giuen **S3**
22   it] *abs.* **S2S4**
27   And the cause] *in red ink* **A**
      the forsaid] th'aforesaide **S2FGS1S4**, this **S3**
29   it can firste be brought] it be first brought **S2GS1S4**, it be brought first **S3**
32   Many other] *in red ink* **A**
35   others] other **S2S1**, *longer pass. diff.* **G**
36   Finis] &c Finis **S2FS1**, *abs.* **GS4S3**
301v:
1   The 3 chapter] *in red ink* **A**
3   Forasmoche] *in red ink* **A**

manifeste the darke places of the same and to open
the palle and obscure wordes by philosofers hidden in the profound
depnes of chaos; the which chaos the wise philosofers found not
onely to perform the elemente workes of kind, but alsoe to hide in
the cavernes therof the same worke beinge performed. 10
For when by great laboure and moste painfull travaill
the aunciente philosofers had attayned to the perfecte mastery therof
and then not willinge that their labours should be buried in
the bottomlesse lake of oblyuion, haue compiled great volums   *
of the same worke, hydinge it vnder shadowe of misticall 15
wordes and names infinite, and haue sealed it in their
writinge from vnworthy people vnder the mistie mantle of
philosofie, fearinge leste the greate wrath of God should com
vpon them for openinge of soe greate a secreate to the
people not worthy therof. For yf the seacreat working 20
of this science were vniuersally knowen, ynnumerable
people wold labour therin. And whoe can tell wher God
wold giue the doinges therof to wicked men to plague the world
therby. For ther can be noe greater plague to the wordle
then wicked men to haue this glorious science, wherby ther might 25
happen subuertion of kingdoms, rebellion of subiectes, dissobe-
dience of seruantes, and innumerable other scourges depending
vpon the same beinge in the handes of wicked men. Therfore,
let noe man entend himselfe to labour in this moste highe science

---

6   *In the left margin*] Chaos **A**
7   hidden] hidden hidden **S2**
9   elemente] *longer pass. diff.* but contains element **G**, excelent **S2S1S4S3**
13 should be buried] to bee buried **S2S1**, *longer pass. diff.* **G**
14 lake] graue **S2FGS1S4S3**
15 same] saide **S2S1S4S3**
    shadowe] the shadowe **S2FGS1S4S3**
17 writinge] writinges **S2S1S4S3**
19 openinge] th'openinge **S2S1S4S3**
    the people] a peopell **G**, people **S2S1S4S3**
22 wher] whether **S2FGS1S4S3**
23 wold] 'l' *written on top of earlier* 'r' **A**
    doinges] doeing **FG**, woorkinge **S2S1S4S3**
26 kingdoms] kindomes **S2**
28 vpon] on **S2S1S4S3**
    Therfore] *in red ink* **A**
29 himselfe] *abs.* **S2GS1S4S3**

vnworthy leste his wicked harte be lifte vp to encrease his dampnation 30
by wickednes, hauinge abundance, and destitute of the grace of God
may be, besides his owne condempnation, the distruction of many other.
But to such as entend to worke this worthie science to the glory of God, to
the relife of those that haue need, and for the maintenance of
the common welle of their country, to such I write; and those, 35
ther is noe doubte, but the God of nature will open vnto them the gate of
the pallace therof, wherby they may enter into the garden of the Ladie *
and Emprise of Phi*losofie*, wherin ther is to be found infinit treasures
wherby the worke may be performed. And here cominge to the theorik
and introduction of this science: As w<h>e haue learned in *
  holy scripture 40
that in the begininge God, the high creator, out of on substance and *
masse created and made all thing*es* that weare made, soe muste
we alsoe knowe that all thing*es* of our secreat workinge in the stone
doth com & is made of on substance and of on masse. Therfore, in this
worke thou shalte need but on thinge, that is to say, argent viue, *
  and on ac- 45
cion, that is, decoction, and but on vessell to the white and to the red.
And forasmoch as o*ur* stone is but on thinge, therfore he hath * 302r
  in himselfe
al that he neadeth to his p*er*fection, and excepte thou canste find in him all y*t*

---

30 vnworthy] vnworthely **S2GS1S4S3**
33 But to] *in red ink* **A**
  entend] entendith **S2S1S4**
  to the relife] the releefe **S2FGS1**, & releife **S4S3**
34 those that] such as **S2FS1S4S3**, *longer pass. diff.* **G**
35 those] to those **S2GS1S4S3**, *longer pass. abs.* **F**
36 the gate] that gate **S2FS1S4S3**
38 of Phi*losofie*] Phi*losophy* **S2GS1S4S3**
39 may be performed] maybee be performed **S2**
41 in the begininge God the high creator] *in red ink* **A**
42 *In the left margin*] On masse & on | substance [*In red ink*] is the content herof i<t> is | [*In black ink*] arg*ent* viue | [*In red ink*] one accion & | on vessell **A**
43 of] in **S2FGS1S4S3**
  in] of **S2FGS1S4S3**
44 of on masse] one masse **S2FGS1S4S3**
302r:
1 And forasmoch] *in red ink* **A**
  he] it **S2S1S4S3**
  in himselfe] in itselfe **S4S3**, in himselfe and with himselfe **S2FGS1**
  In the right margin in red ink] Mercuri hath all that | he ned to his p*er*fection | in himselfe *which* is | our stone **A**

he neadeth to his complmente, thou shalte never make perfecte that thou
seekeste; for all life, soulle, and generation that thing hath in himselfe, which by the industrie of the workman is by crafte drawen out
of him. And soe is it plainly showed that our stone is on substaunce
and on matter, and of himselfe and with himselfe is made, vnto the
which substaunce we neither ad nor take awaie, but remoue
thinges superfluous in his preparation; wherfore our mastery
is made perfecte of on thinge only, one way, by one disposision, with
one deede. Therfore, thou muste learne, before thou begin thy work,
to knowe which be the rotes of the mineralls, for with them only thou shalt
performe our mastrye without any additamente or help of any other thing
save fier, moiste and dry. And soe shall he comm forth buddinge and springinge out of his owne rote without the helpe of any other kind or strange
matter added thervnto. And by thes tokens he is knowen
in his workinge: for first he proceadeth into blacknes, & after that
shall seame innumerable collours, then white, and laste of all red.
And knowe for certain that white and red be foundamentes of all our
masterie, and thuse our worke and waye is moste certain, and
endeth in the same that it beginneth. And noe true tincture is
made but of the red stone. Nor the sprite may not be ioyned with y^e
bodie but by a mean of the soulle. And thinke not that the bodie of
whom the sprite shalbe ioyned is any straunge thinge that should be
added therto, but such a thinge as was hid in the sprite in his first

---

10   of on thinge only] only of one thinge **S2S1S4S3**, of on thinge **G**
11   Therfore] *in red ink* **A**
12   mineralls] minerall **S2FS1**
       thou shalt] shalt thow **S2GS1S4S3**
       In the right margin] Nota de mineralibus et igne **A**
14   shall he] he shall **S2S1S4S3**
       buddinge and springinge] burgeon and springe **S2FGS1S4S3**
16   thervnto] thertoo **S2GS1S4S3**
       And by thes tokens] *in red ink* **A**
17   &] *abs.* **S2FGS1S4S3**
       In the right margin] Coullours in | the stone &c | [*In red ink*] black | whit | red **A**
18   seame] be sene **G**, shine **S2S1S4S3**
19   foundamentes] the foundamentes **S2S1**, the fundamentall **S4S3**
22   but] *abs.* **S2S1S4S3**
23   by] *abs.* **S2S1S4S3**
       of whom] to whome **S2S1S4S3** *
24   any] a **S2S1S4S3**

beinge. And the ende of this worke is likned vnto wax moulton
in the fier, which coaguleth with the qualiti of the could without addic*io*n
or puttinge to of any other thinge. But in the begini*n*ge, the sprite of   *
our mastery is coaguled into his body by the quality of heate, and then
is it made fermente and tincture which ph*ilosofe*rs calleth their gould,   30
which gould is white in deed and red in poware. And white is semi-   *
p*er*fecte and is made p*er*fecte w*i*th rednes without any other thinge then
himselfe save fier. For in this manner is the sonn white in
his appearinge and red by exp*er*ience; and soe is the red man and   *
the white wife maryed together, which ph*ilosofe*rs call by the name of   35
gould. For after their totchinge, ther be too manners of gould: the on
white and the other red. For without the ☉ and the ☽, this crafte of   *
alkamie is not made because they be only the ferment*es* of the ☉ and ☽,
and ioineth the co*m*pound of element*es* together into on fixed   *
    bodye, w*hich* is noe other
thinge after alkami to be vnderstand but moiste and dry, water & earth;   40
and in the water is ayer, and in the earth fier. For after the sep*ar*ation
of element*es*, ther is none of them seene, but earth and water,   *
    and ayer &
fier is in them as qualitie, and cannot be seen nor p*er*ceyued, but worketh
inuisibly. Therfore, yf thou canste turn the element*es*, thou shalte attain   *
to our masterie; and the turninge of the element*es* is nothing ells   45
    but to make
of water earth, and of that that is flyinge fixed. For water, earth, ayer,

---

26 And the ende] *in red ink* **A**
      vnto] to **S2S1S4S3**
      In the right margin in red ink] Thend of the work **A**
31 semip*er*fect] semi-e-p*er*fect **A**
35 the white] white **S2GS1S4S3**
      call] calls **F**, terme ˄{or call} **S2**, call or tearme **S4S3**
      name] names **S2FGS1S4S3**
      In the right margin in red ink] Sol and Luna ar y^e | ferment*es* **A**
36 totchinge] techinge **S2S1S4S3**
      manners of gould] maner of gouldes **S2GS1S4S3**
37 this crafte] the crafte **S2S1S4S3**
39 co*m*pound] compoundes **S2S1**
      In the right margin] Fier earth | water ayer **A**
42 element*es*] th'ellement*es* **S2FS1S4S3**
43 *In the right margin*] What it is to turne | the element*es* **A**
44 Therfore yf thou] *in red ink* **A**

and fier, moiste and dry, fixed and flowinge, white and red, malle and
femalle, sulphur and argent viue be all one and the same that
☿ and waters is; for water is the mother of all element*es*, & of
the grosnes and thicknes of the water is our sulfur procreate and
broughte forth. For when the qualities of the earth overcommethe
the qualities of the water, then is it called earth, for the phi*loso*fers
calle all fluxible thing*es* water, and coaguled or thick thing*es* &
stife or dry, earth. And soe is our stone coaguled or curded, as male
and femalle, which curd muste be taken for the bodie and the fermente,
the venum, and the floware of gould that was hid in ☿ when he
did hould his firste nature called the milke and the curd that
coaguled. For the curd and ☿ be only on in substance soe that
he coaguleth himselfe. ☿ in his firste beinge is white and of
his owne kinde fleinge from the fier, and wi*th* his owne accorde
he is coaguled into sulfur, calcioned, fixed, and houlden;
and this curd or sulfure is the gould of phi*loso*fers and the key
of all their phi*loso*fie, wit*h*out the w*h*ich may noe man enter into the house
of their masterie and co*n*inge; for the curde is gum that
coaguleth our milke, and again our milke dissolueth our
sulfurous gome. Therfore, knowe that run*n*inge water
is the mother of him that is coaguled, and of him he is bound
and broughte forth. Our stone in the begininge is called a femalle
and runinge milke. We calle the woman coaguled by the name,

302v

5

10

15

20

302v:

3   waters] water **S2FGS1S4S3**
    In the left margin in red ink] Water is the mother | of all element*es* **A**
6   the water] water **S2S1**
7   coaguled or thick thing*es* & stife or dry earth] coagulyd thing*es* earth **S2FGS1S4S3** *
9   *In the left margin in red ink; a line in red ink between lines 8 and 9 points to the note*] The curd is taken | for the bodie **A**
11 the milke] milke **S2S1**
12 coaguled] coaguleth **S2S1**, coagulateth **S4G**, coagulats **S3**
    soe] *written on top of earlier* the **A**
13 ☿] *in red ink* **A**
16 curd] crude **S2GS1**
18 gum] o*ur* gum **S2FGS1S4S3**
    that] w*h*ich **S2S1S4S3**
20 Therfore knowe] *in red ink* **A**
21 and of him] and of her **S2FGS1S4**, and for her **S3**
    bound] borne **S2FGS1S4S3**
22 Our stone] *in red ink* **A**
23 name] man **S2GS1S4S3** *

for acction to man is passion to woman. For when our
stone is liquid, flowinge, fleinge, and sufferinge, it is called y̶ᵉ
woman; and the curd that coaguleth is the man, and that
is called stedfaste, permanente, and workinge the coniunction.
And of thes too is our stone medled, perfectly complete and
made. For our stone overcommeth not but when he is simple
and flyinge, and is overcom when he is coaguled and fixed.
And soe he draweth out all life, soulle, and generation of himselfe and none other; for out of his bodie is drawen our flowars
that maketh our stone to burgeon & springe, pourgethe
and entreth soe that the inner parte clenseth the vtter vntill the
rednes appeare, which rednes is hidden vntill the bodie is burned
white in the firste parte of the worke, and the white turned into
red in the laste parte of the worke; wherfore it is spoken of a philosofer
sayinge: Excepte thou canste put away rednes with the whitnes of
☿, and afterwardes turne that whitnes into rednes thou shalt
not com to any true tincture because ther is noe stedfaste
tincture but in the red stone, for that tincture can noe more
be changed. And forasmoch as argent viue accordeth moste
with mettalls, and beste with perfecte mettalls, therfore it is
manifeste that he is the perfection of our worke without commixcion of any other. For ☿ is the rote of everie thinge, and he
only is the tincture of our stone. Therfore, studie in all thy
workes to overcom argent viue, for yf thou canste make an end of

---

24 man] the man **S2GS1S4S3**, *longer pass. abs.* **F**
    woman] the woman **S2GS1S4S3**, *longer pass. abs.* **F**
29 For our stone] *in red ink* **A**
32 none other] of none other **S2FGS1S4**, of noe other **S3**
33 our stone] the stone **S2S1S4S3**
34 the inner parte clenseth the vtter] the iner parte clensith the vtter and the vtter parte clenseth th'iner **S2S1**, the inner part cleanseth the outward and vtter part cleanseth yᵉ inner **F**, iner parte clenseth yᵉ outer & yᵉ ovter yᵉ iner **G**, the inward part cleanseth the outward & the outward parts cleanseth the inwards **S4**, yᵉ inward part cleanseth yᵉ outward & yᵉ outward cleanse yᵉ inward **S3** *
35 hidden] hid **S2FGS1S4S3**
    is burned] bee burned **FG**, bee turned **S2S1S4S3** *
37 wherfore it is spoken of a philosofer sayinge] *in red ink* **A**
42 And] *in red ink* **A**
45 For ☿] *in red ink* **A**

thi worke only with him, thou shalt not be a sercher but a finder of y̆ᵉ   303r
most precious perfection of perfections that passeth all other workes
in the poware of nature. For yf the white fume of ☿ weare
not cold, of it should noe puer gould be engendred by the crafte   *
of alkamy. For without the ascention of the sprite, ther is not nor   *   5
cannot be any generation of the bodie; wherfore we con-
clude that of nothing in the wordle may the philosofers stone be
made but of argent viue. For argent viue that is called ☿
of the philosofers is could and moiste, and God made all mineralls
of him; wherfore he is their begininge, ground, and firste   10
matter. He is alsoe aery, flyinge from the fier; he perseth everie   *
bodie, and is fermente of elixar; he is water permanente,
water of life, and maidens milke; he dryeth and moisteth;
he sleyeth and quickneth. Yet may marcury never be slainte   *
with fier, excepte he be first broughte into perfecte fixation   *   15
craftely and with greate skille by the helpe and benifite of y̆ᵉ
fier. I haue hard of som wise men that followed the
perfecte science and made of ☿ certain workes that abode the fier,   *
and then made with him maruailouse mutations; for as he is
chaunged, he chaungeth, and as he is teyned, he tayneth, and   20
as he is coaguled, he coaguleth, and as he is fixed, he fixeth.
Therfore, he is called homogenius because he hath in himself
all that belongeth to his perfection. He is called by all manner of
names because of the excellency of his workinge. He is found
in everie place because of his elementes. He is wild, and by his   *   25
putrifaction he is made royalle and deare in his vertue;

303r:
6   wherfore we conclude] *in red ink* **A**
8   of argent viue] only of argent viue **S2S1S4S3**
10 their begininge ground and firste matter] ther begening & ground & firste mater **G**,
    y̆ᵉ begining their ground and their first matter **F**, the beginyng the grounde and their
    first matter **S2S1**, the beginning ground & their first matter **S4S3**
    In the right margin in red ink] Argent viue philosoficall | is cold & moist **A**
11 aery] aier **S2FGS1S4S3** *
13 moisteth] moystneth **S2FGS1S4**, moistens **S3**
14 slainte] slaine **S2FGS1S4S3** *
15 into] vnto **S2S1S4**, to **S3**
17 I haue hard of som wise men] *in red ink* **A**
18 *In the right margin*] Workes of ☿ **A**
22 Therfore] *in red ink* **A**
    homogenius] homogenie **S2FGS1S4S3**
26 he is made] is made **S2S1S4S3**

wherfore the contrition of our stone is not with handes but
only with fyer. And the matter of our stone muste be dyuided
into too partes: With the one the bodie is coaguled, and the
other is refeirid with liquifaction and putrifaction and   30
entention is all one thinge. To coagulate is to bringe water into the substance of the earth, and putrifaction is y̅ᵉ
reuolution & turninge of elementes vntill they be confixat
and vnited into a naturall bodie elementally houldinge
the properti of the fier, wherin he encreaseth his quantity and goodnes.   35
Therfore, our putrifaction is not foulle or vncleane, but it is y̅ᵉ
commixtion of wet and dry, that is to say, water and earth. And the
bodie coaguleth the sprite, and the sprite dissolueth the bodie; wherfore it is said fier and ☿ is sufficient for the deed. And
knowe that all the regimente & gouernance of our stone is only in y̅ᵉ   40
fier and in the dyuersities of his degres. Therfore, yf thou canst
measure the fier, as thou oughtest to doe, then hote fier and
water is sufficiente to all the perfection of this worke. For our naturall in himselfe is but water permanent, hauinge all thinges

---

27   *In the right margin*] Contrition of the stone **A**
28   matter] water **S2FGS1S4S3**
29   into] in **S2S1**
      the other is refeirid with liquifaction and putrifaction and entention is all one thinge]
      the other is referred with liquefattion and putrifaction and entencion all is one thing
      F, y̅ᵉ other is reserued with putrifacktion all which is on thinge **G**, the other is rese-
      rued for liquifaccion and putrifaccion and entencion is all one thinge **S2S4S3**, the
      other is reserued for liquificcation and putrifaction and entencion is all one thinge
      S1 *
30   *In the right margin*] Liquifaction | putrifaction | entention on **A** *
31   coagulate] coagule **S2FS1**, congeal **S4S3**
32   *In the right margin*] Coagulatio quid | putrifactio quid **A**
35   the fier] fire **S2S1S4S3**
      quantity and goodnes] quallye & goodnes **G**, quantety quallity and goodnes
      S2S1S4S3
36   Therfore our putrifaction] *in red ink* **A**
39   said] saide that **S2GS1S4S3**
      the deed] thee **S2GS1S4S3**, it **F** *
      And] *in red ink* **A**
40   *In the right margin*] Fier [*In red ink*] & ☿ is sufficient **A**
41   canst] can **S2GS1S4S3**, *abs.* **F**
42   oughtest] ought **S2FGS1S4S3**
43   water] 'w' *written on top of earlier* 'm' **A**
      In the left margin; a line is pointing to the 'w' *in* water] w **A**
      naturall] materiall **S2S1S4S3** *

in himselfe that he neadeth. Therfore, he is vnwise that studieth not * 45
in the gouernau*n*ce of the fier, for the fier dyuideth the water * 303v
   and w*it*h
him the body and the sprite be ioined together, and w*it*h him is made
our coagulac*i*on & co*n*fixation of element*es*; wherfore only water is **
sufficiente for vs, for in water is vapour, whoe houldeth w*it*h
him his brother and co*m*meth out of on belly and is co*n*ceyued & 5
borne vp in the wing*es* of the winde, by the reason wherof he co*n*- *
tinually descendeth vpon the mortified qualities of the eleme*ntes*
vntill all be broughte into a fixed bodie. And then is it called o*u*r *
water that wetteth not the hand. For allthoughe he were in y^e
begininge water, nowe being coaguled into earth, his watry 10
substance cannot be felte. Neyther is he liquified but in
his mothers milke. And when that this sulfur that is of
himselfe white & in poware red is made by the addicion of the **
☉, then is it called the floware of gould and Hermes goulden
tree, and the priuity hid of ph*ilosofe*rs, dry water and dry pouder, ** 15
fier, earth, and the red stone, and the secret of all secret*es* of the
ph*ilosofe*rs. But argent viue is of the nature of the ☽, moist
water, and moiste pouder, the soulle, the stone, the flyinge *
eagle, maydens milke, and the proud woman. And when
thes natures be ioyned together, they shall turn to a white coller * 20
and this is our first p*a*rte of our worke. And this sulphur *
earth is p*er*fecte, and without him this masterie is not done.
Youe muste knowe that it is a longe time to doe the sep*a*ration of

---

303v:
1   w*it*h him the body and the sprite be ioined together] w*it*h him the volatile spirite is
       houlden and with him the bodye and the spirite bee conioyned together **S2S1S4S3**
2   ioined] conioyned **S2S1S4S3**
6   the reason] reason **S2GS1S4S3**
8   is it] it is **S2S4S3**, is **G**
11   felte] 't' *written on top of earlier* 's' **A**
12   when that] when **S2FGS1S4S3**
13   himselfe] 'm' *written on top of earlier* 's' **A**
       In the left margin] Hermes tree **A**
15   ph*ilosofe*rs] all ph*ilosophe*rs **S2S1**, all the ph*ilosophe*rs **S4S3**, ye ph*ilosophe*rs **G**
16   fier] firie **S2S4S3**
       the ph*ilosofe*rs] ph*ilosophe*rs **S2FGS4S3**
17   But argent viue] *in red ink* **A**
18   *In the left margin*] Maidens milk **A**
21   our first p*a*rte] the first parte **S2S1S4S3**
       And] *in red ink* **A**

the sulfur*us* earth, of the w*hich* Hermes was firste finder; wherof in the doinge, thou muste be longe and sweete, for the blacknes
co*m*methe not forth at once, but every dai it goeth forth by lyttle &
lyttle in longe tyme. And the first errour in our crafte
is great haste of fier. And herin co*n*sider howe nature
is broughte forth in the matrix of the mother in temp*er*atnes of
heate, for soe muste our mastery be done. For whatsoever
  ₍{knoweth} o*u*r
secrete will suffer the prolixiti of decoction that he may com
to his purpose. For he that setteth a tree may not presently
locke to haue fruite therof but muste abide his time. And wit*h*out
this vnderstandinge enter not into this crafte. For the certainty
of this worke is the coagula*t*ion of arg*ent* viue; and the sprite of quintaescentia is therin and all our finall secret, and wit*h*out this our
worke may not be done; & beinge thus coaguled into his bodie, he shalbe
collored as white as snowe, & then is it called o*u*r dry water, & not any
vaine water made of corosyues, but a collorike water, hoter then
fier, & is had of our quick water of life, that is, arg*ent* viue coaguled
into sulfur; wherfore we may not make mettell of crud marcury,
as nature doth, except we entend to tarrie natures tyme in the
generatio*n* therof. For nature hath lymited vnto him his tyme in y^e
genera*t*ion of mettalls, w*hich* is 1000 yers, & althoughe it may
  be done in
moch lesse tyme by co*n*tinuall heate, yet noe mans years may attain to
the end therof. Therfore, in this place somwhat to make mention
of thos thing*es* that in this volume I entend to entreate of, I shall first in

---

24 sulfur*us*] sulphurious **S3**, sulphur **S2FS1**, sulphers **G**
    firste finder] the first finder **S2S1S4S3**
26 by] *abs.* **S2FGS1S4**
28 And herin] *in red ink* **A**
29 *In the left margin*] Natur & fier **A**
30 whatsoever] whosoeuer **S2FS1S4S3**, howsoeuer **G** *
32 to his purpose] to ~~the~~ his purpose **A**
    setteth] soweth **S2**, sees **S3**
38 snowe] *otiose(?) mark above* 'o' **A**
    & not any] not any **S2FGS1S4S3**
40 quick water of life] quick water or water of life **S2FGS1S4S3** *
42 entend] meane {entende} **S2**
    tarrie] staye **S2S1S4S3**
45 mettalls] mettall **S2FS1**
47 Therfore in this place] *in red ink* **A**
48 volume] smale volume **S2FGS1S4S3**

his place speake of sublimation, what it is, & what partes of the worke
 it doth 304r
occupie, and then of the elixar minerall & that in diuers manners,
and afterward of the animalle and vegitalle stone, which ar both on
matter, by on order of workinge, and both apertain to the elixar
of life as well as to the transmutation of mettalls, savinge that y{e}   5
animal stone hath in his grosse coniunction more quantity of
matter because he is enduwed with too lightes, that is to
say, the ☉ & the ☽, by the reason wherof nature, workinge   *
in the bowells of the earth, then is brought forth into whitnes, &
afterwardes by longer contrition of the fier and fermentall no-   10
rishmente, the ☽ is made red; & of both beinge ioyned with the quintaescentia vegitall is made the great elixar of life.
The elyxar minerale, of whom I entend first to entreate,
entreth not medison because he is made of the fier againste   *
nature, takinge the greate decoction of the animall and ve-   *   15
gitall stone. Neither is he in their manner fermented but doth
passe his whelle of philosofie by a shorter manner of workinge, as here- *
after in the practice therof shalbe declared. The elixar mi-   *
nerall, as is aforesaid, is elixerate with the fier againste nature, which is a water
drawen from the minerall vaine by addicion of naturall vicinall   *   20

304r:
1   his] this **S2FGS1S4S3**
       what it is] in shewinge what it is **S2FS1S4S3**, shewing what it is **G**
       worke] worlde **S2**, worlde {worke} **S1**
3   which ar both on matter] which are both one & composed of one matter **S2S1S4S3**
       In the right margin in red ink] Animall | vegitall stone
4   apertain] appertaineth **S2S1S4**, appertains **S3**
6   animal stone] *in red ink* **A**
9   then] the sonn **S2S1S4S3**
       forth] *abs.* **S2FGS1S4S3**
11   quintaescentia vegitall] vegetable quyntaecencie **S2FGS1S4S3**
13   The elyxar minerale] *in red ink* **A**
15   takinge] lackinge **S2FGS1S4S3** *
       animall and vegitall stone] animall & vetegable stone **F**, vegetable and animall stone **S2GS1S4S3**
18   The elixar] *in red ink* **A**
20   vaine] vaine **A**, wine **S2FGS4S3**, wine ˄{mine} **S1**
       vicinall] vitruall **S2GS1S4S3** *
       In the right margin] Wine alias vain | as I tak it but y{e} | word was wine **A**

and nyter, very stronge & mortall, drawen wi*th* fier elementalle fro*m*
a stinckinge menstrue made of 4 thinges, vz., of 4 element*es*, & is the
strongest water in the wordle, by whose sprite the tincture of the
fermente is encreased. For wi*th* this minerall sprite is gould teyned
by reason of certain strength of sulfure, moste puer, hauinge abun-   25
dance of erthines, which we calle the sprite of the grene lyon
or otherwise his blod, because that in this vnctuouse sprite is
found the humor of all bodies mete & necessarie for the elixar, whose
child is to vs all that we haue nede of, sufficient for all the
elixars as welle vegitall & animall as minerall, although   30
not by on order of workinge, as shalbe showed in the practiz of the
said elyxars. Therfore, entendinge somwhat to enlarg
the theorike in declaringe the subiecte of the elixar, what
the elixar is, and wherof it oughte to be made, what the fermente is, and howe it oughte to be p*re*pared, for as it is moste   35
certain that the ph*ilosofe*rs stone cannot be made wit*h*out his subiecte,
noe more can the p*e*rfecte forme of the elixar be broughte
forth without his fermente. For as the greate worke of God
together wi*th* reason doth teach vs that the soull of man is y*e*
glorie of the life, the lighte and bewty of the body, soe muste we   40
knowe that fermente, beinge broughte to co*m*plemente, is the soull,
the life, and the tincture of the stone. And here is to be notid
that when we speke of fermente that is the life of the stone, we then mean
the sprite of the water, fermentinge, washing, tayninge, & fixing the
earth wi*th* his water into the whitnes of sulphur, not flyinge from   45

---

21   very] 'v' *written on top of* '&' **A**
25   certain strength] a certaine strengthe **S2S1S4S3**, *abs.* **G**
     In the right margin] Grene lyon **A**
26   erthines] earthlines **S2FS1S4**, eye{a}rlynes **G**
28   the elixar] elixer **S2FGS1S4S3**
29   to] too **A**
32   said] *abs.* **S2S1S4S3**
     Therfore entendinge somwhat to enlarg] *in red ink* **A**
36   the ph*ilosofe*rs stone] the stone of ph*ilosophe*rs **S2S1S4S3**
37   be broughte forth] *abs.* **S2S1S4S3**
39   y*e* glorie of the life] y*e* glory of lyfe **G**, the glorye the life **S2FS1S4S3**
42   the tincture] tincture **S2GS1S4S3**
     And here is to be notid] *in red ink* **A**
43   fermente] th<at> ferment **S2S4**, the ferment **S1**
     In the right margin in red ink] Ferment is the lif | of the stone **A**

the heate of the fier. But when we speake of that fermente that is the 304v
soulle of the stone, then knowe that we speake of another fermente y^t
muste be had of ☉ or ☽, the tincture of the stone as well to the
white as to the red.
And nowe comminge to speke of the subiecte of the elixar, 5
which is a globouse matter, procreate in the bowells of the earth,
& is a thick watri substance with abundance of moisture and
yet not wettinge that it totcheth, and is called the second   *
materia prima. For ther is doble materia prima, and   *
yet but on thinge, that is to say, water from water was   **   10
before heaven and earth was made, as it is said in the
bocke of Genises that the earth was vain & void, and the
sprite of the Lord was borne vpon the waters, and God
made the firmament & departed the waters from water; &
forasmoch as water was before heaven & earth, it is moste 15
manifeste that water was the firste matter of all thinges, wherin
was a beinge in poware, reduciable to the acte that is of noe quality
nor quantitie, of the which matter and substantiall form is
broughte forth the second matter that is a beinge in acte of the
beinge in poware. Soe likewise the second materia prima, 20
which is the first matter of the bodies of the mine, is ingendred
in the bowells of earth, as a viscus water, ronninge in the vains
therof, takinge noe reste but lightly flyeth from the fyer, and
is outwardly white because of his coldnes and moisture, and is
inwardly red because of his drynes & heate, which rednes is 25

304v:
1   But when we speake] *in red ink* **A**
    that fermente] the fermente **S2S1S4S3**
    In the left margin in red ink] The ferment of the | soulle of the stone **A**
5   And nowe comminge] *in red ink* **A**
    the elixar] the ~~stone~~ elixar **A**
    In the left margin] Of the subiecte of | the ~~stone~~ elixar **A**
6   globouse] globeous **FGS4**, cloobouse **S2S1**
    In the left margin in red ink] The elixar is first | made **A**
8   called] called of the philosophers **S2S1S4S3**
10 from] for **S2S1S4S3** *
    In the left margin in red ink] Too materia primaes **A**
11 was] were **S2FS1S4S3**
14 the waters] water **S2FGS1**
17 *In the left margin in red ink*] Water was the first | beginninge of all things **A**
20 Soe] 'S' *partly written in red ink* **A**
22 *In the left margin in red ink*] The second materia | prima is ☿ **A**

hid in his wombe, as the soulle of man is hid in his bodie.
And here is to be vnderstode that ther is difference betwen the
first matter of the bodies of the mine and the first matter of
the stone. For the first bodies of the mine is quicksiluer, but
the first matter of the stone is the ph*iloso*fers ☿, which ☿ must   30
by crafte be drawen from the firste matter of the bodies of
the myne, that is, quicksiluer, wherof must be made a
heavenly coloured water, in the which water the bodies
of ☉ and ☽ oughte to be calcioned and dissolued, and soe coniunction to be made of the said bodies. And althoughe quick-   *   35
siluer of the mine and the ☿ of ph*iloso*fers be on in kind, yet ar
they dyuers in specie, and therfore they said ther be too maners of marcuries, vz., on drawen from quicksiluer of the
myne, and the other from the bodies of ☉ and ☽, and yet but
one quicksiluer of kinde, althoughe diuers in workinge. For   40
when the bodies be dissolued and closed in this man*ner* in the ☿ of
quicksiluer, then is it very ☿ and sulfur of the ph*iloso*fers and
water p*er*manente, the menstrue resoluble, hauinge in him-   *
selfe all that he neadeth to his p*er*fection, fully conteyninge the 4
element*es* in the pryuie bodies, ioyned in the quintaescentia by   *   45
reason of the fatnes of sulphur & y*e* sprite of the water ar putrified   305r
together into a substance, black and stinckinge, which we cale
the menstrue resoluble, wherin is made the matrimony betwen
the body and the sprite, knitinge together the co*n*trarietis & extremeties of the element*es*, as water with fier, earth w*ith* ayer, heat   5
w*ith* cold, moiste with drith, and sulfur w*ith* arg*en*t viue; by the
reason wherof the heaven beinge made darke w*ith* moiste

---

27   And here is to be vnderstode] *in red ink* **A**
     vnderstode] vnderstande **S2**
28   *In the left margin*] The first matter | of the stone [*In red ink*] is | the ph*iloso*fers ☿ *which*
     | is the water of | quicksiluer **A**
29   the first bodies] the first matter of the bodies **S2FGS1S4S3**
31   matter] matters **S2S1**
34   *In the left margin in red ink*] The calcina*tion* | & solution of the | bodies of ☉ & ☽ **A**
37   said] say **FG**, say that **S2S1S4S3**
38   quicksiluer] the quicksiluer **S2FS1**
39   ☉] the sonn **S2S1**
     In the left margin in red ink] Too mann*er* of ☿[ies] **A**
41   closed] loosid **S2FGS1**, loosened **S4S3**
42   is it] is ~~is~~ it **A**
305r:
6   drith] drie **S2FGS1S4S3**

aery drops descendeth vpon the mortified bodyes, & soe
by continuall sublima*t*ion and distilla*t*ion of the sprite, the
matter is made white, and being broughte into co*n*gealed   10
earth and 7 times ymbibed, and then again putrified
and fermented vntill it will flowe in the fier like wax,
w*h*ich is then the elixar white, & is medison vpon ☿ & all imp*er*-
fecte bodies of the mine &c.
And forasmoch as it is necessarie alsoe somwhat to speake of   15
the ferment because it is the form of the elixar, I will therfore
in a shorte mann*er* entreat therof followinge Guido & Raymone
& other ph*iloso*fe*rs because they haue made argument*es* on the
same to declare that w*i*thout ferment can com forth noe   **
true tincture, nor collour of gould nor siluer, be-   20
cause that ☿ vegitable hath not in himselfe sufficient
to forme & to make the elixar. For it may not be graunted
that the same p*r*opertie is apropriate to ☿ vegitable that is
in gould & siluer by nature, although that in naturall   *
adrop be gould & siluer in poware, for that gould and y^t * 25
siluer that is in naturall adrop is only in poware, after
the doctrin of Raymone; wherfore of necessitie in reducing
that powar into acte, it must haue the assistance of gould
and siluer vulgar. And Rhasis saith our gould and our   *
siluer is not vulgar gould & siluer, meani*n*ge therby y^t * 30
it muste be opened & made ayerouse to the entent yt may
be fermented & muste be fixed. For ph*iloso*fe*rs say that   *
adrop in his p*r*ofundity is aerouse gould, meaninge therby   *

---

8   aery] aire **S2FS1S4S3**, *longer pass. abs.* **G**
10   *In the right margin*] Ymbibed | putrified | ferme*n*ted **A**
15   And forasmoch] *in red ink* **A**
     In the right margin in red ink] Of the ferment **A**
17   followinge Guido & Raymone] and followe the sainges of Guidoe Raymonde **S2F-GS1S4S3**
18   on] vppon **S2FGS1S4**
22   to make] make **S2GS1S4S3**
24   &] or **S2FGS1**
25   *In the right margin in red ink*] Adrop **A**
27   of necessitie] in necessity **S2S1**
29   And Rhasis saith] *in red ink* **A**
     Rhasis saith] furth*er*more Racis saith **S2FGS1S4**, further Rasis saith **S3**
30   siluer] volg*er* siluer **S2FGS1S4**
32   say] saith **S2S1S4**, said **F**
33   p*r*ofundity] profunditude **S2**
     aerouse] aireose **S2S1**, airyall **S4S3**

that he houldeth in himselfe the qualitie of gould, by whose vnctuositie
the tincture of gould is encreased. For ph*ilosofe*rs doe agree that 35
adrope in himselfe is leprous gould, hauinge qualitie wit*h*out qua*n*titie;
to the w*hich* qualitie quantitie being added, the stone is therby created.
For Guido, a great ph*ilosofe*r of Greace, speakinge of y*e* ꝫall sprite,
or otherwise of the me*n*struall water that is gotten adrope by arte, w*hich*   *
sprite, saith he, is the soulle of ph*ilosofe*rs, and is called soler water, * 40
arsenick, and Lune. And consequently, in the same place, he
saith the bodie is the fermente; and againe he saith the earth **
is the matter wherin the fier is hid w*hich* doth embibe and fix
the water and ayer. And the water and ayer doth washe,
taine, and p*er*forme the earth into p*er*fecte whitnes, w*hich* is the stone, 45
and is done w*i*thout any strange thinge introduced into him, but
that all his p*ar*ts may be of himselfe coescentiall and co*n*create with- * 305v
out co*n*iunction. For thentent of ph*ilosofe*rs was to make the thinge
complete aboue the earth in a short time, w*hich* nature scarchly
p*er*formeth w*i*thin the earth in a thousand yeares. Therfore,
play they the foolles after ph*ilosofe*rs that from gould and siluer * 5
vulgar seke the ferment of elixar. For Raymond saith *
our tincture is drawen from a ville thinge and is decocted w*i*th another moch more noble. For, saith he,
we fermente with gould vulgar, w*hich* ought to be taken
for gould adrope, w*hich* is more vulgar then gould vulgar. 10
But here note that ferment muste be mad of gould *
altered and of siluer altered, of whom only the clernes
muste be taken both for the elixar and for the ferment.
And yet knowe for certaine that the stone may be p*er*formed
into white and red, w*hich* springe both of one rote, without 15

---

35   For] *in red ink* A
      doe] doth **S2S1**
38   For Guido a great ph*ilosofe*r] *in red ink* **A**
      In the right margin] Guido **A**
39   that is gotten adrope] that is gotten from adropp **F**, that is gotten of natruall adroppe
      S2S1, that is gotten out of naturall adrop **S4S3** *
305v:
2   ph*ilosofe*rs] the ph*ilosophe*rs **S2FGS1**
6   For Raymond saith] *in red ink* **A**
11   But here note] *in red ink* **A**
      In the left margin in red ink] Wherof the ferment | is made **A**
13   for the ferment] ferment **S2FGS1S4S3**
14   And yet knowe for certaine] *in red ink* **A**

gould & siluer vulgar, but never may be made the
elyxar of the stone but by addition of gould and siluer
vulgar, which ought with the ☿ of the stone to be altered, reuegitated,   *
and elyuated into sulfur, and to be fixed, and partly to be rubi-   *
fied, and partly by the white oille of argent viue and sulfur   20
to be kepte in a mean of whitnes, and both the sulfurs of
gould and siluer to be olefied. And soe of them with the ☿   *
vegitable, by continuall circulation and rotation of the
elementes, is made the greate elixare both of white and red,
whose ferment is not vulgar but philosofacall fermente of   25
fermente. And thus ought it to be vnderstode of ferment
not to be vulgare but philosofically altered into newe
qallities soe that in fermente, well nere all men ar   *
deceyued, fermentinge with oills and waters drawen
from the bodies not altered at all, not perceyuinge the true   *   30
doctrine of philosofers sayinge of nothinge white nor red which
nature hath created can be made the elixar white or red
vntill it hath gon through the whelle of philosofie soe that
the first qualities beinge destroyed, the second qualities
may be introduced; wherfore in this sorte ought we to vn-   35
derstand the philosofers, which after the ignorant seameth to disagre,
sayinge owr gould is not ☉ vulgar, and yet with ☉
vulgar we fermente the tincture. Neyther can ☉ or ☽
com forth without ☉ or ☽. And the sprite of our ☿ is ☉ of philosofers.
And therfore knowe that ☿ vegitable cannot make the elyxar   40
of sufficient form without gould and siluer vulgar because   *
we entend by this arte to make ☉ & ☽. And wheras the philosofers   *
speaketh of ☉ & ☽ vulgar, it muste thus be vnderstode not to be
gould & siluer vulgar, but ☉ & ☽ altered, which must be done before   306r
it can be made fermente. And the ☉ of philosofers is ☿ vegitable,
without the which the stone is not nor cannot be engendred

---

16 the elyxar] of th'elixer **S2S1S4S3**
31 philosofers] the philosophers **S2S1**
32 the elixar] elixer **S2FGS1S4S3**
35 Wherfore in this sorte] *in red ink* **A**
40 ☿] the mercury **S2S4S3**
42 And wheras] *in red ink* **A**
43 it muste thus be vnderstode] it is thus to bee vnderstanded **S2**, it is thus to be vnderstood **S4S3**, it is this to be vnderstode **S1**
306r:
1 *In the right margin in red ink*] What is ment by ☉ & ☽ | vulgar &c **A**
3 cannot] can **S2S4S3**

because the said ☿ vegitable muste be fixed with his ferment,
which, after Raymond, is said with gould vulgar, yet not as it is   5
vulgar but altered into a spirituall quallity, and soe likwise
with ☽ vulgar, altered, which before their alteration ar accounted
to be vulgar but not after. Therfore, Raymond saithe   *
with gould vulgar, to declare wherof should be made the moste
true fermente. And another philosofer saith our gould is not   *   10
☉ vulgar to signifie for fermente it must be taken for ☉   *
altered. And in this manner spake Guido, when he said
that ☿ vegitable is the ☉ of philosofers to declare therby what   *
was the subiecte of the elixar, with whose sprit ☉ & ☽ ought
to be altered, wherby their tincturs ar to be encreased.   15
For Avycen saith ☉ taineth not, excepte it be first tayned.
But howe ☉ is tayned, saith Reply, is a secrete, and soe it shall remaine.
But this knowe for certain that ☉ is not tayned but with the ☿ all
sprite, wherin it muste be dissolued, and his clernes separated
from his grose earth, and then putrified as well in the ferment   20
as in the stone &c. And here is declar<e>d sufficiently the
moste nedfull partes of the theorik, both for the elixar and fermente.
And nowe sumwhat of sublimation, and then of the elixeration
of the minerall stone, dyuydinge it into seuerall partes for yᵉ
better vnderstanding of the diu<e>rsity of this worke and the   25
abreuiation of the same in his elixeration. And althoughe I haue
not in this my simple labour set forth all thinges soe
plainly as I haue knowen them, yet I haue more largly opened

---

8   Therfore Raymond saithe] *in red ink* A
10 And another philosofer saith] *in red ink* A
11 ☉ vulgar] volger S2S1S4S3
12 Guido] *in red ink* A
16 For Avycen saith] *in red ink* A
17 Reply] *in red ink* A, George Rippley S2FGS1
18 but with] otherwise but with S2S1S4, otherwise yⁿ with S3, without G
    In the right margin] The tayninge | of ☉ A
19 separated] seperate S2S1
21 And here is declar<e>d sufficiently the] *in red ink* A
    declar<e>d sufficiently] sufficiently declared S2FGS1S4S3
23 And nowe sumwhat of sublimation] wherfore I will God permittinge speake somwhat of sublimation S2FS1S4S3, wherfor I will specke God wiling somthing of subblimaton G
25 this worke] his worke F, his workinge S2S1S4S3
26 abreuiation] abreuiations S2FS1
27 labour] *abs.* S2, woorke S1, sort S3, ~~sort~~ S4

the hid secretes of this science then I haue seen them by any
other set forth; and for that it is not necessary to reuell   30
vnto the wordle the meruailouse workinge of nature in the knittinge
together of the 4 complections, in ioyninge water with fier, earth with ayer,
cold with heate, and drith with moisture, in the which coniunction nature
showeth himselfe more meruailouse then the harte of man
can thinke before the eye haue seen; wherfore the greate &   35
wonderfull apparances then & after showinge, I humblye
commite them to the moste holly will of him by whom they
weare firste showed to dispose by his godly wisdom to whom
it pleaseth him. Soe be it. Amen.
      Finis theorika. 1590. Noviter scriptum * 40
      per Simonem Forman. Iuni i9.
          ☿

Hereafter followeth the declaration of sublimatione ** 307r
in the practike parte, what it is, & what parte & tymes of the worke it
dothe occupie, and howe and after what manner the workman
oughte to vnderstand therof, accordinge to the meaninge
of the philosofers &c. Capitulum 4um 5

For bycause y^t sublimation dothe occupie every parte of the workinge of y^e
philosofers stone soe that the whole worke is comprehended
vnder sublimation, I wil therfore speke somwhate therof
as a thinge necessary for the better vnderstandinge of the same.
For in the workinge of the stone ar many sublimations, wherof * 10

---

32 in ioyninge] by ioyninge **S2S1S4S3**
33 cold with heate and drith with moisture] heate with could & moyst with drought **G**,
  heat with coulde and moyst with drie **S2S1S4S3**, hott with cold and moyst with dry **F**
  coniunction] commixtion **G**, *abs.* **S2S1S4S3**
34 himselfe] herselfe **S2FGS1S4S3**
  the harte] in the hart **S2S1S3**
39 Soe be it] *abs.* **S2GS1S4S3**
40 theorika] theorica &c **S2S1**
  1590 Noviter scriptum per Simonem Forman. Iuni i9 ☿] *abs.* **S2FGS1S4S3**
[*306v blank*]
307r:
1 Hereafter] *in red ink* **A**
2 in the practike parte] *abs.* **S2FGS1S4S3**
5 Capitulum 4um] Cap 4 **FS1**, Capitula 4 **S2**, Chap 4th **S4**, Chap ye 4th **S3**
6 For bycause] *in red ink* **A**

the firste is the sublimation of quicksiluer and is done in this man*er*
and forme followinge: from dry thing*es* not to him accordinge sublyme quicksiluer, vz., from vitriall and salte let him be well sublymed. But youe muste beware that, in the changinge of him, his
actiue poware be not loste. For yf the actiue poware of quicksiluer   15
be loste in sublimation, then is he not able for the worke of the stone.
Also yf the sprite of quicksiluer be not well kepte in his sublima*t*ion
but that he respire and fly away, then will the ☿ be freshe and of noe
strength to dissolue. Furthermore, yf quicksiluer be not strongly
commixed before his sublimation w*ith* those things he muste be   20
sublimed, then in the sublima*t*ion will the quicksiluer arise
with the ☿ sublimed. For when the crud ariseth with the dry, then
is the ☿ combred and will wax blacke in his grindinge. And her is
to be noted that the very cause wherfore we sublime quicksiluer
is to change his co*m*plection and to giue him his working vertue   25
and actiue poware. Therfor, knowe that yf quicksiluer, w*hich* is cold &
drye, should be mingled with that w*hich* is hote & moiste, it should therby
lose his working vertue; wherfore quicksiluer may in noe wise
be suffred to arise w*ith* that that is sublymed. For this knowe
that yf quicksiluer be well kepte in in his sublimation, he putteth   30
away his quallity of cold and dry, and receyueth in the place therof his working vertue, that is hote and moiste, throughe the *p*articipa-
tinge and ioyninge him with the sprite of vitriall. Therfore, beware
that noe quicksiluer arise w*ith* that that sublimeth. For yf any
arise, then will the matter wax blacke — but with stronge fyer   35
yt may be made white — and then it will be freshe as water
or the stannes and is note able for the elixar. But yf it be well

---

11   the sublimation] sublimac*i*on **S2S1S4S3**
       In the right margin] The sublimaon of | quicksiluer from | vitriall & salte **A**
13   from vitriall and salte] *in red ink* **A**
16   sublimation] his sublymac*i*on **S2FGS1**
17   Also] *in red ink* **A**
       kepte in] kepte in yn **S2G**
18   of noe strength] is of no streng{t}he **S2S1S4S3**
19   Furthermore] *in red ink* **A**
26   Therfor knowe] *in red ink* **A**
       In the right margin] Quicksilu*er* is cold & dry **A**
29   For this knowe] *in red ink* **A**
33   vitriall] the vitreall **S2S1**
       Therfore beware] *in red ink* **A**
36   will be freshe] will fresh **S2**
37   the stannes] other stones **S2S1S4S3**, *abs.* **G** *
       But yf] *in red ink* **A**

wroughte it will be hard as roche allam, cleare & clean, bytter
and sha<pt>, and hath receyued his actiue poware & workinge   *
vertue, and is able for the worke of the stone. And in this
manner may youe knowe the goodnes of sublimation &c. Finis.   40

      Another sublimation   307v
      of quicksiluer
            Capitulum 5um
Take i^{li} of quicksiluer & grind it strongly with on pound of common salte. **
Then put all together into a pan of seathinge water, and ther let it   5
take residence vntill the salte be all resolued. Afterward powre
out the saulte water & salte, and put to other warme water, &
soe washe it 2 or 3 tymes. Then dry it well by the heate of y^e
fier. Then take 2^{li} of vitriall finely ground and dryed a-
gainste the fier. Then take i^{li} of salte alsoe well dryed,   10
and grind the vitriall and salte together strongely, and again
dry them. Then take your quicksiluer and put it to the vitriall
and salte by lyttle and lyttle, contynually grindinge them   *
together soe longe as youe can perceiue any quicksiluer to
appeare. And here is to be noted that to the compound of vitriall   * 15
and salte beinge 3^{li}, youe muste put on pound of quicksiluer.
And when they ar throughly encorporated together, put them   *
in certain potes of sublimation, set one their covers and
lute them faste. The covers muste haue, every on of y^m,
a lyttle hole in the top wherby the moistnes may pass out;   20
and when the lute is dry, set your sublimatoryes
vpon a furnace of sublimation, and vpon the hole of
every cover lay a peace of glasse that youe may knowe
when the moistnes is vapored away. For soe longe as any

---

39   sha<pt>] sharpe **S2FGS1S4S3** *

307v:
1   Another sublimation] *in red ink* **A**
3   Capitulum 5um] Cap 5 **FS1**, Cap 5th **G**, Capitula 5 **S2**, Chap 5th **S4S3**
4   Take] *in red ink* **A**
6   resolued] desolued **S2FGS1S4S3**
7   & salte] *abs.* **S2FGS1S4S3**
14   to appeare] appeare **S2S1S4S3**
15   And here is to be noted] *in red ink* **A**
17   encorporated] encorporate **S2S1**
18   in] into **S2S1**
22   vpon a furnace] in a furnes **S2S1S4S3**, *longer pass. abs.* **G**
24   away] *abs.* **S2FS1S4S3**

moistnes remayneth in the substance, soe longe shall youe
find drops of sweete hange vpon the peeces of glasse, &
when the peaces of glasse shalbe found dry, then stop faste the
holes, and encrease the fier. For before the moistnes is clean
gone out, youe muste haue a softe fier, w*hich* fier muste be
encreased from the first 6 howars vnto 24. Then let
your furnace {be} coulde, and warely take of the covers of the
sublimatories, & youe shall find the ☿ sublimed faire &
white. Take out then the ☿ sublimed, and grind him again
with his feces, and sublime him as before, and this doe
6 tymes. Then tak him awai from his feces, and in sublimatories
of glasse sublime himselfe 3 or 4 tymes, and then youe shall haue
him melte in his sublimation into a cake like vnto roch allam.
And the firste feces sublime again w*ith* stronge fier, and the ☿ that
sublimeth put to the other sublimed, and sublime them together and yᵉ
feces that is of the vitril & salte that remayneth after the sublima*tion*
reserue, for ther is noe better corrosyue to cleare a wond of ded
fleshe or to clense an olde sore. Finis.

   The seconde sublimation is the sublimaon
    of ph*ilosof*ers mercurie
     Capi*tul*um 6ᵘᵐ
Knowe that in this sublimaon the heauy is made lighte for because
that in this sublimaon the bodies of ☉ & ☽ ar dissolued and broughte a-
gain to their center of their begininge. For when the said bodies ar in this
manner dissolued into ☿, then we saie that they be sublimed into
ayer and be broughte the one with the other into vapour. For the soulle
beinge dissolued from the body is made spiritualle and flyinge into

---

26 hange] *abs.* **S2S1S4S3**, *longer pass. diff.* **G**
29 youe] it **S2FGS1S4S3**
31 {be} coulde] coole **S2FGS1S4S3** *
33 then] *abs.* **S2FGS1S4S3**
36 himselfe] him by himselfe **S2FS1S4S3**, him **G**
42 Finis] &c finis **S2S1**, &c **GS4S3**
308r:
1 The seconde] *in red ink* **A**
2 of ph*ilosof*ers mercurie] of the mercury of ph*ilosoph*ers **S2FS1S4S3**, of yᵉ ☿ of yᵉ philociphers **G**
3 Capi*tul*um 6ᵘᵐ] Cap 6 **G**, Capitula 6 **S2**, Chap 6 **S1S4S3**
4 Knowe that in] *in red ink* **A**
  Knowe] Therfore knowe **S2FGS1S4S3**
6 their center] the centre **S2S1**

the ayer. For the water, openinge the flyers gate, maketh the bodye spi- 10
rituall like vnto itselfe, and by reason therof they arise bothe
together into the ayer, takinge ther the inspiration of life of their   *
owne moisture, growinge and multipliinge in their kind, as doth all other
growinge things in their kinde. For vapoure conteyneth vapoure forasmoch as they be ioyned together by decoction; wherfore   *
    the stainge   15
of kind in this place is lefte in seruiage to flighte to the entente
they may be ioyned by sublimacion; wherfore they muste with softe fyer
be reared vp into vapoure and soe be norished in the ayer vntill they
receyue life of every stedfaste kinde. And therfore al the entente of this*
mastery is only in the workinge of nature and in the sublimation   20
of his water, and this is the sublimation of the ☿ of phi*losofe*rs. Finis.

The thirde sublimation

This thirde sublimation showeth as well the workinge of   *
nature after putrifaction as before; wherfor I entend to showe howe that by
continuall sublimation the earth doth drawe downe into his bodie his   25
water and ayer. For when the watry substaunce is turned into
earth by dissention, then we say that all his strength and vertue is
wholy enclosed in himselfe. For this must we alwais knowe that

---

11   by reason] by the reason **S2FS1**
       arise] fly arise **A**, rise **S2FS1**
19   receyue] may receue **S2S1S4S3**
       of every stedfaste kinde] for th'inspiraci*o*n of th'ayre is the life of euery stedfast kinde **S2S1S4S3** *
       And therfore] *in red ink* **A**, wherfore **S2FGS1S4S3**
21   of the ☿ of phi*losofe*rs] of y^e ☿ of ye philo*sophe*rs **G**, of phi*losophe*rs **S2S1S4S3** *
22   The thirde sublimation] The thirde sublimation **A**, The 3^d sublimaci*o*n Cap 7^m **F**, The thurd subblimati*o*n Cap 7 **G**, The thirde sublymaci*o*n Capitula 7 **S2**, The thirde sublimation Chap 7 **S1S4S3**
       The thirde] *in red ink* **A**
23   This thirde] *in red ink* **A**
       This] The **S2GS1S4S3**
24   I entend to showe howe that by co*n*tinuall sublimation] I entende somwhat to speake therof to shewe therby the more plainly howe that by continuance of sublymaci*o*n **S2FGS1**, I intend somewhat to speak thereof to shew thereby ye more plainly that by continuance of sublimation **S4S3**
28   alwais] also **S2S1S4S3**
       that it is not] *in red ink* **A**

it is not ynoughe to resolue, but that we muste after solution bringe
it again to a naturall bodie, vz., into an oyly substaunce, pure and
clean, white, shin*n*inge. And this sublimation is nothing ells but the
workinge of kind in the knittinge together of the element*es*, and
 is done w*ith*-
out any manuall workinge at all. For when we see the quallity of the
water overcome the earth soe that all is made water, we calle that solutio*n*
because the water hath made the body like vnto himselfe; and again
when we see that the quallities of the earth hath overcom the quallitis
of {the} water soe that all is made earth, then we say that the water
is co*n*gealed firmly to abide with his body. For as the water busieth
himselfe in the begininge to dissolue the earth and to bringe it into
subtill kinde, soe lykewise doth the earth afterward co*n*gealle his
water and ayer into a fixed bodie that it may be able together, the
on with the other, to suffer the sharpnes of the fier soe that by this
means it plainly appeareth that all is but only on workinge. For howe
moch soever is by the on dissolued, soe moch again is by the other co*n*geled;
by the reason wherof the heavy bodyes in this sublima*ti*on
 ar returned
again to the center of their substanciall forme, beinge made
moch better then they weare before when they did hould their
fixed nature in gould and siluer. For beinge thus altered
by solution, sep*a*ration, putrifaction, sublima*ti*on, and co*n*gelation of the sprite, they ar broughte into the beste fixation y*t*
can be, wh*er*by they ar made able to chang all imp*er*fect
bodies of the mine into most p*er*fecte fixation and cullour,
accordinge to the quallity of the fermente. For as solutio*n*
is the sublima*ti*on of the sprite, soe in like manner is the
congela*ti*on of the sprite the sublima*ti*on of the body, for soe
is it reared vp into worthines &c. Finis.

---

29 to resolue] for to solue **S2FS1**, to solue **GS4S3**
  muste] must also **S2S1S4S3**
30 vz] that is to bringe it **S2FGS1S4S3**
31 shin*n*inge] and shininge **S2FGS1S4S3**
33 For when] *in red ink* **A**
37 {the} water] th<at> {the} water **A**
42 soe that by this means] *in red ink* **A**
308v:
1 the reason] reason **S2S1S4S3**
3 then] then then **S2S1**
7 imp*er*fect] vnperfect **S2FGS1**
12 it] *abs.* **S2S1S4S3**

The 4th chapter of the 4th
sublimation

The iiiith sublimation is after putrifaction when the
blacknes passeth away, and the corrupte humiditie is drawen out. For
putrifaction is nothing ells but the mortification of moiste with
drie; which drines we calle the bodie, and the moiste humiditie
we call the sprite; which sprite in the beginninge, beinge cold and moiste,
doth kindly fly from the heate of the fier, but after putrifaction y^e
bodie soe conteyneth him that he may not fly because that then
the heavie may not be parted from the lighte, neither can y^e
lighte be drawen from the heavy. For compassion is the cause
therof and ixer is the begininge and endinge of all our worke.
And reason aproueth that it is needfull in putrifaction that it
be made blacke: for as by corruption the first form is destroid,
soe in like cause corruption is the generation of the second
forme. And putrifaction is nothing ells but that heate, workinge in
a moiste bodie, engendreth blacknes, which in the begininge of putrifaction
is called the ravens heed; wherfore putrifaction is needfull that
the bodies may be corrupted, without the which corruption can com
forth noe encrese. Witnes our sauiour Christe, which saith: Excepte
the grain die and rote in the ground, it bringeth forth noe fruite.
Soe lykwise in this worke, excepte the medison doe putrifie and
rot, he may not be made fluxible & melting. For yf he will
not melte before ♃ fly, then turneth it to naughte and is not
profitable for this worke; wherfore knowe that our putrifaction
is not foulle & filthy but faire and clean; for it is none other thing
but the medlinge of earth till all be broughte into on body of ye fier.

---

13 The 4th chapter of the 4th sublimation] The 4th sublimac*i*on Cap 8^m **F**, The forth
    subblimation Cap 8 **G**, The fourth sublymation Capitula 8 **S2**, The fourth sublima-
    tion Chap 8 **S1S4**, Chap 8 The fourth sublimation **S3**
    The 4th chapter] *in red ink* **A**
15 The iiiith sublimation] *in red ink* **A**
    In the left margin] Putrifaction | quid est **A**
17 of moiste] wth m of moiste **A**
19 which] the which **S2GS1S4S3**
28 And putrifaction] *in red ink* **A**
30 *In the left margin*] Ravens | heed **A**
32 which saith] where he saith in his gospell **S2FGS1S4**, where he say's in his gospell **S3** *
34 Soe lykwise] *in red ink* **A**
38 none] no **S2FGS1S4S3**
39 of earth] of earth with water and water with earth **S2S1S4S3** *

Therfore saith the ph*ilosofe*rs that vapoure, co*n*tinually movinge the   *   309r
matter, ceaseth not vntill all together be congealed into glorius
earth. For yf the matter should not die, the dyuers collours
should not appeare. Therfore, let the vessell be well closed
till all be made earth. For then it is called the dragon eating   *   5
his owne wing*es*, showinge dyuers collours, in diuers
manners, chauninge from collour to collour, vntill it
com to stedfaste white. For all the cullours of the wordle
shall seam vpon him when the black humidity is
dryed away. Of the variable coullours take noe heed,   *   10
for blacke and white be p*er*fecte coullours, and all the other
be mean collours, p*ar*taking of the other too leess or more
accordinge to the tyme and place of the worke soe that
ther is noe way from the on end to the other but by the
blacke and the white. And knowe for certain that the sprite   15
fixeth not but in white collour, for then is he abill to abid
the fier for the first p*ar*te, and this is the matter that all
ph*ilosofe*rs resteth one, which, after he is norished, turneth
argent viue and all other bodies of the mine into p*er*fecte
siluer &c. Finis.   20

    The 5th sublimation   309v
    Cap*itulum* 9^m

The 5th sublima*t*ion is of the medison w*hich* turneth vn-
p*er*fecte bodyes of mettalls into siluer, and that is to rear it into
gould, w*hich* we calle alsoe sublimation, forsoemoch that gould   *   5
    is brought

309r:
3   die] drie **S2S1S4S3**
4   Therfore let the vessell be] *in red ink except for the initial* 'T' **A**
5   *In the right margin*] The dragon | eating his | wing*es* **A**
6   in diuers manners] and in diuers maners **S2FGS1S4S3**
8   stedfaste] perfect **S2S1S4S3**
9   seam] bee seene **F**, shewe **S2S1S4S3**
     In the right margin] Diuers | collours **A**
10   Of the variable coullours] *in red ink* **A**
15   *In the right margin*] The fixing of | the sprit is in | whit collour **A**
309v:
1   The 5th sublimation] *in red ink* **A**
2   Cap*itulum* 9^m] Cap 9 **G**, Capitula 9 **S2**, Chap 9 **S1S4S3**
3   of] *abs.* **S2S1S4S3**
5   that] as **S2FGS1S4S3**

therby into a dignity and worthines aboue siluer. For this sublim{a}*tion*
is not to reare it from the loweste of the vessell to the highest, but to
make kind worke w*i*th kind, yeldinge his vertue without number.
For at every tyme it is solued and congeled, we say that it is sublimed;
and therfor solution, congelation, & multiplica*t*ion is the sub- 10
limation of this p*ar*te of the worke, subliminge it both into goodnes &
quantity. And therfore, yf our stone be made of simple kind, as
of earth, ayer, water, and fier, then is it made of the beste thing*es*
that can be, for then he gyueth everlastinge life to all those of
his kinde, wh*ich* as yet w*i*thout him remayneth dead, w*hich* life may 15
not afterward*es* be taken from them. Therfore, Hermes saith
as the worlde is made, soe is our stone, vz., of heavie & lighte, of softe
and of harde, of thicke and of thin, of hote & of moiste, of cold and of dry;
wherfore thes quallities once made equally to accord, beinge peaceably bound and p*er*petually knite together, is made the greate 20
elixar of life. For when they be eyther w*i*th other co*n*uenably
espoused, ther is engendred of them a temp*er*ate substau*n*ce, the
wh*i*ch noe fier by violence may overcom, nor noe filth of earth
defille, nor noe slyme of water bind, nor noe breath of ayer duske,
beinge thuse broughte into a spirituall body. Therfore, saith the ph*iloso*fer 25
of al thing*es* in the wordle our gould is moste worthy, for it is
leavene to the elixar, w*i*thout whom noe gould can be made. For
gould soe made gladdeth mans harte, conserueth youth,
reuyueth age, and healeth all bodyly malladies; for why he
is leaven to our paste, as reninge of milke in chease, as fier 30

---

6   into] to **S2FGS1S4S3**
10   and therfor solution congelation & multiplica*t*ion] *in red ink* **A**
16   Therfore Hermes saith] *in red ink* **A**
18   of harde] harde **S2FS1S4S3**
     of thin] thin **S2FGS1S4S3**
     of dry] drie **S2FGS1S4S3**
23   noe fier by violence] no violence of fire **S2FGS1S4S3** *
24   bind] bemud **S2S1S4S3**
25   Therfore saith the ph*iloso*fer] *in red ink* **A**
     ph*iloso*fer] ph*iloso*phers **S2FGS1S4S3**
26   gould] stone **S2S1S4S3**
27   whom] 'w' *written on top of earlier letter(?)* **A**
29   reuyueth] reneweth **S2FS1S4S3**
30   as reninge] renninge **S2S1**

in colles, as barme to alle, & muske in good electuaryes, the which
soferantly yeldeth delectable sauours. Finis.

    The dissolution of ☿ sublimed   310r
      Cap*itulum* 10^m

First make a malgam of ii^li of quicksiluer and of ii^li of puer
tyne in this manner: Melte your tyne in a meltynge pot
that shalbe large and greate ynoughe for your malgam. Then
haue another crucible, wherin youe muste make your
quicksiluer hote, and when he is hote, put him into the tynne
that is moulton, and stur them strongly together. Then
take the crucible from the fier, & stir it contynually vntill it
be somwhat could. Then caste the malgam into a pan
of clean water, and washe the malgam vntill the water
becom all blacke. Then pour out the same water, and put
to newe, and soe contynue washinge and powringe out
the water vntill it com cleare from the malgam without
any more blacknes. Then dry the said malgam, and
grind your ☿ sublimed and your malgam together. And
her note that to your malgam of ii^li of quicksiluer and ii^li of
tynn, youe muste put ii^li of sublimatum, and grind them
well together vpon a marble stone, and when they ar well
ground, lay them vpon a table of glasse that muste be made
for the purpose, hauinge a phillet about it to kepe in the
water that it haue noe way out but at the spoute, which muste be
made for the yssue of the said water as it dissolueth, which
table muste be set in a cold place or in som seller, for
otherwise it will not dissolue to any good purpose. And
when it is wholly dissolued and distilled from the plate of
glasse into som great still body or other vessell of earth, well

---

31   as barme to alle] *abs.* **S2FGS1S4S3**
32   sauours] sauoure **S2S1S4S3**
      Finis] &c finis **S2FS1**, &c **S4S3**
310r:
1   The dissolution of ☿ sublimed] *in red ink* **A**
2   Cap*itulum* 10^m] Cap 10 **G**, Capitula 10 **S2**, Chap 10 **S1S4S3**
3   First make] *in red ink* **A**
7   and when he is hote] but not so hott that ˏ{it} ~~you~~ fume and when he is thus hott **S2S1S4S3**, *abs.* **F**
16 And her note] *in red ink* **A**
17 note] you must note **S2FGS1S4**
25 And] 'A' *first written in red ink but later on filled in with black ink* **A**

leaded, then let it be filtered dyuers tymes, for soe shall y*ᵉ*
filth remain at every filtringe behind in the bottom of the
vessell. Then take the malgam and the quicksiluer that 30
remayneth after the filteringe, and put them together in a pan
of clean water, and washe them soe longe as any filth will
com from it and that the matter be clean and clear. Then
sep*a*rate your quicksiluer from the tynn by strayninge it through
a stronge lynnen cloth, w*hich* quicksiluer will serue youe again 35
for to make another malgam. Afterward*es* take youre tyn
that remayneth after the straninge, and lay it vpon the plate of
glasse somwhat thine. And yf ther be any quiksiluer remani*n*ge a-
monge the tyne, it will com ron from it. Then enkitch yo*u*r **
malgam again, and soe dissolue from tyme to tyme at y*o*ur 40
pleasure vntill youe haue soe moch as will suffice for youre
worke &c. Finis.

    The distillation of the forsaide 310v
    dissolued water of m*e*rcurie
        Cap*itulum* 11ᵐ

Thus muste the dissolued water of ☿ be distilled: First take
a still of glas somwhat greate, and fill the third p*a*rte therof w*ith* 5
the said dissolued water, and set theron his limbick, showerly luted *
vnto the bodie soe substancially that noe breath may goe out
throughe the lute; and when the lute is well dryed, eyther
againste the fier or in the sonn, set the forsaid still in balneo,
firste with a verie softe heate vntill the fainte water be com forth; 10
and when youe shall feelle the water com sharpe and byting
vpon your tonge — for soe muste youe proue it — then chang

---

30   Then] 'T' *first written in red ink but later on filled in with black ink* **A**
34   it] *abs.* **S2FS1S4S3**
36   Afterward*es*] *in red ink* **A**
39   enkitch] enrich **S2S1S4S3**, *longer pass. diff.* **G** *
41   as will] as will as will **S2**
42   Finis] *abs.* **S2GS4S3**
310v:
1   The distillation of the forsaide] *in red ink* **A**
     forsaide] afforesaide **S2FS1**, sayd **G**
3   Cap*itulum* 11ᵐ] Cap 11 **G**, Capitula 11 **S2**, Chap 11 **S1S4S3**
4   Thus muste] *in red ink* **A**
5   fill] still **S2S1**, destill **S4S3**
6   theron] on **S2FGS1S4S3**
     showerly] showerly {surely} **A** *

the receptorie and put to another, and lute him showarly
to the pipe of the limbick, and make your balneo soe hote
as youe may suffer your hand vnder the bottom of the glass;   15
and co*n*tynue it in the same heate soe longe as any heate will   *
distill; and when it leaveth dro{p}pinge by the same heate, suffer
your balneo to {be} colde, and take awaie the receptory, and stope
him suerly with wax. For with this water youe muste dissolue
your calce of Lune, as it is taught in the 21 chapter of this bock;   20
and with this water muste ☉ be calcioned, as it is taught in
the 23 chapter folowinge. For the office of this water is to dissolu
Lune, to calcin Sol, and to make fermente for the white
elixar; and the matter that remayneth in the glass, wherin
is the fier of nature, may not be distilled before your ☉ be calci-   25
oned; and p*r*osed with the dissolution of your gould as it is said   *
in the 23 chapter. And here is to be vnderstode that we
mean it to be don in this m*er*curiall water and to be precipitated in the sam, and not wi*th* rawe ☿ as he is mercuried,   *
for the rawe m*er*cury cleaveth not to the gould as the ☿   30
water doth because of his sulfurus drynes, w*hich* is taken from
the water in his sublima*t*ion, solution, and distillation of

---

14   soe] as **S2FGS1S4S3**
16   heate will] water will **S2FGS1S4S3** *
     In the left margin in Napier's hand] Calx of ☾ | calcind | y^e ☾ must be dissol|ved in this water **A**
18   to {be} colde] to coole **S2FGS1S4S3**
19   For with this water youe muste dissolue your calce] <u>For with this water youe muste dissolue your calce</u> *underlining in Napier's hand* **A**
     For with] *in red ink* **A**
     youe muste] must you **S2S1S4S3**
20   *In the left margin in Napier's hand*] Chapt*er* 21 **A**
21   calcioned as it is taught in the 23 chapter folowinge] <u>calcioned as it is taught in the 23 chapter folowinge</u> *underlining in Napier's hand* **A**
     taught] shewed **S2FGS1S4S3**
23   for] of **S2S1S4S3**
     In the left margin in Napier's hand] & ye calx of ☉ | must be dissolvd | in this water | c. 23 **A**
26   p*r*osed] p*r*osed {proceed} **A**, then proceede **S2FS1S4S3** *
27   the 23 chapter] the same 23th chapter **S2S1S4S3**
     And here is to be vnderstode] *in red ink* **A**
     vnderstode] vnderstande **S2**
28   this] the **S2S1S4S3**
29   mercuried] mercuryzed **S2S1S4S3**, mercurie **FG** *

the sprite; wherfore when thou readest this my bocke,
estem it not to co*n*teyne in every chapter therof, from the first
minerall vnto the laste animall, the p*er*fecte worke of phi*los*ofie, 35
but yf thou canst confer on place w*i*th another and labour
rightly, as I haue taughte, thou shalte moste showerly & *
certainly find the neste of the bird of humanity, who, imbrasing
his beloued, bredeth vp her yonge ons in the wing*es* of wynd,
& norisheth them in the bowells of the earth vntill they com 40
to p*er*fecte age to change as they be chaunged &c.
                              Finis.

  The fermentation of Luna 311r
  Cap*itulum* 12ᵐ

Fermente for ☽ᵃ muste be done and pr*e*pared in this
manner followinge: First youe muste make readie yo*u*r calce of
Lune, and ioine him in equall weighte w*i*th the m*er*curiall water, 5
well rectified, and put them together into a body of glasse, and set
one his limbicke, suerly luted and dryed againste the fier, and set
the still in balneo with a verie softe h<e>ate, and ther he will dissolue
and breath vp his fainte water; and when he is dissolued and all
his fainte water com from him soe that the whole matter is be- 10
com a gren oille, then let your balneo co{o}le, and take forth the
stille, and power the matter into a glas of circulation, and
close the said glasse by meltinge of the pipe, and set him in
balneo, and ther let him stand in soe temp*er*ate a heat as youe
may alwais suffer to hould your hand vnder the bottom 15
of the glasse vntill the matter be becom blacke as pitch by
rottinge and putrifiyinge the quallity of every eleme*n*te, with

---

33 wherfore when] *in red ink* A
34 to co*n*teyne in every chapter therof] in euery chapter therof to contayne **S2S1S4S3**
39 wynd] the winde **S2FGS1S4S3**
311r:
1 The fermentation of Luna] *in red ink* **A**
  of Luna] for Luna **FG**, for Lune &c **S2S1**
2 Cap*itulum* 12ᵐ] Cap 12 **G**, Capitula 12 **S2**, Chap 12 **S1S4S3**
3 Fermente] *in red ink* **A**
6 into] in **S2S1S4S3**
7 the fier] {the soonne} **S2S1S4S3**
15 alwais] alwaie **S2F**, *abs.* **G**
16 be becom] becom **S2FGS1S4S3**

troblinge, rysinge, and setlinge, showinge ynnumerable collors
& chaunging from cullor to cullor, vntill the matter shall caste
vp bubles white like vnto perrells. Then after 3 daies let
your fier cease, for then haue youe fermente readie for
the white elixar, which youe may encrease in this manner
followinge from tyme to tyme at your pleasure: Take
the calce of Lune and the water of ☿ in equall portions,
as is showed afore, and ioine them together in a glass, and set
them in balneo to dissolue for 7 dais. Afterwardes waygh
your fermente, and put vnto him his weighte of the said dissolution
in a circulatori vessell, and the wholle matter wil becom
black. Then closse vp your glasse as is aforsaid, and set him
in balneo with an indifferent strong heate, and ther let
him stand vntill it becom all white, which wilbe done in
shorter tyme then at the firste. And thus may youe multiply
your fermente at your pleasure, and soe shall youe encrese
his goodnes, vertue, and sapoure. Then ferment your
medison as is said in the 21 chapter followinge, and continue his fermentation as it is said in the same chapter, alwais
reseruinge a portion of your fermente for his multiplication;
for as youe multiply your medison by multiplication of ferment,
soe may youe multipli your fermente with the white oylle
of ☿ and the calce of ☽ᵃ, as ofte and to what quantity youe
will &c. Finis.

---

18   troblinge rysinge and setlinge] troubling riseing and setting **F**, burblinge risinge and saltinge **S2**, bublinge risinge and saltinge {setlinge} **S1**, riseing & setting **S4S3** *
     ynnumerable] inumeberable **S2**
19   *In the right margin*] Then make | oill of him & | ferment your | medison **A** *
20   bubles] burbles **S2S4**, brobles **G**
22   *In the right margin in red ink*] Multiplication | of the elixar **A**
23   take the calce of] *in red ink* **A**
24   portions] proportion **S2S4S3**, proorsion **S1**
25   afore] before **S2FGS1S4S3**
26   for 7 dais] 7 dayes **S2FGS1S4S3**
27   dissolution] solucion **S2S1S4S3**
31   him] it **S2S1S4S3**
32   *In the right margin*] But nevere | ferment lest | it be made | oylle **A** *
34   Then ferment your medison] *in red ink* **A**
35   *In the right margin in Napier's hand*] Chapt 21 **A**
36   fermentation] fermacion **S2S1**
39   the white oylle] the white of oylle **A**

The fermentation 311v
of ☉ Cap*itulum* 13ᵐ

The preparation of the fermente for the red elixar
muste be done as followeth: Let your gould be first m*e*rcurized,
and when it is thuse m*e*rcurized, let the ☿ be vapoured from the   5
gould in a crucible; and when the gould shalbe found w*ith*out fvminge,
take yo*u*r crucible from the fier, and caste your gould into an
earthen pan, beinge well leaded; wherv̇pon poure warm
water, and washe the said calce after the same manner 3 or 4
tymes, and then dry it vpon som place of veri cleare yron,   10
& when it is dry, put the said calce into a glasse with a streight
mouth, and poware vpon him his weighte of the oyll of ☿,
w*hich* we call our naturall fier. Then stop the glasse well &
showerly, and set him in balneo vntill all the whole matter
be dissolued into an oylle of gould collour; & when it is p*e*rfectly   15
dissolued, take out your glasse, yo*u*r furnace being first cold, &
poware the matter into a circulatori vessel, and set {one} yᵉ
pipe therof a small still hed, well luted to the crowne of the glass,
& set the glass in balneo that the fainte water may breath vp
into the heed; and when youe feelle that all the fainte water   20
is com forth, then nip your glasse as afor in the p*re*paration of yᵉ
ferment for Luna, and set the glasse again in balneo
w*ith* such a heate as youe may alwais suffer your hand vnder
the bottom therof, and let it stand vntill the matter becom
browne and then blacke, & co*n*tynue the same heate alwais,   25
and youe shall see the matter chaunge and showe forth inum*e*rable
collours in his seathinge, and when he hath lefte his boyling
soe that the bobles stand firmly vpon the earthe, then after
3 dais let your furnace coole and take out your glasse, and

311v:
1   The fermentation of ☉] *in red ink* A
2   of ☉] for goulde &c S2FS1
    Cap*itulum* 13ᵐ] Cap 13 G, Capitula 13 S2, Chap 13 S1S4S3
3   The preparation] *in red ink* A
10 place] plate S2FGS1S4S3 *
    cleare] cleane S2FS1S4S3, *abs.* G *
12 *In the left margin*] Naturall | fier is yᵉ | oyll of ☿ A
15 gould] good S2
17 {one}] ~~vnder~~ {one} A, ouer S2FGS1S4S3
28 bobles] brobles FG, burbles S2S1S4 *
29 and turne it into oylle] *abs.* S2FGS1S4S3 *

Edition 193

turne it into oylle. And then is your ferment made for the 30
red elyxar, which youe may alsoe encrease at your pleasur
with the calce of ☉ & the oylle of ☿, as is aforesaid in the ferment
for Lune. And note that this ferment of ☉ is moste fragra*n*t
of smelle, w*hich* sapoure encreaseth in every multiplication 35
as well as his vertue & goodnes. Then with this said sulfur
& oill ferment your medison as it is taught in the chapters **
of this bocke vntill it will flowe in the fier lix waxe. Finis. *

      To fixe a malgam * 312r
      Cap*itulum* 14

Make a malgame of xii oz. of ☿ and ʒ i of ☉, & put it
into a peare glasse. Alsoe make another malgam of ʒ xii of ☿
and ʒ i of Lune, and put it likewise in another pear glasse, 5
and put vpon eyther of the malgams 4 penni weighte of the elixars,
and stop the mouthes of your glasses, and put them in a furnace,
and cover them with ashes 4 fingers thicke, and make aboue them
a fier, neither to greate nor to smalle but a mean fier, and soe
let it be kepte 3 dais in the same heate; and when your ashes be cold, * 10

---

30 And then is your ferment made] for then yo*ur* ferment is made **S2FGS1S4S3**
32 aforesaid] saide afore **S2S1**, afore **F**, taught **G**
33 for] 'f' *written on top of earlier* 'ff'(?) **A**
      And note] *in red ink* **A**
36 & oill] *abs.* **S2FGS1S4S3** *
37 lix] like **S2FGS1S4S3** *
38 Finis] &c finis **F**, &c **G**, &c finis fermentum **S2S1**, &c the end of fermentation **S4S3**

312r:
1 To fixe a malgam] *in red ink* **A**, The sixte amalgame **S2S1**, The 6*th* amalgamation **S4S3**
2 Cap*itulum* 14] Cap 14 **G**, Capitula 14 **S2**, Chap 14 **S1S4S3**
3 Make a malgame] *in red ink* **A**
    ʒ i] i ʒ **S2FGS1**
    In the right margin] xii to on **A**
4 ʒ xii] xii ʒ **S2FS1**, 12 **G**
5 ʒ i] i ʒ **S2FS1**, on **G**
6 eyther] each **S2S1S4S3**
7 stop] nipp **S2FGS1S4S3**
8 cover] kiuer **S2S1**
9 neither] not **S2S1S4S3**
10 your ashes] the ashes **S2GS1S4S3**

take out your glasses, and youe shall find your malgam congealed and
fixed. Then put vpon each of your malgams 4 penny weighte of his
medison as at the firste tyme, and make the like fier over them for
the space of 3 dais, and doe soe 3 times, and the malgams will be
fixed to melte and flowe vpon a hote plate without fewming.   15
Then is it ready to caste vpon ☿ crud, on parte therof vpon 10   *
    partes, &
yt shalbe fine, and ☿ congealed therby as medison on vpon a hundred.
And yf youe will encrease your medison, dissolue the elixars in
too glasses with the ☿ vegitable, called the quintaescentia, into an oylle,
and drawe the water from it oftentymes vntill it com to an oyll   20
white or red. And then yf youe caste on drop therof on a thousand
drops of ☿ made hote in a crusible, it turneth it to perfecte medison,
wherof on parte changeth 4 partes of vnperfecte bodyes into ☉   *
or ☽, accor-
dinge to the quallity of the medison. And yf youe will make coa-   *
gulations of the said quintaescentia vntill it be turned into   25
    a thick   *
oylle cauled aurum potabile, it then healeth all infirmityes of
mans body, both within and without &c. Finis.

    The prologe to the mineral   ** 312v
        stone Capitulum 15

Forasmoch as in this first parte of the body of our worke of
alkamie it is necessarie that we begin with the introduction
of the minerall stone, I will therfore that noe man rashly   5

---

14 doe soe] so doe **S2S1S4S3**
17 as] is **S2FS1S4S3**, *longer pass. diff.* **G**
18 And yf] *in red ink* **A**
    In the right margin] To turn the | elixars to | oille **A**
21 on a thousand] vpon a thowsande **S2FGS1S4**
    In the right margin] To encrese | your med|ison **A**
26 *In the right margin*] Aurum | potabile **A**
312v:
1   The prologe to the mineral stone] *in red ink* **A**
2   Capitulum 15] Cap 15 **G**, Capitula 15 **S2**, Chap i5 **S1S4S3**
3   Forasmoch] *in red ink* **A**
5   I will therfore that noe man rashly take in hand the said minerall stone] I haue therfore
    thought it good as by the waye of a shorte admonicion to giue warninge vnto such as
    shall hereafter entre into the practise of this most worthie science not rashlye to take in
    hande the saide minerall stone **S2FGS1**, I have therefore thought it good & by the way
    of a short admonition to give warning vnto such as shall hereafter enter into the practice
    of this most worthy science not rashly to take in hand the said minerall stone **S4S3**

take in hand the said minerall stone but vpon good aduice,
and the practice to be as well prouided for of char*ges* for
errours as for the thing itself. For as it seameth to the
vnskilfull to be a way moste easy, leaste of coste, and sone
done, soe is it found by practice to be a way full of daungers
wherby the worke may sone be loste. And the cause therof is
in this minerall way the water of the stone is often drawen
from his bodie and ofte poured again vpon his earth; by
reason wherof w*i*thout great takinge heed, the sprite may
fly away, wherby may com the death of the stone; the w*h*ich beinge well kepte in, the stone shall therby be made lyuinge &
soe shall attain to his vertue op*er*atiue. And when the matter is broughte into an oylle that will not fly nor congell
in stronge fyer, then is it a p*er*fecte medison for all gould
alkamy, sufficient good for the gouldsmith. But it entreth
not medison. For noe gould alkamy is cordiall althoughe
it be never soe well & p*er*fectly made because he passeth not
into his p*er*fection by the gates of putrifaction & fermenta*t*ion,
but is elixerate only with the fier againste nature, w*h*ich
fier of himselfe is verie mortall. And by the reason of the
shortnes of his workinge, he is not purged to the purity of his
p*er*fection; wherfore he is not multiplicable, as the vegitall
and animall stones, of whom is made the goulden drinke
and the greate elixar of life, as herafter more at large
shalbe declared in the 25, 26, & 27 chapters of this treatis.
Finis.

---

6   good aduice] good aduise taken **S2FGS1S4S3**
7   prouided for] prouided **S2FGS1S4S3**
8   as for the thing itself] that maye happen in the workinge of the saide stone as also for the ordenary charges beforehande appoynted to the same to be passyd w*i*thout the defect of errou*res* **S2S1S4**, y*t* may happen in y*e* working of y*e* s*ai*d stone as also for y*e* ordnary charges beforehand appointed for y*e* same & passed w*i*thout any defect of errors **S3**, *abs.* **FG** *
9   sone] soonest **S2S1S4S3**
11   is] is that **S2S1S4S3**
14   reason] the reason **S2FS1S4**
27   is] is is **S2**
       vegitall] vegetable **FGS1S4S3**, vegitably **S2**
28   stones] stone **S2FGS1S4S3**
29   and the greate] & the great and the great **S2**, *abs.* **S3**
31   Finis] &c Finis **S2FS1**, *abs.* **S4S3**

The first parte of our stone
called the mineralle stone Cap*itulu*m 16

In this firste p*a*rte of the min*e*rall stone, let ☉ be m*e*rcurized w*i*th 24
pr*o*portions of naturall adrop into such a {thinnes} that it may together with that it is ioyned be passed thorowe the middell of a fine
lynnen cloth without any globe remayninge; for then may it be p*re*ci-
pitated in a longe vessel with a straighte mouth and stronge,
beinge luted over all, savinge the top. The calce, once mercurized, muste be vapored clean from the adrop; and when he is
vapoured that he fume not, let the calce be washed 3 or 4
tymes, and at the laste tyme dry it again the fyer vpon a clean
plate of yron, and, beinge made dry, put him into the said glasse,
& put to 3 pr*o*portions of the first water of mercurie, and set the
said glass in a dry furnace, first with a mean fier for the
space of 24 howars. Afterward*es* encrese yo*u*r fier, and soe
contynue it vnto the full end of 21 dais; wherin let the matter bruble
and calcin into a red pouder by violence of the fier; after w*hi*ch tyme
let the said pouder be dissolued in the fier againste nature, w*hi*ch
fier must thus be made: Take of the beste and cleaneste vitriall &
vermilion of equall weighte, grind them well together, and dry
them in the ☉ or against the fier. Then drawe your ardente
water and powre it vpon your vitriall and vermilion, and

*   35

*

40

45

*   313r

5

---

32   The first parte of our stone called the mineralle stone] *in red ink* **A**
     our] the **S2FGS1S4S3**
33   stone] stone &c **S2S1S4S3**
     Cap*itulu*m 16] Cap 16 **G**, Capitula 16 **S2**, Chap 16 **S1S4S3**
34   In this] *in red ink* **A**
     firste] *written on top of earlier* parte **A**
35   {thinnes}] <u>thicknes</u>; thinnes *inserted in the right margin as a correction* **A**, thines **S2F-GS1S4S3**
36   that] that that **S2S1S4S3**, *longer pass. abs.* **G**
42   again] againste **S2FGS1S4**, before **S3**
313r:
1   vnto] vntill **S2FGS1S4**, till **S3**
     bruble and calcin] calcyne & burble **S2S1S4**, calcine & buble **S3**
2   w*hi*ch] the w*hi*ch **S2S1S4**
3   *In the right margin*] Fier against | nature quid **A**
4   take of the] *in red ink* **A**
     beste] finest **S2S1S4S3**
5   grind] and grinde **S2FS1S4S3**

put it into a still of glas, & set one his limbeke, suerly luted, &
set the still in balneo, and drawe vp all the weke water. Then
remoue the still into dry fier, first with a mean heate, afterward with
a stronge fier as the mannor is with the sharpe water of
philosofers, and the sprite of quintaescentia of vitriall &
vermilion, which
principally maketh the minerall stone, shalbe minerales
& ioyned with the quintessence of the ardent water, and this distillation
must be continued vnto the fifte tyme; and let the matter of vitriall
& vermilion be well dryed after every distillation, and at every
distillation put to newe stufe of the vitriall & vermilion, dryed
as at the first; & when youe haue performed your forsaid distillations, powre vpon your calce his weighte of the said water,
& set your glasse in balneo 7 dais; and then substracte the
water from the calce with a lente fier, and again calcin y^e
matter into a dry pouder. Then rectifie the water, and pour
it again vpon the calce, and set it in balneo to dissolue as
before 7 dais; after the which tyme drawe vp his watter again
with the said heate of balneo, rectifie the water, and again calcine
the earth, and contynue this order of imbybinge, distillinge, and
calcioninge vntill the water haue noe sharpnes, and the earth
will congealle noe more but remain in the bottom of the vessell
moulten like wax. Then put the said substance in a vessell of
calcination, and continue it in a fornace with a convenient fier,
made both vnder and aboue, by the space of 7 dais; and yf
it continue in the same heate his oyly substaunce without congeling,
then proue his fixation vpon a fiery plate of clean yron. Yf
it fewm not vpon the plate but melte and runn as wax
without makinge any noise or hissinge, it is well & fixed, and is
able to change all vnperfect bodies into gould alkamy, sufficient
good for all workes of the gouldsmith. But yf youe perceiue y^t

---

8   into] in S2S1S3
    & set] set S2FGS1
11 stronge] stronger S2G
    water] water~~s~~ A *
12 quintaescentia] the quintaecency S2S1S4S3
20 substracte] subtrahe S2FS1
24 7 dais] by 7 dayes S2S1, for 7 dais S4S3
    the which] which S2S1S4S3
33 then proue] then proue then proue S2
35 is able] able S2GS1S4S3
37 But yf] *in red ink* A

it be not fully fixed by the first water, then ad vnto it a newe
water, wel rectified, and in equall quantity of the said matter, &
procead in like manner as youe did with the first water con- 40
tynuinge vntill all the sharpnes be gone out of the watter; and
when ther shall remain any more strength in the water, then let   *
the matter be calcioned in a stronge fier by 3 dais. Yf it ther congeall not but stand moulten in the same heate, proue his fixation
again as is aforsaid. Yf it moulte & ron vpon the plate without [fewming], 45
colouringe the plate with a depe colloure of gould, then is he fixed & is a 313v
good elixar of all gould alkamy. But this elixar of the minerall
stone entreth not medison because he is elixerate only with the
fier againste nature, and is mad elixar without putrifaction;
wherfore he is of nature to mans bodie veri mortalle. But of 5
the minerall bodies he healeth their leprouse diseas & maketh
of an ymperfecte bodie a perfecte bodie, abydinge every examination and
tryall of the fier. I haue in the first parte of the mineralle
stone in the chapter folowinge made the ardent water of
the oille & water of ☿, distilled both together, because that the oill 10
of ☿ is his vnctuosity, soner fixed, and is moch agreinge with   *
the nature of gould then the mercuriall water. And yf youe
will haue your medison multiplied and encresed in
goodnes, then ferment him with such a fermentation as
we haue spoken of. And soe shaull youe haue your desier for 15
all gould alkamy of the firste order. Finis.

---

38   a newe water] newe water **S2S1S4S3**
42   any] no **S2GS1S4S3** *
45   moulte & ron] runne & molte **S2S1S4S3**
       [fewming]] *catchword* **A**
313v:
2   of all gould alkamy] for all goulde alkamye **S2FS1S4S3**, of ye munerall stone **G**
      But this elixar] *in red ink* **A**
8   I haue in the] *in red ink* **A**
      the first] this first **S2FGS1**
9   in the chapter] as in the chapter **S2S1S4S3**
11   ☿ is his vnctuosity] mercurie is by his onctuositie **S2GS1**, mercurie is for his victuositie
      F, ☿s vnctuositie **S4S3**
      moch] more **S2FGS1S4S3** *

The seconde parte of the minerall stone
   Cap*itulum* 17
In the second p*ar*te of the min*e*rall stone, gould is veri excelle*n*tly
elixeratid with the fier againste nature compounded w*i*th the
fier of nature, beinge distilled both together; and this water beinge minerall & vegitall, it worketh co*n*trary thing*es*. For so moch
as it dissolueth, soe moch again it congealeth and lifteth in glorious
earth, w*hich* of phi*lo*s*ofe*rs is called the secrete dissolution of the stone,
w*hich* is not without the co*n*gelation of his water. Therfore, in the
name of Christe, let vs goe to the practice vpon the calce of the
bodie, beinge first prep*ar*ard and made into verie fine pouder, as it is
said in the chapter nexte before. Then let the same pouder be
put into an apte vessell for the purpose, and poure vpon it soe moch
of the co*m*pounded water as may cover the calce halfe the thicknes
of a thumb, and the water will straightway boill in his vessell w*ith*-
out any externall fier dissoluinge and liftinge it vp into the
forme of yce, with dryinge vp of itselfe, which after muste
be remoued by the hand of the workeman. And the calce remai-
ninge muste again be well dryed by the fier; and then put to soe
moch water as before, and soe continue the same order of workinge vntill all the calce be wholy dissolued, and then againe
dryed & made into fine pouder, and let it be put into a good qua*n*titi
of the fier of nature, and well rectified, and, the vessell beinge
well stopped by the means of heate outwardly administred, yt
may be dissolued into oylle, w*hich* will sone be done by the reson
of the simplicity of the matter; the w*hich* thinge beinge done, substracte the water from it by distillinge in balneo vntill ther
remain in the bottom of the vessell a thin oylle; the w*hich* oille the

---

17 The seconde parte] *in red ink* **A**
    stone] stone &c **S2S1**
18 Cap*itulum* 17] Cap 17 **G**, Capitula 17 **S2**, Chap i7 **S1S4S3**
19 In the] *in red ink* **A**
23 in] into **S2S1S4S3** *
27 prep*ar*ard] prepared **S2FGS1S4S3** *
30 cover] kiuer **S2**
31 water] mater **S2S4S3**
32 into] to **S2S1**, in **GS4S3**
38 and let] then let **G**, let **S2GS1S4**, *abs.* **S3**
39 and well rectified] well rectified **S2GS1S4S3**
42 substracte] subtrahe **S2S1**

more oftner it shalbe resolued and the water taken from it, soe
moch the more shal it be made thine and subtill. And this
be calcioned into dry pouder w*ith* softe fier, imbibed, and afterward
  fermented,
& at every ferme*nta*tion strongly calcioned vntill it will not dry but stand
in the fier fixed & flowinge as wax; for then is he elixar that will
change arg*ent* viue and all imp*er*fecte bodies of the mine into puer gould alkamy,
good for all work*es* savinge medison. The same oill, beinge put into
kymia sealed to putrifie in fier of the firste degre, may be made y*e*
greate elixar both for white and red, according to the quality of the
fermente, which may be auugmented w*ith* fermentall oille infinetly.
And when it is cast vpon a malgam and fixed vpon the same, it turneth it into elixar, w*hich* muste be caste vpon vnp*er*fecte bodies. Moreover, the white elixar muste be augmented with m*er*curiall water,
and the red with m*er*curiall oille, w*hich* both muste be had of ☿ dissolued
only by himselfe. When we say by himselfe, we mean another
dissolution that muste be had of his owne substaunce, and not
of tyne or sal armoniak, but as it is said in the vegitall
stone. Finis.

Cap*itulum* 18   The thirde parte of the mineral stone

This third p*ar*te of the minerall stone showethe howe the bodie of
gould is elixerate with the fier againste nature, holpen afterward*es* with
the fier of nature, as hearafter shall be showed, and the manner is

---

45 more oftner] oftner **S2S1S4S3**
  resolued] dissolued **S2FGS1S4S3**
46 the more] more **S2S1**
314r:
1 be] beinge **S2GS1S4S3** *
  softe fier] a softe fire **S2FGS1S4S3**
4 gould alkamy] alkamy **S2S1**
5 The same oill] *abs.* **S2S1S4S3**
  into] yn **S2GS1**
13 we say by himselfe] wee say of himselfe **S2S1S4S3**
16 Finis] Finnis &c **G**, {&c} Finis **S2S1**, &c **S4S3**
17 Cap*itulum* 18] *partly in the left margin* **A**, Cap 18 **G**, Capitula 18 **S2**, Chap 18 **S1S4S3**
  The thirde parte of the mineral stone] *in red ink* **A**
  stone] stone &c **S2S1**
18 This] *in red ink* **A**

this: Dissolue the bodye of gould in the fier againste nature,
    which bodie oughte
firste to be mercurized and precipitated into verie fine pouder, which pouder
muste afterwardes be calcioned in the first liquor that distilleth out of
Lunarie, which we take to be the firste parte of our ardente
    water, wher-
with the fier againste natuer is made. And the said water beinge a-
gaine taken from it in balneo, let the matter be calcioned into dry
pouder with softe fyer, and then put him againe into the fier against
nature to dissolue in balneo. And when it is dissolued into citrin,
cleare, and veri fair water after 7 dais, let the same water
be drawen from yt by the same heate of balneo, and ther will
remain in the bottom of the stillatorie a thick oyll; whervpon let
the water, well rectified, be poured again, and set in balneo
other 7 dais, and again drawe vp his water from him, and soe
contynue the same order of workinge vntill the water be wekned
as well water; and doe in like manner as is said of the first
water with newe water, well rectified, by powringe one and
drawinge vp vntill all the sharpnes of the water remaine in
the substance of gould. And here is to be noted that after
everie distillation the matter muste be calcioned into dry pouder
with softe fier, and after the laste calcionation let ther be poured vpon yt fyer of nature, perfectly rectified, and let the quantity
be doble to the substaunce of the bodye soe dissolued, and let the
vessel be wel stopped & set in balneo 7 dais, and then the water shalbe
found verie citrine, which by his attractiue vertue hath taken from
the fier againste nature his tincture. Then after the said 7
daies, let the water of nature soe tayned be warely emptid
into another vessel, hauinge a straighte mouth wherby he
may be the more surely clossed. Then let other water of

---

21 this] thus **S2S1**
      dissolue] *in red ink* **A**
22 verie fine pouder] fine pouder **S4S3**, a very fine powder **S2S1**
27 softe fyer] a softe fire **S2FS1**
29 same] saide **S2S1S4S3**
38 And here is to be noted] *in red ink* **A**
40 softe fier] a softe fier **S2S1S4S3**
314v:
1 set] put **S2S1S4S3**

nature be poured vpon the oille of the bodie of gould, and let ther
be made balination as before; and when the water is tayned
with the oill of gould, then let it be as before warely emptyed into the forsaid vessell, and again surely clos<e>d vp the glasse,   * 10
and power newe water of natuer vpon the said substaunce
of gould, and soe contynewe vntill the oill of gould will noe
more tain the water of natuer, for then yt is a signe yᵗ   *
all the tyncture is drawen out from the fier againste nature of the body dissolued. Then let the said evacuations be put   15
into a stillatory vessell with his limbick, and with an easy fier
lifte vp the water of nature vntill in the bottom ther remain
an oille; whervnto put newe fier of nature, verie well rectified, and after the matter hath stand in ballneo 7 daies,   *
let the same water be substracted from yt again by dist-   * 20
illinge, and soe contynue the worke with powring one and
drawinge vp vntill ther be made a subtill oille without any feces or grosnes remayninge therin. And be suer
that nothinge of the water of nature be lefte behind but only yᵉ
substaunce of gould remayninge in the bottom of the vessell, turned   25
into a moste puer oille. This oille, being put into kymia
& sealed to putrifie, may be made the great elixar by the
same order of workinge as is said in the chapter going before.
Nowe to com to the practice as we entend vpon the said oille: *
Sublyme argent viue with Romain vitriall and coulbuste salte;   * 30
and afterward let it be sublimed 3 or 4 tymes by itselfe, partly   *
from his grosse feces; the which thinge beinge done, and the
☿ made into pouder, let it be put into his vessell fixatorie,
and put to it a certain quantity of the forsaid oill of goulde,
wherwith it may be scarcly couered. And put the vessell,   35

---

7 the oille] oyle **S2S1S4S3**
and let ther be made balination as before and when the water is tayned with the oill of gould] *abs.* **S2S1S4S3**
10 clos<e>d] close̶d̶ **S4**, close **S2FS1S3** *
19 hath] haue **S2S4**, has **S3**
20 substracted] subtrahed **S2FS1**
26 into kymia] into a kymea **S3**, in kymea **S2S1**, in fymo **G**
27 & sealed] sealed **S2FGS1S4S3**
29 Nowe to com] *in red ink* **A**
entend] entended **S2S1S4S3**, *abs.* **G**
30 coulbuste] combust **S2GS1**, combustable **S4S3** *
31 partly] purely **S2FGS1S4S3** *
35 be scarcly] scarcly be **S2S1S4S3**

well closed, vnder a softe fier vntill the matter be well ioyned together. Then afterward let it coolle, and then bruse it on a
clean marble stone, & incerate it with the said oill of gould,
and let it be warely put again into his vessell, and set vn-
der the fier as before; and let the vessell stand doble the tyme as   40
at the firste that the matter may be the more strongly compacte
together. And this ruell in this worke is generally to be obserued that the vessell with the matter in it to be fixed be alwais
set vnder the fier, & the fier augmented from tyme to tyme; & let   315r
  yᵉ inceration
be continued vntill the argent viue be perfectly fixed with the oill of ☉,
  which muste be
proved vpon a fiery plate of siluer. And yf it shalbe found fyxed, yet
let it haue another inceration, and put it vnder a stronge fyer 3
dais. Then grind it on a stone, and encere it ther vntill yt be thick   5
as ointmente, and after it is dryed with a softe fier vntill yt will
noe more dry but that it be made to the fier as flowinge wax.
For then it transmueth siluer substancially into gould
alkamy, veri good for all workes of the gouldsmith. But yt entreth
not medison but work<e>th in poware according to his elixe-   10
ration &c. Finis

---

36 matter] matters **S2FS1**
37 then bruse it on] when its cold bruise it on **S3**, when it is coulde let it be brused vpon
  S2FGS1S4
38 incerate] increse **S2FGS4S3**, increse {incere} **S1**, infues **G**
41 matter] matters **S2FS1**
42 generally] geneally **S2**
315r:
3 yet] *abs.* **S2S1S4S3**
4 *In the right margin*] 3 **A**
5 encere] increace **S2S4S3**, increase {incere} **S1**, grind **G**
6 and after it is dryed with a softe fier] and after it is dried with a softe fire let it be
  calcined with a stronge fire which thinge don increace it oftentimes & dry it with a
  softe fire **S2S1S4**, & after its dry'd with a soft △ calcine it with a strong △ which
  done increase it often & dry it with a soft △ **S3** *
8 For then] soe the For then **A**
  transmueth] trancemuteth **S2GS1S4**, transmutes **S3**, tranfumeth **F**

The 4ᵗʰ part of the minerall stone
    Capitulum 19

In this 4ᵗʰ part of the mynerall stone is showed howe to elixerat
the bodie of gould with ☿, is in this manner to be vnderstode
    that kibrik
beinge dewly prepared, that is, mercurized & precipitated, ought to be dissolued into 12 proportions of the water of the sea; and when it is
dissolued, strain the matter thorowe a clean lynnen cloth, and the balle
that remayneth in the cloth put in kymia, and set it over an easy fier,
as it weare the heate of the sonn, vntill the red culler begin to appere
in the vtter face therof. Then ad vnto yt the 4ᵗʰ parte of the forsaid
water, which oughte to be kepte cleane and close in a vessell apte for the
purpose; and as before, set fyer to yt, and let yt dry with easy heate.
For in this worke, howe moch leess is put of the sprite and more of yᵉ
bodie, soe moch the better solution shalbe made, which is ever with the
congelation of his water. Youe must therfore beware that the belly be
not made over moiste, for then shall not the matter receyue drynes;
& this manner of imbibition muste be contynued vntill all the water be
dryed vp into a bodie. Then let the vessell be sealed after the
manner of philosofie, and aptly place the said glasse in his furnes, putting
vnder yt a mean fier, which may not wax cold from the fi\<r>st howar it
is set in the furnes vntill the accomplishmente of the wholle worke. And
when the matter is sublimed, let it again be compelled downe with
fier conveniently set over yt to dissend by lyttle and lyttell without violence. That done, let yt again {be} sublimed, and soe let the
    sole of the
sonn vulger, which is our mean oyntment, without the which ☉ the sprite
is not ioyned with the body. Therfore, let yt ascend into heaven,
    and a-
gaine let yt descend from heaven into earth and that ther be made a
substaunce not soe hard as the body, nor soe softe as the sprite, but houldinge
a mean dispotition, standinge fyxed and abydinge in the fyer like a

12 The 4ᵗʰ part of] *in red ink* **A**
    stone] stone &c **S2FGS1**
13 Capitulum 19] Cap 19 **G**, Capitula 19 **S2**, Chap i9 **S1S4S3**
14 In this 4ᵗʰ] *in red ink* **A**
15 vnderstode] vnderstande **S2**
17 into] in **S2S1**, *abs.* **S4S3**
19 in kymia] in fymoe **G**, ₐ{in}to kymea **S2S1S4S3**
31 mean] newe **S2S4S3**, newe {meane} **S1**
33 again be compelled] bee agayne compelled **S2S1S4S3**

peace of whyte wax moulten in the bottom of the vessell, which
    must be no-
rished with milke & meate vntil the quantity therof be encreased at thy
pleasure. This medison fermented into red with a portion of Soll
dissolued in the reddishe water of the sea & putrified into
sulfur, separatinge the first forme from the matter, and soe
may it be broughte into a purple colloure with stronge fyer
contynually administred vnto yt. And thus in the minerall
way of the white into the red shalbe made the true elixar, &
this elixar white or red shalbe encreased a hundredfold
in the vertue of his quintescens beinge fixed vnto yt, yf it be dedused
with the fier of nature into a thine oylle, which muste be don
in a circulatory vessell. For then the leste drop doth congeall a hundred dropes of ☿ into medison, which turneth
all other bodies into Sol or Lune accordinge to the fermente. Finis.

  The laste parte of the minerall stone
    Cap*itulum* 20

This laste parte of the minerall stone showeth howe ☿ only
is elixerate; wherof I entend not moch to entreate because
of the tediousnes of his workinge. But yf he be ioyned with
fermente, putrified, and made meltynge, he wilbe a good elixar,
able to change argent viue into medison and all vnperfecte
bodies of the mine into siluer, the beste that can be. For by such
an order of workinge of ☿, ioyned with his fermente, may be
made the complete elyxar, both of white and red, and is
multiplicable as is said in the other chapters &c.
And forasmoch as the 4 chapters hearafter folowynge be
not wroughte after the manner of the minerall stone, nor all-

---

41 peace of whyte wax] wh˰{i}te pece of wax **S2FS1S4S3** *
315v:
2 reddishe] reddest **S2FGS1S4S3**
13 Finis] &c Finis **S2FS1**, &c **G**, *abs.* **S4S3**
14 The laste parte of the minerall stone] *in red ink* **A**
15 Cap*itulum* 20] Cap 20 **FG**, Capitula 20 **S2**, Chap 20 **S1S4S3**
16 This laste] *in red ink* **A**
  ☿] *our* mercurie **S2S1S4S3**
20 vnperfecte] imperfecte **S2S1S4S3**
25 And forasmoch] *in red ink* **A**

together after the manner of the vegitall and animall stone,
I haue therfore set them as a mean betwen both to a farther
relation, as shalbe showed hearafter, not terminge them the
on nor the other, because of their diuersitie in workinge &c.
Finis cap*itulo* 20^(mo). 30

  Cap*itulu*m 21 The preparation of the white   *   316r
  oille for the animalle stone

In this 21 chapter, beinge different from the way of the minerall stone, shalbe showed the manner and way howe to p*r*epare
the white oille for the animall stone, w*hich* must be done as folowethe: 5
Let the bodie of ☾^a be m*er*curized and made into verie fine pouder,
w*hich* pouder of som is called the calce of ☾^a, and of other som the
ashes of Lune. For Moryen saith: Set not at naughte the ashes y^t   *
remaine in the bottom of the vessele, for they, saith he, ar the diadem
of the harte because they ar of things everlastinge. But the calce   10
cominge in this p*ar*te of the worke is not the ashes that Morien
speaketh of, for he meaneth a calce or ashes that commeth in
another place of the worke; for in those ashes, co*m*inge in that p*ar*t
of the worke, is the sulfur found that wise men seaketh; wherfore
knowe that this calce of himselfe is not a p*er*fecte worke but a good be-   15
gininge when he is suerly p*r*epared, w*hich*, after his p*r*epara*tion*, must be put into
the first liquor that distilleth from Lunarie, beinge firste taken   *
from the roote, that is, adrop, w*hich* we muste vnderstand to be ☿ dissolued.

---

30 &c] *abs.* S2GS1
316r:
1 Cap*itulu*m 21] *in red ink* A, Cap 21 FG, Capitula 21 S2, Chap 21 S1S4S3
 The preparation of the white oille for the animalle stone] Of the annymall stone G,
 abs. S2FS1S4S3 *
3 In this] *in red ink* A
4 the manner and way] the waye and man*n*er S2S1S4S3, ye waye G
 In the right margin] The whit oill | for the animall | stone A
6 let] *in red ink* A
8 Moryen saith] *in red ink* A
 Moryen] Morien the ph*iloso*pher S2FGS1S4S3
12 meaneth] meanes S2S4S3, *longer pass. diff.* G
13 place] parte S2S1S4S3
15 *In the right margin*] Adrop quid A
16 p*r*epared] preperate S2F
18 the roote] his rote S2S1S4S3
 that] w*hich* S2S1S4S3, *longer pass. diff.* G

For therin may the said calce be broughte into puer watter, which will
in his dissolution cloth himselfe with a gren vesture, and therfore
it is called the gren lyon, for the body soe dissolued in the said water
wilbe of the collour of an emerald. This water, before the dissolution
of the bodie, conteyneth but 3 elementes, vz., water, ayer, and fier,
and the elemente of water philosofers calle his moistnes. The element
of ayer they calle the kind of water that maketh the body to melte.
The element of fyer they calle the kind of water that burneth, calsioneth,
and dissolueth the bodie into form of a sprite. For when the body is
thus dissolued, then is ther elixar engendred and matrimony
betwen the bodie and the sprite, as in the firste parte. For Lune in
this manner dissolued and returned again into ₍{the}₎ simplicity of his
    first beinge
before he was fixed into a body of the myne may be slainte in the
wombe of his mother, and putrified in sighte to a filthy matter, wherof afterward may be made a bodie naturalle, sufferinge the violence of the fyer. Therfore, when ☽ is thus dissolued, and the vessell
well closed, let it be set in balneo 7 dais; after the which tyme let the water
be substracted from the body by the strongest heate of balneo. The vessell
beinge well stopped wherin the body remayneth, poure the water into a
vessell with a lymbicke, and distill the water dyuers tymes vntill the
weaknes be drawen out and the calce somwhat dryed. Againe
power on the said water vpon the body of Lune, and set him in balneo
as before; & when he hath stand in balneo certain daies, drawe vp his
water as at the first tyme, and let the water be distilled soe that ther
remain behind noe parte of his flegma or fatnes; & thus doe from tyme to
tyme vntill the water haue noe sharpnes, which may be proued by the

---

19   watter] 'w' *changed from earlier* 'm' **A**
20   himselfe] 'm' *changed from earlier* 's' **A**
       In the right margin] Gren lyon **A**
26   water] the water **S2S1S3**
       burneth] burnes **S3**, brenneth **S2S1**, brineth **G**
27   form] the forme **S2FGS1S4S3**
31   slainte] slayne **S2FGS1S4S3** *
32   in] *written on top of earlier* to **A**
34   Therfore] *in red ink* **A**
       In the right margin] The ☽ dissol|ued must be | set in balneo | 7 dais **A**
35   the which] which **S2S1S4**, *abs.* **S3**
36   substracted] subtrahed **S2FS1**
43   fatnes] feyntnes **S2FGS1S4S3**

taste of the tonge; and when all the strength of the water shall remaine 316v
in the bodie, not ascendinge by the heate of balneo, then take
newe water, well rectified, to the full quantity of the matter,
& power it vpon the substance of Lune, and set it in balneo as
is said of the other water, and followe the aforsaid order in   *   5
drawinge vp, rectifiynge, and powringe one vntill the water
haue noe taste of sharpnes but as a common water, and the
matter in the bottom of the vessell remain a cleare, subtill
oylle. Then let the said oyll be put in a circulatorie vessell, philosofically sealed, and set in fier of the firste degre to putri-   10
fi vntill it turn blacke as moulton pitch; for then is ther con-
<iu>nction matrimoniall betwen the body and the sprite, as in the
second p*a*rte. And of this coniunction speketh Morien the   *
ph*iloso*fer, sayinge: But yf the vnclean bodie be putrified, clensed, and
made white and ioyned w*i*th the soulle, thou haste not don right   *   15
in this worke. Nowe, therfore, when the black collour is p*er*fect
soe that he burble, settell, and putrifie blacke as pitch moulton
on the fyer, w*hich* is brought to passe by reason of ayer ascendinge &
dropes co*n*tynually fallinge vpon the mortified bodyes,
the water decreasinge and the water begini*n*ge to dry, then   20

316v:
5   *In the left margin*] This is the wife set | her not to putrifie | without her hus|band for a woma*n* | ca*n*not bring forth | a child w*i*thout a | man **A** *
7   a common water] co*m*mon water **S2GS1S4S3**
8   cleare subtill] subtill clear **S4S3**, cleare and subtile **S2FS1**, subtill **G**
9   in] into **S2GS1S4S3**
12 matrimoniall] matrimonyallie **S2S1S4S3**, matremony **G**
13 And of this coniunction speketh Morien the ph*iloso*fer sayinge] *in red ink* **A**
14 sayinge] when he saith **S2FGS1S4**, when he sai's **S3**
      But yf the vnclean bodie be putrified] but if y*e* vnclea<n> bodye be ‸{not} putrifyed **G**, this is the wife set her not to putryfie w*i*thoute her husbande for a woman cannot bringe foorth a childe w*i*thoute a man but if th'vnclene body be putrified **S2S1S4**, this is y*e* wife set her not to putrifie without y*e* husband for a woman ca*n*not bring forth a childe without a man but if y*e* vnclean body be putrifi'd **S3** *
      and made white] made white **S2S4S3**
15 *In the left margin*] The black collour | will com in the | on as well as in the | other but never | p*er*fectly except | they be together **A** *
18 reason] the reason **S2S1**
20 the water begini*n*ge] y*e* mater begining **G**, that the water begin **S2S1S4S3**, the matter beginn **F**

somwhat strength the fyer, and in that mannor co*n*tynue it vntill the water be dryed vp and that ther appeare brubles
vpon the earth in the forme of p*er*elles or the eyes of fishes. And
in this order let the fier be co*n*tynued vntill all the matter be dryed
vp into a whit pouder or clear and shin*n*inge, w*hich* is our whit sulfur 25
ready to be made elixar. Then without all doubte may the
glass be opened, and caste heaven vpon earth, that is to saye, **
water vpon sulfur viue; wherfore take water the 4th p*ar*te,
& power it vpon his sulfur, and the matter wil becom blacke.
Then close vp the vessell, and set it again in balneo vntill 30
the water be conioyned
with the body soe that all the matter be congealed into whit pouder. Then poure on again as before his 4th p*ar*te of water, close
vp the vessell, and set it in balneo vntill it be wholy dryed
vp, as in the other imbibition; and thus doe with the water vnto 35
the 4th tyme, the 5th, and the 6th tyme, and put to too p*ar*tes
of his water, and at the 7th time put to 5 partes, and then
in a circulatory vessell putrifie the wholle matter a-
gain in balneo vntill the darknes of his ecclips be paste. *
For when the said darknes is gon out, he will showe himselfe 40
lyke moste puer christall, w*hich* matter must be fermented w*i*th
his ferment made and p*re*pared of the body p*er*fect, putrified in
the bodie vnp*er*fecte; wherof Moryen speaketh, saying: Let ther be *
made co*n*iunction of ferment p*re*pared & made subtill with the vnp*er*fect body
cleansed, for in this manner ought ferment to be p*re*pared. Therfore, 45
put into the wholl matter the 4th p*ar*te of fermente, and set him
again in balneo vntill all be well co*n*ioyned together. 317r

---

22 brubles vpon the earth] vppon the earth burbles **S2S1**, vpon the ▽ bubles **S4**, on y^e ▽ bubles **S3**
25 or] *abs.* **S2FGS1S4S3**
26 Then without all doubte] *in red ink* **A**
27 *In the left margin*] This teacheth the | imbibition but | make no sulfur | of on alone except | in the vegitall stone | for when I compi|led it I ment to | haue sent it into | Ingland as a pr*e*sent | & mediator to help | me home out of | Russia wherfore | I made it the more | darke that I | might the sonner | be sente for home | for to doe it my|selfe **A** *
29 power] powre and powre **S2**, cast **G**
30 vntill the water] vntill ~~it be wholy dryed vp as in the other imbybytion~~ the water **A**
36 and put] put **S2GS1S4S3**
43 Moryen speaketh] *in red ink* **A**
46 into] vnto **S2S1S4**, to **S3**

Then ad vnto yt as be-
fore the 4th p*ar*te of fermente, and set yt again in balneo, and thus contynewe thy fermentation vnto the 4th tyme, and at the fifte tyme
let the matter fermented again be set in dry fier of ashes, and in y^t
order of fyer, let the fermentation be co*n*tynued vntill the matter 5
will congealle noe more but stand in the fyer liquid and moulton
as wax, haueinge the collour of burnished siluer or moste puer
christalle. For then ys it the p*er*fecte elixar white, able to turn arg*ent*
viue into medison and all other vnp*er*fecte bodyes of the mine into
p*er*fecte siluer. And this elixar, w*i*th longer co*n*trytion of the fier, 10
the goulden oyll, and ferment of red m*er*cury, may be made the red elixar
for gould. And for the declaration of the ferment, what it is and
howe it oughte to be p*re*pared, I haue largly spoken therof before
in the chapters of fermentation. Therfore, such as shall year after **
be p*ar*takers of this my simple labour, I farther co*m*mit them to ye 15
p*er*vsinge of such bockes as ar writen of this science, and specially
the worke of George Reply in English miter called his XII Gates. *
For I haue found more fruite in that on bocke for the p*er*fecte
vnderstandinge of this science then in all other bock*es* that haue com to
my hand*es*. For he hath truly set forth the straighte way to all those 20
that the God of glory hath ordeyned to the vnderstanding therof; *
wherby I haue lerned to knowe the croked way and the matter
that the ph*ilosofe*rs hid and coverid vnder the mantill of phi*losofie*; for the
w*h*ich to the eternall God of poware, in his sonn Iesus Christe and the
Holly Ghoste, be everlasting praise wordle w*i*thout <e>nd. Amen. Finis. 25

Cap*itulu*m 22 Of ☽ and ♀ *

Forasmoch as this 22 chapter is somwhat different from the chapter *

317r:
1   vnto] to **S2FS1**, *longer pass. diff.* **S4S3**, *longer pass. diff.* **G**
4   fermented again] beinge fermented **S2FS1S4S3**, *longer pass. abs.* **G**
7   as wax] *abs.* **S2S1S4S3**
8   ys] 's' *written on top of earlier* 't' **A**
9   vnp*er*fecte] imperfect **S2S1S4S3**
10 p*er*fecte] ˄{moste} perfect **S2S1**
    In the right margin] Contrition **A**
14 year after] hearafter **S2FS1S4S3**, herof **G** *
16 specially] esspetially **S2FGS1S4S3**
25 <e>nd] 'e' *written on top of earlier letter* **A**
26 Cap*itulu*m 22 of ☽ and ♀] *in red ink* **A**, Cap 22 **FG**, Capitula 22 **S2**, Chap 22 **S1S4**, *abs.* **S3** *
27 Forasmoch] *in red ink* **A**

next before, I haue therfore dyuided them into too p*ar*tes for because
y*ᵗ* in this chapter the bodie of Lune is ioyned w*i*th another body,
     wh*ich* body
of som is called the mean of transmutation, and of som the vnp*er*fect
body feminine, hauinge a collour aduste, wherv*n*der is hid gren, whit,
and yellowe. Thes too bodies beinge ioyned by liquefaction ought to be
mercurized and calcioned into very fine pouder, as is said in the
other chapter, which pouder must afterward be put into a portion of
the whitteste water of the sea, and then being set in balneo, yt
will be straightway decked with a grene collour like vnto an
emerauld; and when the body is thus dissolued and moulten in y*ᵉ*
sayd water, let the water be distilled from the oill in balneo, w*h*ich oill
wilbe like oille oliue, enclyninge somwhat to blacknes. And the water
being well rectified, let it be poured again vpon the said oylle, and set
it in balneo as at the first, and after certain daies drawe vp his
water, as is in the other chapter; and when all the sprit of the water shall
remaine in the bodye of Lune and Venus, then caste away the said
water & begin again with newe water, powringe one and drawing
vp, vntill the sprite of the water remain with the oylle in the bottom of
the vessell, which oylle will then be cleare, thin, and subtill. And
furthermore, sheurly set again the whole matter again in balneo
w*i*th a verie softe heate that he may breath vp his faintnes; and
when all the faintnes is lifte vp from the said oylly substance,
let the matter be put in his vessell kymia, sealed to putrifie.
When we say sealed, we mean the stopple and the glasse to
be moulten together. And being thus well closed, let it be set in y*ᵉ*
fier of the first degree, and ther let it stand a longe season in
the same moiste, temp*er*ate heate, vntill it be turned blacke as
moulton pitch, w*h*ich after will coagulate into dry pouder after
mani brublin*ges*, wherin shalbe sene ynumerable collours.

---

32   ioyned] 'i' *written on top of earlier* 'y' **A**
36   be straight way] streight way be **S2S1S4**, straight be **S3**
42   as is] as **S2GS1S4S3**
317v:
4   furthermore sheurly] for the more surety **S2FS1S4**, to be more sure **S3**, *abs.* **G**
      set again] set **S2FGS1S4S3**
9ˑ   in y*ᵉ* fier] in fire **S2FS1**
11   moiste] most **S2FS1S4S3**, *abs.* **G**
      vntill] till **S3**, *at the end of the line* vn- **S2**
12   coagulate] coagule **S2**, congeal **FGS1S4S3**
13   brublin*ges*] burblinges **S2S1S4**, bubleings **S3**
      shalbe sene] shall shewe **S2S1**, will shew **S4S3**

And when that that is fine shall ascend vpward in the vessell
white as perrells or the eyes of fishes, then somwhat strength 15
the fyer that all the said calce may dry into a white pouder,
for then is it verie sulfur and thend of our firste p*a*rte of our
worke and the begininge of the second. For Raymon saith *
the first p*a*rte of our worke is the creation of our sulfur, and the
second p*a*rte the makinge of the medison; soe that Raymon deuyded 20
the wholle worke into too p*a*rtes. But other some haue dyuided
it into 3 p*a*rtes. For, sai they, the first is p*r*eparation, the second p*a*rte
creation, and the third norishmente. And althoughe Raymon
and the other doe not agree in one order of teachinge, yet in
effecte ther is noe difference betwen them. And when the said 25
sulfur is created into moste white pouder, then open the
glasse and giue it drinke 7 tymes of the milke of his mother, & *
at the laste tyme of the 7, let the wholle matter be circulate
vntill the blacknes be gone out and that the substance in the vessell
showe like moste fair, burnished yron or a sword of puer met- 30
talle, newely made clean. For then is it called the arcenike of
ph*ilosofe*rs, and the yonge man wayned from his mothers brestes, w*hich* *
muste be fed and norished vntill he com to p*er*fecte age; w*hich*
norishment we cal soware dowghe or ferment, wherwith
he muste be fede vntill he dred not the fyer but will stand 35
therin fluxible and flowinge as wax. For then is he elixar
and medison vpon mercurie and vpon all other bodies of
the myne. Finis.

   Cap*itulu*m 23 Of the dissolution of ☉, and of the red * 318r
    pouder, red sulfur, red ☿, & red oille

This 23 chapter showeth howe the body of gould ought to be m*e*rcurized,

---

17 our firste p*a*rte] the first parte **S2GS4S3**
18 Raymon saith] *in red ink* **A**
  In the left margin] The first p*a*rte | of the worke | S For*m*an **A**
24 the other] other **S2S1**, others **GS4S3**
31 is it] is he **G**, it is **S2S1**
37 vpon all] all **S2FGS1S4S3**
38 Finis] Finnes &c **G**, &c Finis **S2FGS1**, &c **S4S3**
318r:
1 Cap*itulu*m 23] *in red ink* **A**, Cap 23^m **FG**, Capitula 23 **S2**, Chap 23 **S1S4**, *abs.* **S3**
  Of the dissolution of ☉ and of the red pouder red sulfur red ☿ & red oille] *abs.* **S2F-GS1S4S3** *
3 This 23 chapter] *in red ink except for the final* 'r' **A**

as it is taughte in the chapter goinge before. Therfore, in the name of God, take the
bodie of gould, welle mercurized into fine pouder, and calcin the said pouder 5
into the first parte of the body of alkamy, & set it in balneo 7 dais. Afterward
power the said water into a strong glasse, well luted over all, saving
the top, & set it in dry fier, & ther let it calcin into a red pouder by
the space of 21 dais by violence of fyer; and when the said pouder is
soe calcioned, let it be put into the second parte of the body of alkamy,
    and set 10
it in balneo vntill the pouder be wholly dissolued. Then in the same
heate drawe vp his water, rectifie it, and power it on again, and
thus contynue vntill the sprite of of the water remain in the substance   *
of gould. Then begine again with newe water, and doe with it
as with the firste vntill all the sprite shalbe found to remain in the substance 15
of gould. For then yt will be a puer oille, which muste be put into his circulatorie vessell, philosofically closed; and set it in fyer of the first degree,
& ther let it stand vntill it be made blacke as moulton pitch, & soe contynue
the same heate vntill the vpper face therof showe with a   *
    yellowe collour
mengled with white. Then somwhat strengthen the fyer vntill 20
ther appeare vpon the earth brubles enclyninge to rednes. But yf
they be all white, it is noe matter, for the ymbibition and fermentation
will bringe again the red collour. And when the earth is well dryed
into powder, gyue it drinke 7 tymes by imbybition, as it is showed
in the other chapters; and then again putrify the wholle matter. 25
But here thou muste vnderstand this drinke to be noe other thinge

---

4   it is] is **S2GS1S4S3**
       chapter] chapters **S2FS1**, *abs.* **S3**
5   welle] 'll' *written on top of earlier* 'r' **A**
       calcin the said pouder] which must be calsined **G**, when it is thus made into fine powder let the saide powder be calcined **S2S1S4**, when it is thus made into powder let ye said powder be calcyned **F**, when its thus made let it be calcin'd **S3**
6   into] in **S2GS1S4S3**
       body of alkamy] body alkamy **S2FGS1S4S3**
10   body of alkamy] body alkamy **S2FGS1S4S3**
12   his water] his saide water **S2S1**, the said ▽ **S4S3**
13   of of] of **S2FGS1S4S3** *
15   *In the right margin*] Oille **A**
16   into] in **S2S1S4**
17   set it] set ~~in~~ it **A**
20   strengthen] strength **S2**
21   brubles] burbles **S2S1S4**, bubles **S3**
26   But here] *in red ink* **A**
       noe] none **S2S1**, nought **S3**

but only the blod of the lyon, from the which blod the watrines muste   *
be taken away. And when the earth is thus purged by this laste
putrifaction, then is it very sulfur of nature, whervnto must
be added ferment made of the perfecte body, drowned and slainte   *   30
in the blod of the said lyon, and philosofically circulate into sulphur; for
with such fermente it is necessary that our sulfur be fermented. Then
encres<e>th the said sulphur with the 4th parte of fermente, and set it in balneo
7 daies that the matters may the better be incorporated together. Then
remove it into dry fyer, and ther let it stand till yt be congealed, and then en- 35
crease it again, and thuse contynue vntill the matter will dry noe more
but stand in the fyer fluxuble & flowinge as wax, for then it is medison and able
to change argent viue into elixar and imperfecte bodies of the myn into
gould better then the naturall gould of the myne. Nowe, when thou wilte
augmente thy medison, take as moche of the ferment as of the elixar, &
      ioin them   40
together into a circulatory vessell, and set them in balneo certain dais.
Then calcyne it in stronge fyer vntill it stand in the glasse moulton as wax
without any brublinge. And thus mayst thou multiply thy medison   *
      infnitly,
& in every fermentation the medison shall encrese in goodnes to a greater elixar &c.
Or otherwis<e> youe may multiply the elixar by lyttle and lyttle, and calcin
      him as   45
before, and soe encrese him at thy pleasur &c. Finis.

    Capitulum 24   318v

The 24 chapter is of ☿ and the red bodie ioyned with the   *
grene bodie of feminine kinde called the mean of transmutation,

---

30 slainte] slain **S2FGS1S4S3** *
33 encres<e>th] increse **S2FGS1S4S3**
34 may the better be] may bee the better **S2FS1S4S3**
35 till] vntill **S2FGS1S4**
37 it is] is it **S2S1S4**
39 Nowe] *in red ink* **A**
41 into] in **S2GS1S4S3**
      In the right margin] Multipliing | of the medi|son **A**
42 stronge fyer] a stronge fire **S2FGS1S4S3**
43 brublinge] 'r' *written on top of earlier* 'l' *or* 'b' **A**, burblynge **S2S1**, bubleing **S4S3**
318v:
1 Capitulum 24] *in red ink* **A**, Cap 24 **G**, Capitula 24 **S2**, Chap 24 **S1S4**, *abs.* **S3**
2 The 24 chapter] *in red ink* **A**
3 transmutation] transmentacion **S2**

or the mean betwen the ☉ and the ☽, the w*hich* bodies be ioyned together
by liquefaction. Let them be m*er*curized into subtill pouder
by drynge in a clean crusible, w*hich* pouder must be calcioned
in a certain white water aptly choson for the purpose, wherin
it must bruble by the space of xxi daies into a red pouder,
w*hich* pouder muste afterward be put in a portion of the laste
water that distilleth from Lunarie, wher it muste be brought
into the thicknes of oylle by often takinge his water from
him. That beinge done, let it be put in his kymia, surely
sealed, and set it in his furnace, as in the other chapter,
vntill it be chaunged blacke as moulton pitch. And soe
contynewe forthe the same heate vntill ther appeare in
the vessell a yollowe collour mixed with white and
grene. And then doe as before in the other chapter. Strength
the fyer, & soe let it stand vntill ther appeare vpon the
earth brubles in the form of perrells or fishes eyes.
Then embibe the sulfure as it is said in the other chapter,
and ferment it with vulgar gould, altered and purified
in the laste liquor of Lunarie, phi*losofi*cally sirculate and
made able to be ioyned with the sulfur of nature &c.
And whatsoever is needfull more for the p*er*forminge of this
chapter is sufficiently declared in the nexte chapter before.
Therfore, p*ro*cead in everie order of workinge as is showed
in the same chapter, and soe shall youe make alsoe of
this laste a p*er*fecte elixar. Finis.
         Vince et viue. SF

 25 Chapter

Wheras in the 4 chapters goinge before I haue dyuided
the elixeration of gould and siluer into 4 p*ar*tes, I entend

---

4 be] beinge **S2S1S4S3**
8 bruble] burble **S2S1S4**, buble **S3**
13 it] *abs.* **S2FGS1S4S3**
17 doe] *abs.* **S2FS1S4S3**, *longer pass. diff.* **G**
19 brubles] burbles **S2S1S4**, bubles **S3**
21 purified] putrified **S2FGS1S4S3**
27 alsoe of this laste a p*er*fecte elixar] of this also the perfect elixer **S2FS1S4S3**, of it also ye p*er*fit elixer **G**
28 Finis] Finis &c **S2**, &c Fines **G**, &c **FS4S3**
29 Vince et viue SF] 'e' *written on top of earlier* 'e' *and another letter* **A**, *abs.* **S2FGS1S4S3** *
30 25 Chapter] *in red ink* **A**, Cap 25 **FG**, Capitula 25 **S2**, Chap 25 **S1S4**, *abs.* **S3**
31 Wheras] *in red ink* **A**

by the grace of God here to includ them in on chapter into too p*ar*tes, by one order of workinge, as shalbe showed,
sauing that in the first p*ar*te of this chapter, I shall somwhat   35
entreat of the animall stone, of his order of workinge, &
howe of him may be made the greate medison cauled y^e
goulden drinke and aurum potabile, the w*hic*h stone, yf God
will, shalbe co*m*posed out of the 21 & 23 chapters heare
nexte before. For of that matter, that is to saie, of the white   40
oylle & of the red, beinge ioyned together, moste certainly may
be made the great elixar of life, the w*hi*ch elixar or goulden oyll
is appropriat and may be had abundantly out of the said
animall stone; of the which stone I will alsoe in the end of
this bocke entreate, for soe moch as shalbe necessary for   45
the same, because the animall stone is the wholle and finall
end of all our workinge as co*n*cerni*ng*e this worke of the
loware astronomy &c. And in the second and laste p*ar*t
of this chapter shalbe showed howe the oylles that ar made
in the 21 & 24 chapters, being ioyned together, may be made   50
a stone of greate vertue and shall be nothinge inferiour to the   319r
vegitall stone.
For of this stone alsoe may be made an excellente medison for the health
of mans bodie. Wherfore in this chapter shalbe declared the greatest treasure of all worldly treasures; for in this treasure is all that the harte
can desier, for in it is found health, welth, and longe life. For   5
what greater treasure can ther be vpon earth then to haue health
and wealthe & therwithall to enioye longe life, w*hi*ch ar all to be had
in the said animall stone, as a moste excellente medison for the pre-
seruation of mans health & life? Wherfore Moryen saithe: He y^t
hath it hath all and all may doe co*n*serninge wordly things. And ther-   10
fore, it is giuen vnder highe co*m*maundemente to be kepte secret fro*m*
vngodly people. For from such hath God alwais w*i*thdrawen his be-

---

38   aurum potabile] aru*m* potabyle **S2S1S4**
40   of that matter] ˏ{if} that matter **S2S1S4S3**
41   of the red] the red **S2FGS1S4S3**
44   animall] ayall **S2**
50   21] 22 **S2FGS1S4S3**
319r:
2   alsoe] *abs.* **S2S1S4S3**
4   treasures] 't' *written on top of earlier letter* **A**
7   longe life] a longe life **S2S1S4**
8   animall] ayall **S2**
12   benignyty] 'ny' *written on top of earlier* 'hty' **A**

nignyty. But vnto good men hath he gyuen it at tymes accordinge
to his pleasure to doe this often, as he did to Noye when he made
the arke, to Moyses when he made the tabernacle, to Sallomon
when he builded the Temple, and to Machabeus when he reedyfiid
Ierusalem. For vnto thes, it appeareth, God gaue abundance of the
workes of kind to perform his will. And as it pleased his almighty
maiesty in the first age of the wordle to giue vnto Noye his excellente scyence, as to a man notably beloued that he mighte therby
performe that that was by his God commaunded him, so alsoe in the
second age of the wordle his pleasure was to giue it to a holly
woman, the syster of Moyses, that the Lordes apointed busynes might
be by Moyses plentifully performed according as it was prescribed to
him. Soe lykewise in this laste age, he hath gyuene it to many
godly and well-disposed men, who hath lefte in writing the
testimony therof, wherby we that nowe followe in penultima
etate mundi doe suck a certain kind of swetnes out of their laborrs,
wherby we attain and com thorowe his godly permission to knowe
the highe secretes of nature, composed and hid in a globus masse
called of the philosofers the leess wordle because of the eavennes of his
elementes, and the course of the ☉, ☽, & plannetes, the complection of
humors, signs, and stars that ar raygninge in him; wherby
he is brought by manuall workinge and nature to a merveylouse
aptation and an excellent form of soe highe vertue that ther is
nothing in the wordle conserninge earthly thinges may be compared thervnto. Therfore, whoesoever will worke in this

---

15 arke] ship **S2FS1S4S3**
19 his] this **S2S1S4S3**
22 giue it to] gaue it to **S3**, giue it **S2S1S4**
24 as it was prescribed to him] as it was by his almighty maiestye prescribed vnto him **S2FGS1S4**, as it was by his maiesty prescribed vnto him **S3**
25 laste age] last age of the worlde **S2FGS1S4**
  he hath gyuene] it hath pleased him to giue **S2FGS1S4**, it has pleasd him to giue **S3**
  to] *written on top of earlier* in*(?)* **A**, vnto **S2FS1S4**, *longer pass. diff.* **S3**
26 who hath lefte in writing] who hath lefte in who hath left in writinge **S2**
27 therof] of the blessinge giuen them of God **S2FGS1S4**, of his blessing given of God **S3**
  nowe] *abs.* **S2S1S4S3**
  in penultima etate mundi] in perultimo etas mundi **S2S1S4S3** *
29 attain and com] com and attaine **S2S1S4**, attaine **S3**
  permission] perpmission **A**
30 composed and hid] compased & hide **G**, compased and hidden **S2S1S4S3**

science and entendeth to seke out the perfecte worke of kind, let
him first endeuour himselfe to knowe the nature of this
excellent creature, and what is conteyned in him, and after-   40
ward his aptation and the mannour howe he is changed and
broughte into a naturall body, which body must be engendred
& brought forth in the fier; wherfore it is of Morien the   *
philosofer said in Turba Philosoforum: Honor our kinge comminge
from battaille, vz., from the fier, crowned with a dyadem and clothed   319v
    with a kinges
clothinge. And he saith moreover: Let him be norished vntill he com to
perfecte age, whose father is ☉ & whose mother is ☽ soe that our stone muste
be brought forth with the norishmente of his father and mother, vz., with
the ferment of ☉ and ☽, which ferment, as is declared in the chapters therof, 5
muste be broughte to complemente before the stone can be fermented.
And here referringe this parte to the theorica, I will procead to the practice
of the said elixars. Take therfore the oylle of ☽, as it is made in
the 21 chapter, and the oill of ☉, as it is made in the 23 chapter,
and let them be puerly rectified from their flegma with a softe   10
fier soe longe as any fainte water will arise; and when they ar well
rectified, let them be ioyned together by eaquall waighte, and put them in
kymia, surely sealed, as it is showed in the other chapters, and then
set him in fier of the first degree, vz., 40 daies to his blacknes,   **
    and ther

---

38   worke] woorkes **S2FGS1S4S3**
319v:
2   he saith moreover] moreouer he saith **S2S1S4**, moreover he sai's **S3**
3   ☉] the sonn **S2FGS1S4S3**
    whose mother is ☽] & his mother is the moone **S4**, & his mother yᵉ ☽ **S3**, the moone his mother **S2FGS1**
    soe that our stone muste be brought forth] wherfore it behooueth our stone in this manner to be brought foorth **S2FGS1S4**, so it behou's our stone thus to be brought forth **S3**
7   to the theorica] to the theory **S3**, vnto the theoryca **S2S1**, vnto the theory **S4**
    I will procead to the practice] I will by the grace of God proceede foorth vppon the practice **S2FGS1**, I will by the grace of God proceede forth to the practice **S4**, I will God permit proceed to yᵉ practice **S3**
8   Take therfore] *in red ink* **A**, Therfore in the name of Christe take **S2FGS1S4S3**
    it is made] it made **S2**
14   vz 40 daies to his blacknes] *abs.* **S2FGS1S4S3** *
    In the left margin] 40 dais in his blaknis | 340 dais goinge | out of his blacknes | and 30 dais in his | siccasion or | dryinge **A** *

let him stand even vnto the terme of 5 monethes, in the which tyme youe
shall see the matter becom blacke, and with the longe contynuance of his
seathinge and brublinge, he will dry and drinke vp all his moistur
& becom a whit substance like vnto froson snowe. And in the drying
therof soe many collours shall appeare as may be thought to be in
the wordle, which collours will pass awaie on after another, &
when they ar all gone out soe that the water hath made white
the earth by conuerting his moisture into drynes, ther will stand
vpon the earth brubles as it were perrelles, which we call the
diadem or thron of our sulfur, our tella foliata, and the
sprite of our sublimed bodies, purified and lifte vp into a most
excellent substance. And when thes signs shall appeare, then somwhat
strength the fyer, and let it be contynued vnto the full end of 5 mounthes;
for soe shall he receyue his perfecte drynes and a naturall appetite to
drinke vp his imbibition, which muste be gyuen him as is taughte before in the 21 chapter. And when the 5 monethes ar compleate and
that he be broughte as I haue said into dry pouder, whit and shininge,
then open the glasse and power vpon the sulfur his 4th parte of
water, and set him in balneo vntill all be congeled into sulfur.
Then power one again his 4th parte of water, and procead with
his imbibitions as it is declared in the forsaid 21 chapter;
and when the imbibitions be compleate soe that he hath passed
his second putrifaction, let the sulfur be dyuided into too parts:
the one parte for the white, and the other for the red. And that reserued
for the white elixar, let it be fermented with white fermente, y^t

---

15 even vnto the terme] vnto th'ende and terme **S2FS1**, to the end & tearme **S4**, ye end **S3**, for ye space **G**
 the which] which **S2S1S4S3**
16 the longe contynuance] ye long continoinge **G**, long continewence **S2S1S4S3**
17 brublinge] burblinge **S2S4**, purblinge **S1**, bubleing **S3**
18 becom] so becom **S2S1S4**
21 white the earth] th'earth white **S2S1S4S3**
23 brubles] burbles **S2S1S4**, bubles **S3**
24 thron] crowne **S2S1S4S3**
 tella foliata] stella foliata **F**, terra folliata **S2S1S4S3** *
25 purified] putrifyed **S2FGS1S4S3**
26 then] *abs.* **S2FGS1S4S3**
29 imbibition] inbybycions **S2FGS1S4S3**
 is taughte before] it is taught before **F**, it is taught afore **S2S1**, it is tought **G**, is taught **S4S3**
35 his imbibitions] the inbybycions **S2S1S4S3**
 the forsaid] th'aforesaide **S2GS4**, *abs.* **S3**
 In the left margin] Parte the sulfur | before the | first imbibi|tions **A**

is to saie, with ferment made with Lune, as it is showed before. And let the ferme*n*tations be conty*n*ued vntill the wholle matter stand in the fier moulten without co*n*gealinge; for then it is the white elixar for siluer. Then take the other pa*r*te reserued for the red, and imbibe the said sulfur 7 tymes with the red oylle of ☿, and let him pass his order of imbibition in everie degree of workinge as was done with the sulfur of the white elixar; & when yt is congeled into gloryouse earth, let him be fermented with red mercurye, vz., red fermente, made and p*re*pared of the body of Sol, putrified in the sole of the volatiue sprite. And let his ferme*n*ta*ti*on be co*n*tynued vntill he will not congealle in a stronge fyer but will melte before ☿ fly, for then is he the very elixar for gould.

Then tak of the too elixars by equall waighte, as moch of the one as of the other, & vpon the white elixar poware his weighte of water of ☿, purely rectified, and vpon the red elixar his waighte of the fier of nature, p*er*fectly rectified, & ioine them together in a circulatory vessell, that is to say, the too elixars, that they may worke the more strongly. The vessell being well closed, let them be set in balneo, and ther let them stand vntill the matters be well co*n*ioyned and that the elixars haue co*n*geled their waters & oille soe that all be made elixar. Then let it be remoued into dry fier, first wit*h* a mean heate encreasinge the fyer vntill the matter be broughte into a p*er*fecte purple collour and stand moulton in the fyer as it were oylle. Then let the vessell cole, and poware vpon the same elixar the quintaessencia of the vegitable stone, called the vegitall quintessence, and let it be done in equall p*ro*portion of the said elixar, and set him in balneo to circulate, the vessell beinge well closed. And when it hath stand in balneo certain dais, remoue it into dry fyer, and giue it first easy fyer and afterward stronge fyer. Yf it therin remain a puer oille wit*h*out co*n*gealinge, then the leste drop therof shall change 1000 drops of ☿ into medison. And again

40

45

50

320r

5

10

15

---

40   with Lune] of Lune S2GS4
      showed] saide S2S1S4S3, *longer pass. abs.* G
43   Then take] *in red ink* A
47   *In the left margin*] Elixar for gould | .i. in the water ⌃{of} ☿ A
49   volatiue] volatille S2FS1S4S3
320r:
1   Then tak] *in red ink* A
     tak of] take S2FGS1S4S3
2   water] the water S2FGS1, his water S4S3
     purely rectified] rectified purely S2S1S4, well rectifi'd S3
7   waters] water S2FS1S4S3

let him be fixed into the said quintessence, and doe soe 3 tymes, and then
is he made the highest medison in the wordle for mans health, of the
greateste transmutation that can be, and aurum potabile, to whose
vertue noe treasure in the wordle may be compared &c.
In this second parte of this chapter shalbe showed howe the oilles, as they
ar to be made in the 22 and 24 chapters, oughte to be elixerate; wherby
may alsoe be made of them an excellent medison, as I haue said in the
first parte of this chapter; the practice wherof beginneth in this maner
folowinge: Take the said oylles as they ar made in the said chapters, and
ioine them together in equall waighte, and put them into a circulatorie
vesselle, and vpon the top therof a small styll heed prouided for the purpose,
& let him be luted to the crowne of the vessell, and set the glasse with the
oyles in balneo, and ther let it stand soe longe as any fainte ayer
will arise; and when thou fealeste the water com sharpe, take out y^e
glasse, & close him after the mannor of philosofie, and when he is thus closed,
set him in fyer of the firste degre, and ther let him remain vntill it becom all black as moulten pitch; and contynue it forth in every order of
workinge as it is said in the other elixar vntill it be broughte to a perfecte
medison vpon ☿ and all other bodies of the myne. And yf thou wilte
afterward ioyne the elixars together as it is showed in the other parte
of this chapter and fix them vnto the quintessence of the vegitall stone, he
then shall be an excellente medison for bodyly health &c. Finis. 1590

---

18   into] vnto **S2S1S4**, to **S3**
19   of] and of **S2GS1S4S3**
22   In this second parte] *in red ink* **A**
      In the right margin] Oilles **A**
25   the practice] *in red ink* **A**
26   take the] *in red ink* **A**
39   1590] *abs.* **S2FGS1S4S3**

The 26 chapter of the 320v
vegital stone &c.

Nowe of the vegitall stone, whose vertue is brought to pass
only by the fyer of nature, and howe that fyer is had and wherof,
I will somwhat declare; for the matter wherof is an vnctuous * 5
moisture and is the nereste matter of o*ur* ☿ vegitall and
philosoficall, w*hich* Raymon in his Op*er*tory called black, blacker *
then blacke, the menstrue resoluble engendred of the body *
dissolued; the w*hich* ph*iloso*f*er* in another place saith that the men- *
strue resolutiue springeth of wyne, or the water therof; but * 10
in another place, he saith our water is a mettaline water
because it is engendred of mettaline kinde only and is a simple *
substance, coescentiall of his *proper pa*rtes, concrete and wrought to-
gether, beinge broughte to yt by the dissolutiue menstrue; by ver- *
tue wherof they doe multiply their similitud*es*; wherfore 15
our true water mettaline is our vnctuous humidity of the
body dissolued, lyke vnto black pitch, and then is it called the
menstrue resoluble. For when we haue attayned to that same *
water resoluble, then we say our secrete coniunction is be-
gone wherby co*n*trarie natures be ioyned together, vz., earth & 20
ayer, water and fyer, w*hich* before this coniunction were repugna*n*t,

---

320v:
1   The 26 chapter of the vegital stone &c] *in red ink* **A**, The 26th chapter is of the veg-
    itable stone &c Capitula 26 **S2**, The 26th chapter is of the vegetable stone Cap 26^m
    F, The 26 chapter is of ye vigetable Cap 26 **G**, The 26th chapter of the vigetable stone
    &c Chap 26 **S1**, Chap 26th **S4S3**
3   Nowe of the vegitall stone whose vertue is brought to pass] Forasmuch as I haue
    promised somwhat to speake of the vegitable and a*ni*mall stone I will therfore in this
    place entreate of the vegetable stone for bycause his whole vertue is brought to passe
    S2S1S4, Forasmuch as I haue promissed somewhat to speake of the vegetable stone
    & animall stone I will therefore in this place intreat of ye vegetable stone for because
    his whole vertue is brought to passe **F**, Forasmuch as I haue p*ro*mysed somwhat to
    speacke of ye vegitable stone & anymall I will therfor in this place intreat of ye veg-
    itable for becaues his holle vertu is brought to passe **G**, Forasmuch as I haue p*ro*mis'd
    to speak of ye vegitable & animall stone I'le therfore treat of ye vegitable stone be-
    cause his vertue is bro*ugh*t to pass **S3**
    Nowe of the] *in red ink* **A**
4   fyer] force **S2S1S4S3**
8   body] bodies **S2S1S4S3**
10 water] tarter **S2S1S4S3** *
16 our vnctuous] an ounctuose **S2S1S4S3**

the one againste the other. For when we see the ravens hed
appeare, which say that the peace is made by ioyninge of
    too extremitis
together, vz., sulfur & argent viue, which in this place ar con-
strayned to remain as ded bodies, mortified into a filthie 25
matter. And because we cannot perform the worke of the stone without
the menstrue resolutyue, in the which the bodies of ☉ and ☽ ought to be calcioned and dissolued, therfore it is necessarie that such a menstrue
be hade, and as I haue spoken before of a secrete coniunction wrought
and brought to passe by the menstrue resoluble, soe muste we haue 30
alsoe a menstrue resolutyue, for in this place is the ioyning together
of bodies made clean, the which bodies in their clensing ar retorned
again to the center of their begininge and soe ar made cleane
inwardly and outwardly; by the means wherof this coniunction
is the cause of the philosofers worke. For whatsoever is done be- 35
fore this coniunction apperteyneth to the gross worke and is done
by manuall workinge, as in subliminge, dissoluinge,
distillinge, calcioninge, and dissoluinge of the bodies. After
this coniunction we leaue of all manuall workinge and commit
the wholle state of our worke begone to natuer and fyer vnto 40
the full accomplishmente of the same; wherfore the philosofers
worke and nature beginneth together after the testimony of
an old writer, which saith the philosophers worke doth not begine till all be cleane without & within; for all the partes beinge
made cleane, then nature by force of workinge, by his o<w>ne 45
strength, ioyneth together contrary thinges, as heat with cold, and moist
with dry, and sulfur with argent viue, which being closed together in on vessell will not leave stryuinge vntill by their vehe-
ment wrastelinge ther appeare in the vessell drops of sweete
descending vpon the mortifyed body black and stinckinge, [and] 50
afterward the said drops, being changed into the form of perells or the eyes   321r
of fishes firmly standing on the earth, is the sign that, after that longe

---

23   which] we S2GS1S4, *longer pass. abs.* S3 *
        that the peace] tha{t} peace S2S1S4, peace S3
31   alsoe] *abs.* S2S1S4S3
38   After] And after S2S1S4S3
45   o<w>ne] 'w' *written on top of an earlier* 'v'(?) A
46   and moist] moyste S2FGS1S4S3
50   [and]] *catchword* A
321r:
2   on] vpon S2FGS1S4
       that longe] there longe S2FGS1S4S3

strife, they ar made frind*es* into such an vnity as cannot be sep*a*rated. But
betwen thes co*n*iunctions ther will appeare a meruailouse kind of working,
for nature will change the body dissolued into all collours of the wordle   5
before the co*m*inge of the white sulfur; and when the said sulfure
shall showe him drye without brublinge, then must he be fed forth w*i*th
milke and meate vntill he be encreased in quallity and quantity   *
& soe made able to striue with the fyer and overcom it.
Nowe to p*r*ocead somwhat to speake of the too forsaid menstrues,   10
howe and after what mannour they muste be soughte, I will rehers
the word*es* of Raymond wher he saith that dissolution shall be made
with the sprites of wyne. The same ph*iloso*fer entendeth that in this be had
another menstrue resoluble, w*hi*ch was first the menstrue resolutyue,
without the w*hi*ch menstrue resolution can never be made. For our   15
menstrue resoluble is engendred only of the kind of mettalls & is
a potentiall vapour of the bodyes dissolued, ioyninge together too extremities: sulfur and argent viue; wherfore it appeareth plain that
our water is a mettaline water and not drawen from wyn as
som suppose, and for that it sauoreth of every extreme, our menstrue   20
resolutiue is therfore broughte to acte. And howe that menstrue may
be hade, w*hi*ch is vnctuous, moiste, sulfurous, and mercuryall, well
agreinge with the nature of mettalls, wherwith our bodyes muste be
artificially dissolued, I will here followinge put the practice &c.
Therfore, in the name of God, take the sharpeste humidity of y*e*   25
grape, which is not the grape of the vine, but the black grape of   *
Kotolory, w*hi*ch in this worke of the vegitall stone must be made   *
   into the
manner of a gum, w*hi*ch gum ought to be made of ☿ and water, and beinge made, yt will haue a sharpe taste like allam and is cauled
of Raymond vitreall azoc. This gum must be made, as I haue said,   30

---

4 co*n*iunctions] 2 coniunctions **S2FGS1**, two contraries **S3**, 2 contrays **S4**
7 brublinge] burblying **S2S1S4**, bubling **S3**
8 quallity] e-quallity **A**
9 made] bee made **S2S1S4S3**
21 therfore] therby **S2FGS1S4S3**
24 the practice] the cleare practice **S2FGS1S4S3**
  In the right margin] Practise **A**
26 *In the right margin*] M*r* sb aqua m*er* **A** *
27 Kotolory] Katalony **S2FGS1S4S3** *
28 a gum] gum **S2GS1S4S3**
30 vitreall azoc] vitreall &c azoc **A**
  must be made as I haue said] as I haue saide must be made **S2GS1S4**, as is s*a*id must
  be made **S3**

of ☿ and water and is a puer viniger, as Antazagurras said. On time,
said he, I soughte in bock*es* that I might com to the knowledg of this science.
Afterward I praid vnto God that he should teach it me, and my prayer
was hard that ther was showed me a cleane water, w*hi*ch I knewe to
be a clean viniger, and howe moch more I red in bockes, soe moch
more they weare made open vnto me. This viniger must be ioyned
with argall in this manner: Take puer argall made into fine
pouder on pound, & put vpon it on pinte of the said viniger distilled, and stop
the vessell verie close from taking of ayer, and soe let it stand 24 howars.
Then distill the water from the earth by the heate of balneo, hauinge a
verie softe fyer; which water then will haue noe taste of sharpnes or otherwise verie lyttle because all his hole fier will remain in the earth. Afterward let the earth be well dryed, and when it is congealed, suffer the
vessell to be cold, and then begine againe with newe aquauite or
viniger, distilled after the same mannor as before; and this doe
orderly vntill youe haue such quantity as shalbe sufficient
for your worke. For Avicen saith in his Epistle to his
Sonn: Sone, it behoueth the to haue 60^{li} of gould adrope
for the accomplishment of the elixar. And when the gum
is made by the same order of workinge as is aforsaid, yt
wilbe lacke of collour and taste like allam. Then cut
the said gum into smalle peaces, and put as moch therof
into a stillatorie as shalbe sufficiente for on distilla*ti*on,
and drawe the water from the earth, first with a softe
fyer encreassing the fyer vnto the end, which muste be
done with stronge fyer, soe longe as any drop will com;
w*hi*ch water muste be closslye stopped from takinge of ayer.

---

33   vnto] to **S2S1S4**, *abs.* **S3**
        and my prayer] my praiers **S3**, forsooth saith hee my prayer **S2FGS1**, forsooth said
        he my prayers **S4**
34   that] and **S2GS1S4S3**
36   more] the more **S2S1S4S3**
        vnto] to **S2FS1S4S3**
39   of ayer] ~~from~~ of ayer **A**
41   fyer] heate **S2S1S4S3**
321v:
3   Avicen saith] *in red ink* **A**
        saith] saide **S2FGS4S3**
        In the left margin] Viniger poured | on argole **A**
4   Sone] So*nn* saide he **S2FGS1**, Sonne saith he **S4**
        gould adrope] adrop golde **S2S1S4S3**
7   lacke] black **S2FGS1S4S3** *

And within 2 howars after it is drawen, it oughte to be pow-
red vpon the calce of the bodie to be dissolued, and the earth
that remayneth in the bottom of the stillatory calcin in
a stronge fyer, and kepe it well, for w*ith* yt thou maiste
dissolue ☿ infinitly. And this water soe drawen from
the gum is our ardent water and aquavite, the
menstrue resolatiue, w*hich*, before he was drawen from
his earth, was resoluble, a potentiall vapour, able to
dissolue bodyes, to putrifie and purifie, then to dyuid the
elemen*tes* and to exhale the earth into a m*er*veilouse soulle
by his vertue attractiue. Therfore, he that thinketh it is any
other water then this, he erreth as totchinge this worke. This
water hath a sharp taste and p*ar*tly stincking savoure, &
therfore it is called the stincking menstrue, wherin dwelleth our beaste againste nature, w*hich* ap*er*teyneth not vnto
this worke of the vegitall stone but in his primer dissolution. And because the vegitall stone is elixerate only
w*i*th gould and the fyer of nature, youe shall take of the
said water dissolued with a malgam, as is taughte in the
chapter of dissolution, and put therof soe moch into a still
bodie as shall suffice for on distillation, and lute on his
lymbicke, & set him in balneo, and with a softe fyer lifte vp
the faint water; and when the water beginneth to com sharp,
put to another receptorie and receyue the water soe longe as
any will cum in balneo; in the w*hich* water youe muste calcyne your ☉, and when your ☉ is calcioned, then put vpon
the said calce a quantitie of water that was drawen
from the forsaid gum, w*hich*, because of the ayerynes of the water,
muste be done within too howars after it is drawen; and as

15   *In the left margin*] The dissolutio*n* | of ☉ & ☿ A
20   from his earth] from y̓ᵉ g̶u̶m̶ his earth A
23   exhale] exhalte S2FGS1S4S3 *
24   Therfore] *in red ink* A
     thinketh] thinks S3, thinketh that S2FS1
28   vnto] to S2FGS1S4S3
33   therof soe moch] so much therof S2S1S4S3
36   com] wax {(com)} S2
38   cum] *changed from earlier* runn A
40   said calce] said w̶a̶t̶e̶r̶ calce A
     water] the water S2GS1S4S3

sone as it is put vpon the calce, it wil begin to boill without any externall
fyer. Then yf the vessell be wel shut, it shall not cease to worke, without outward heate added vnto yt, vntill all be dryed into calce; wherfore ther ought not to be put of it in greater quantity then may scarcly
suffice to cover the said calce, and after the calce that is lefte shalbe
again well dryed by the fyer, then put soe moche more water vpon it
as before, and procead as is aforsaid, and contynue the selfesame
worke till all the substance be wholly dissolued and again dryed.
Let it be put into a good quantity of the fyer of nature, and the vessell,
beinge well stopped, by meanes of heate vttwardly administred it may
be dissolued into oylle, which will sone be done by reason of the
simplicity of the matter; the which thinge beinge done, substract
the water from it distillinge by the same moste temperat heate
vntill ther remain in the bottom of the glass a thin oylle, which the more
ofter it shalbe dissolued and the water taken from it, soe moch
the more it shalbe thine and subtill. This oylle, thus made and
put in kymia, sealed to putrifie, in fier of the first degre being
moiste, it waxeth black as moulten pitch; for then may the fyer
haue his accion in the bodie, soe opened, to corrupt it, and therfore
it waxeth black. For heate workinge in a moiste bodie engendreth blacknes, which is the sign of corruption. And because y^t
the corruption of the first form is the generation of the second,
therfore of it is engendred a bodie naturall, which bodie is certainly
inclinable to what fermente youe will. Yet must the ferment
together with the bodie alkamy be altered, for it behoueth that y^e
wholle substaunce of our elixar should hould of the nature of quintaessentiae; otherwise it should nothing availl. And betwen black and
white that shall com ther shall appeare a grene colloure with soe
many collours afterward as may be thoughte by mans wit. But
when the white collour shal begin to appeare, like eyes of fishes, a
man may say that sommer is nighe at hand, whom followeth attume
luckely with longe lokinge for rednes, beinge first discended

322r:
1 it is put] it put **S2**
   externall] externall ~~water~~ **A**
5 cover] kiuer **S2**
12 substract] subtragh **S2S1**
17 fier] the fire **S2S1S4S3**
24 inclinable] enclined **S2GS1S4S3**
26 quintaessentiae] the quintaencie **S2GS1S4S3**
31 say] se **G**, know **FS1**, then knowe **S2S4S3** *
32 lokinge] looked **S2FGS1S4S3** *

from his south seate by dewe course, perfecte, laterall,
    and shin*n*inge,
by grosse solution into the weste p*a*rte and imp*er*fecte, and thence
    into y^e
north, blacke, darke, & purgatoryall, alterate and watry, and here-   *   35
hence into the este, laterall, lighte, in his first p*er*fection christalyne,
paradiscicall, and somerly, and laste of all takinge his fiery
chariot, compassinge into the south, imp*er*iall, fieri, sheninge,
heavenly, autumnalle, highest, and of the beste p*er*fection, for then it
hath gone throwe the whell of phi*lo*so*fi*e. And when the oyll   **
    is simplified   40
of the body altered and is disposed by the fyer in his vessell dewly
sealled, what other thinge is ther then on w*hi*ch nature hath made   *
    to be
engendred of sulfur & argent viue in the bowells of the earth? Wherfore
o*u*r stone is nothinge ells but malle and female, ☉ & ☽, heat and cold,   322v
    sulfur & ☿. For
whatsoever names be rehearsed in this arte ar put to & named colorably to de-
ceyue the vnworthy of this science. For our worke is a worke of 3 thing*es*, vz., a
bodie, a soull, & a sprite. The body is the substance of the matter, which is earth.
The sprite is the vertue, vz., of quintaescentie, w*hi*ch raiseth vp nature   5
    from deth
and is water. The soulle muste be taken fo\<r\> the fermente, w*hi*ch may
not be had but of the moste p*er*fecte body. In sulfur is erthlynes
for the body; in ☿ is aerynes for the sprite; and in them both   *
    a naturall

---

33   laterall] latall **S2FGS4S3**, latall {naturall} **S1** *
34   p*a*rte] pa\<i\>l **S2GS1S4S3**, peele **F** *
35   alterate] alterable **S2FGS1S4S3** *
36   his] this **S2**, this {his} **S1**, the **S4S3**
322v:
3   For our worke] *in red ink* **A**
      a bodie] of ˄{a} body **S2S1S4**, *longer pass. abs.* **S3**
4   a sprite] of ˄{a} spirit **S2S1S4**, *longer pass. abs.* **S3**
5   The sprite] The *in red ink* **A**
6   The soulle] *in red ink* **A**
      fo\<r\>] *changed from earlier* from **A**
7   In sulfur] In ~~earthy~~ sulfur **A**
8   a naturall vnctuosity of the fermentall body] a naturall vnctuositie for the sowle to-
      gether w*i*th the ounctuosity of the fermentall body **S2FGS1S4S3** *

vnctuosity of the fermentall body commixed, vnsep*a*rable vnited by the
leste p*a*rtes, wherhence the stone is formed, who yet beinge absente may
in noe wise be formed, for it is the peculiar prop*er*ty of ☉ & ☽, wh*ich*
app*er*teyneth not to the stone ˄{of} itselfe, to giue the form of gould and siluer.
And this elixar, whether it be white or red, may be augmented
with fermentall oille, white or red, infinitely; and when it is caste
vpon ☿, it turneth it into elixar, which muste be caste vpon vnp*er*fecte
bodies. Moreover, the white elixar is <a>ugmented w*i*th mercuriall
water, and the red elixar w*i*th mercuriall oylle, w*hich* both muste be
had of ☿ dissolued by himselfe. And when the elixar shalbe brought
to a purple coller, let it be dissolued with his owne menstrue, being
first rectifyed into thine oylle, wherevpon it muste be fyxed by circulation of the sprite of our water; and then hath it powar to
turne all bodies into puer gould and to heall all dizeases in
man, for this is the potable gould and noe other, which is
made of gould soe coningly eleme*n*tated and circumduced by the
whell of phi*losofie* that ther is nothinge in the wordle more p*re*cious
sauinge the potable gould of the animall stone, w*hich* is for health
& life lyvely, and is the highest medison in the wordle, as is aforesaid and shalbe showed in the chapter followinge &c. Finis.

Cap*itulu*m 27 Of the animall stone

Here muste we com again in this chapter to the gretest secret and
moste noble, that is to saie, of the animall stone, w*hich* riche &
poore haue. We saie rich and poore haue the matter of our
stone because every man & creature is enduwed with y*e*
4 element*es* wherof our animall stone muste neades be made. For because the animall stone is evenly co*m*pounded and co*m*pacte together

---

9   vnsep*a*rable] inceperable **S1**, inceperably **S2FGS3**, insep*e*rablely **GS4**
13 And this elixar] *in red ink* **A**
15 be caste] cast **S2**
20 it muste be fyxed] let it be fixed **S2FGS1S4S3**
23 noe] none **S2FGS1S4S3**
323r:
1   Cap*itulu*m 27] *in red ink* **A**, Captm 27 **G**, Cap 27 **FS3**, Capitula 27 **S2**, Chap 27 **S1S4**
    Of the animall stone] *abs.* **S2FGS1S4S3** *
2   Here] *in red ink* **A**
6   animall] ayall **S2S1**
7   animall] ayall **S2S1**

in element*es*, therfore of the element*es* he muste be elixerate &
not of wyne forsoemoch as it is wine, nor of eggs, here or blod
forsoemoch as they ar eggs, hear, & blod, but must be made of the elem*entes*; 10
wherfore is is said that of everie pore, sick, and neady ph*ilosofe*r   *
    this worke
may be made by his reason and wisdom that God hath gyuen
him; wherfore we muste seke to haue the element*es* in the excele*n*cy
of their simplicity and rectification because the eleme*ntes* ar   *
    the mother
and root*es* of all thing*es*. Yet the element*es* of the moste   *   15
    eccellent creatur,
in whom they ar soe evenly co*m*pounded, entreth not into the worke
of the elixar but by vertue and co*m*mixtion of the sprit*es* and bodies
of mettalls, and soe they enter in and accomplishe the elixar of life.
For amonge all those that ar animalls, ther is found on   *
    more excelle*n*t
and more naturall then any other, w*hich* ph*ilosofe*rs haue taken for   *   20
    the nea-
rest matter of our stone, for the singular p*er*fection gyuen of God
to the lesse wordle, wherin ar the figures, courses and effect*es* of the   **
plannet*es*, signs and stars, complection of humors, sprit*es*, and naturall vertues; the w*hich* bodie after Aristotle muste be dyuided   *
into 4 p*ar*tes, which p*ar*tes muste again be co*m*passed together vntill   25
they repugn not; wherby too vertues op*er*atiue ar ioyned together,
vz., sulfur and argent viue, alias water & fier; wherby is made   **
a naturall bodie of an excellent form. And here we muste vnderstand that this diuision may not be manuall but must be   *
brought to passe by the only workinge of nature by himselfe, for   30
in this place is covered the greateste secrete of this worke. Yet to y*e*   *
experte workeman all the gross worke or manuall working
is sertainly knowen, but because this secret resteth not in the gross

---

10 ar] bee **S2S1S4S3**
11 is is] it is **S2FGS1S4S3** *
12 his] the **S2S1**, *abs.* **S4S3**
14 ar] be **S2S1S4S3**
      the mother and root*es*] y*e* rottes & mother **GS4S3**, the rootes and mothers **S2FS1** *
15 Yet] and yet **S2S1S4S3**
      the moste] that most **S2S1S4S3**
22 effect*es*] effect **S2GS1**
23 *Unidentified symbol in the right hand margin, resembling a capital* 'H' *with a crossbar on top* A

worke, I will not in this place altogether open the matter therof, because in the other chapter before I haue declared howe the oylls should
be made and ioyned together, and how the body alkamy should be
pr*e*pared for the calcination and dissolution of ☉ & ☽, wherof must
be made white oille and red oylle, puerly rectified as is said before;
w*hich* oylls must be put together in kymia, phi*loso*fically sealed, to putrifi
in balneo a longe season into a blak thicknes. Then by the second
degre of fyer, let it be coaguled into dry pouder after many brubling*es*
which it shall make; in the which shall showe inumerable collours;
and when that which is fine shall ascend vpward in the vessell white as
eyes of fishes, the worke is complet in the first p*ar*te, for the same
white then hath fullie in himselfe the nature of the white
sulfur not breninge, and as the verie sulfure of nature,
p*r*ocreat and made of arg*ent* viue. Then let the said sulfure be
imbibed and again putrified. Then dep*ar*te the sulfur in
too p*ar*tes, the one for the white and the other for the red, as
is said for in the chapter of the animall stone. Then ferm*en*t
the white with the white m*er*curie vntill it will stand in
the fyer fluxible and flowinge as wax. Then set the
other p*ar*te in fyer as it is taughte in the chapter of the animall
stone, and embibe it w*ith* the oill of arg*ent* viue 7 times, &
let it be putrifyed as is said afore in the white stone, and
afterward fermente it with red ☿ vntill it be broughte
to a purple collour and will melte in the fyer like
wax and shine in a darke place like a burninge cole.
Then let the second sulphur named the white and red

---

35   chapter] chapters **S2FS1S4S3**
39   *In the right margin*] Oille **A**
41   brubling*es*] burblinges **S2S1S4**, bublings **S3**
42   the which] which **S2S1S4S3**
       showe] shine **S2S1S4S3** *
323v:
2   the white sulfur] the white ~~tinctur~~ sulfur **A**
3   as] is **S2FS1S4S3** *
4   the said] saide **S2**
5   in] into **S2S1S4S3**
7   for in the] *in red ink* **A**
       for] afore **S2FGS1S4**, afores*ai*d **S3** *
8   the white m*er*curie] w‸{h}ite mercury **S2S1**
10   it is] is **S2FGS1S4S3**
       animall] ayall **S2**
11   it] 't' *written on top of earlier* 'r'(?) **A**
16   red] ‸{the} red **S2GS1**

sulfur be ioyned together with the oille of the white elixar
and the oylle of the red elixar that they may worke the more
strongly; vpon the which yf the quintaessentie of the vegi-
tall stone be fixed, youe shall haue the highest medisons   20
both for the health of mans body and alsoe to transmute in-
to puer gould and siluer everie bodie of the myne.
And here in this place seameth well with this *pre*ciouse   *
stone to end, whose noble secret*es* of many wise men hath
byn depely hid and covered with the obscurity of phi*lo*sofie &c.   25
And forasmoch as the lesse worke of alkamy is not
reconed amonge the ph*ilosofe*rs to be a stone minerall, animall
or vegitall, but a medicine transmutable into
sonn and moone, yet of moch greater *pe*rfection then
the minerall stone, I haue therfore placed the said   30
worke in the end of my bock, as a necessary member
vnto the same. Althoughe his principall p*art*es be declared
in the former chapters, yet they shalbe moste largly
spoken of again in thes fewe chapters folowinge &c.

    In this 28 chapter begynneth   324r
    the leesse worke of alkamy   *

---

20 medisons] medicines of all medicines **F**, medicyne of all medicines **S2S1S4S3**, med-
disone in ye world **G**
23 And here] *in red ink* **A**
seameth] it seemeth **S2FGS1S4**, it seems **S3**
24 to end] to ende the treaties of oure precious stones **S2FS1S4S3**, to ende w*it*h this
p*re*shious stone ye treates of *ou*r philo*sop*hers stones **G**
noble] notable **S1**, on*er*able; '*er*' *in a later hand(?)* **S2**, honorable **S4S3**
many] diuers **S2S1S4S3**, *abs*. **G**
25 depely hid and covered] hid and couered deepely **S2S1S4**, hid & deeply coverd **S3**,
hedden deppely covered **G**
26 And forasmoch] *in red ink* **A**
27 animall or vegitall] vegitable or a*nym*all **S2FGS1S4S3**
31 my] this my **S2FGS1S4S3**
33 moste] more **S2S1S4S3**
34 &c] &c finis **S2FS1**, *abs*. **S4S3**
324r:
1 In this 28 chapter begynneth the leesse worke of alkamy] *in red ink* **A**
2 alkamy] alchimie &c **S4**, alkumie Cap 28 **F**, alkamy &c Capitula 28th **S2**, alkimye
&c Cap 28 **G**, alkemy Chap 28 **S1**
Between title and first line of the chapter in Napier's hand] distill a pottle of spirit*es* of
wyne fr*om* {a} pond<e> of christallyne tartar & draw it of & then put fresh christalline
tartar {lib s} & poure on it thy spirites of wyne & still apply ye fresh chryst tarta<r> lib
s & soe doe ten tymes & then resive it close[?] suppe[?] **A** *

Take therfore the argull of wyne, grind it to smalle duste, and put it
in a bodi of glasse, and vnto on pound therof put a pinte of aquauite
pouring it vpon the pouder of argull. Then stop the vesselle
    closse from
taking of ayer, and let it stand soe 24 howars. Then distill y^e
water from the earth with the heate of the fyer; the which water will
haue noe taste of sharpnes because all his wholle fyer doth remain
in the earth. And this is it that Raymond doth say: Hid fyer in the
loweste partes. Morever, Hermes doth saie: Pace the substance of the
erth by whotnes and moistnes vntill they com into daye. Put fyer
in fyer, vz., put sal combust in aquavite vntill fyer in fyer doe
liquifie, and when the earth is congeled, suffer the vessell to be
cold, and then begin again with newe aquavite after the same ma-
nore as youe did before, and doe this 15 times, and when thou
haste done thes things, then thou haste the fyer of nature. Then
distill fier from fyer, vz., water from fyery earth, because it
burneth the body more then common fyer. The liquor being thus dis-
stilled is a moste subtill oylle, white like whay or milke, and it is
cauled our eagle, our ☿, our sal armoniak,
our heaven, our quintaessencia, and the key of all our secretes, by which
thou maist enter into the chamber of the Ladie and Emprise of Philosofie,
wher thou shalte find infinit treasures, and fewe doe find
this bird althoughe many doe hunt after her. Kepe it well &c.

---

3   to smalle duste] small to dust **S2FGS1S4S3**
4   in] into **S2S1S4**
       put a pinte of aquauite pouring it vpon the pouder of argull] powre vpon it aqua vite vnto one pownde of powder put a pinte of aquavite **S2FGS1**, pour vpon it aqua vite vnto i^{li} of pouder a pint of aqua vitae **S4**, pour on it aqua vitae to i^{lb} of pouder a pint of aqua vitae **S3** *
5   Then] and **S2FGS1S4**, *abs.* **S3**
9   this is it] this is **S2FGS1S4S3**
       is] 's' *written on top of earlier* 't' **A**
10   Pace] place **S2GS1S4S3** *
11   daye] a˄{i}re **S2FGS1S4S3** *
12   put] *abs.* **S2S1S4S3**
16   *In the right margin*] Fier of | nature **A**
19   *In the right margin*] Our eagle | our ☿ | our sal armo **A**
21   which] the which **S2S1S4**
22   of Philosofie] Philosophy **S2FGS1S4S3**
23   doe find] doth finde **S2GS1S4**, finds **S3**
24   her] it **S4S3**, *abs.* **S2S1**
       &c] &c Finis **F**, Finis **G**, Finis Ca 28th **S2**, Finis Chapter 28th **S1**

Capitulum 29 25

Nowe, farthermore, let vs com to the practice that we may haue
water of ☿, which is our tincture, and to haue of ☉ and ☽ white oille
and red oylle, which is our fermentation. Out of the same is made precious
stones, the goulden drinke, and the great elixar of life &c. Finis.

Capitulum 30 30

℞ ergo in nomine domine the forsaid distilled water, and put into it thin
plates of gould and siluer seuerally, and it will be consumed straight-
waie into oylle without the heate of the fyer, and this is our ☿ mettalin
called the vegitable sprite of the menerall wyne by his proper place of
generation of leuing thinges doe com by the influence of the ☉, without ye 35
which neither bodie nor soulle hath life. Soe it is to saie of 3 fold nature, that is to saie, of water, ayer, and fyer; wherfore Raymon doth
saye that our water is made of 3 natures, and is the water wherof mettells ar procreated, and thes natures is ☿, which is the menstrue of
Raymond, which is resoluble, by the which too extremityes be 40
    ioyned together,
that is to say, sulphur and argent viue, fastened in the bodies of 324v
Sol and Lune but yet not fixed, and therfore it is called the grene
lyone vegitable for because by quietnes he doth reste in his proper

---

25 Capitulum 29] *in red ink* A, Captitm 29 G, Cap 29 F, Capitula 29th S2, Chap 29 S1S4S3
26 Nowe] *in red ink* A
30 Capitulum 30] *in red ink* A, Cap 30 F, Capitula 30 S2, Chap 30 S1S4S3
31 ℞ ergo in nomine domine] Take therfore in the name of Christ S2FGS1S4, In ye name of Xst ℞ S3 *
    ℞ ergo] *in red ink* A
    into] in S2S1S4S3
32 seuerally] {into severall glasses} *gloss in Napier's hand* A
35 leuing thinges] liuinge thinges bycause liuinge thinges S2S1S4S3 *
36 it is to saie] is it saide S2, it is said S1, it is said to be S4, its said to be S3, that it is G
37 Raymon doth saye] *in red ink* A
38 3 natures] 3 ~~wat~~ natures A
    wherof mettells ar procreated] of metteles procreate G, of mettalls and are procre-
ated F, of mettales for bycause of one only nature all mettals are procreated S2S1, of
mettalls because of one onely nature all mettalls are procreated S4S3 *
40 be] are S2S1S4S3
324v:
3 proper and naturall] proper naturall S2FGS1, proper S4S3

and naturall place from whence he was taken out of the matrix of the erth by vertue of handlinge and the heate of the sonn.
Many doe speake of the gren lyon w*hich* doe not knowe what
it meanneth. I giue thank*es* to God, w*hich* hath gyuen me this
vnderstandinge. And soe doe youe giue thank*es* to God, for because when the bodies be dissolued into the moste p*re*ciouse
☿, circulatinge thes thing*es* till it be made oylle, and then
youe shall haue gould and siluer to drinke, and the chefest
ferment of elixar white and red. The goulden oylle doth
helpe leprosie and all infirmities, and p*re*serueth youth
vnto the tyme apointed. Finis.

  Cap*itulu*m 3i

And here to com to the true knowledg of the elixar and to haue
the transmutation of mettalls in this less worke of alkamy. Thou
shalte knowe that ☿ is the mean of ioyninge the tyncturs betwen
the ☉ and the ☽, and soe it is called the body of alkamy in this lesse
worke but not in the greatest worke. The great worke is in y^e
☉ and ☽, with whose natures the quintaessentia is fastened
by circulation, hauinge vertue to change ☿ and all other imp*er*fect
bodies vnto p*er*fecte ☉ and ☽ &c.

---

5 *In the left margin in red ink*] Gren lyon **A**
6 Many doe] *in red ink* **A**
  doe speake] doth speake **S2S1**, speak **S3**
7 to] vnto **S2S1S4**
8 to] vnto **S2FGS1S4**
13 helpe] heale **S2S1S4**, healeth **G**, heals **S3**
  leprosie] leapryes **S2S1S4S3**, leporis **G**
  all] also **S2S1S4S3**
14 Finis] &c Finis **F**, &c Finis Capitula 30th **S2**, &c Finis Chapter 30th **S1**, &c **S4S3**
15 Cap*itulu*m 3i] *in red ink* **A**, Cap 31 **F**, Capitula 31th **S2**, Chap 31 **S1S4S3**
16 And here] *in red ink* **A**
  In the left margin] ☿ | the body of | alkamy q*u*id **A**
17 less] last **S2S1S4S3**
19 body of alkamy] body alkamy **S2GS1S4S3**
20 greatest worke] greater **S2S1S4S3**, more **G**
21 ☽] the moone **S2FS1S4S3**
22 other] *abs.* **S2S1S4S3**
23 vnto] into **S2GS1S4S3**
  &c] &c Finis **S2FG**, &c Finis 31 Chapter **S1**, *abs.* **S4S3**

Capitulum 32

This lesse worke of alkamy did Alkamus the philosofer invent
for the chaunging of ☿ and all other bodyes into the ☉ and ☽. And
the mean of transmuptation is Venus, which is the corruptible
bodie and therfore to neither body alterable, vz., into a newe
bodie masculine and feminin of the ☉ and ☽, the which after it
doth putrifie in his mothers milke, and this is ☿ alkamistik
and the aquavite of mettalls and the small ferment of ☉ and ☽,
and shall com to perfecte elixar &c.

    * 25
    *
    *
    *
    30

Capitulum 33 showeth the composision of
    mercury alkamistick

In the name of Christ, Amen, ℞ i ℈ of ☿, and grind him with
℈ iii of the vegitable salte. The salte is made in the first chapter
from the which our vegitable ☿ was drawen by great heate of
the fyer. Grind the ☿ and the salte together vpon a marble stone
vntill it be well incorporat; the which being don, put it in a glass
vessell, well stopped, and put it in balneo with a softe fyer, and the ☿
will dissolue into whit milke. Then put the same milke vpon

  35
    *
  40

---

24 Capitulum 32] *in red ink* A, Cap 32 F, Capitula 32ᵗʰ S2, Chap 32 S1S4S3
25 This lesse] *in red ink* A
     worke of alkamy] worke of ~~philosofie~~ alkamy A
     did Alkamus] did I Alkamus S2FGS1S4S3 *
26 *In the left margin*] ☿ A
30 doth putrifie] doth putryfie and is putryfied S2S1S4, doth putrifye as is putrifyed F, putrifies & is putrified S3
31 ☉] the sonn S2FS1S4S3
     In the left margin] Aquavite of | mettalls A
32 &c] &c Finis FG, &c Finis Capitula 32 &c S2S1
33 Capitulum 33 showeth the composision of mercury Alkamistick] Chap 33 shews yᵉ composition of yᵉ ☿ alkamistick &c S3, Cap 33 sheweth for yᵉ composition of the mercurie alkumistick Cap 33 F, This 33ᵗʰ chapter sheweth foorth the composition of yᵉ mercury alkamistick &c Capitula 33 S2, Chap 33 This 33ᵈ chap sheweth forth the composition of the ☿ alkamistick &c S4, *m.* G
     Capitulum 33] *in red ink* A
     In the left margin in Napier's hand] ☿ purifyed A
35 In the] *in red ink* A
     In the left margin in Napier's hand] The vegetille salt of christalline tartar A
36 ℈ iii] 3 ³ᵉˢ S2FS1, *m.* G
40 vessell] *abs.* S2FS1S4S3, *m.* G
41 same] saide S2S1S4S3, *m.* G

a pound of ☿ in balneo and the same pound vpon 10^li and that *
ten pound vpon 100^li, and all will dissolue into milke; and when all * 325r
is dissolued by putrifaction and made thine, then still it in
ashes with with a softe fyer first, and caste away the first water **
and change your receptory, and still it with a more strong fyer, *
and ther will com a thick water that will dissolue all bodyes, putrifi 5
them, fyx them, and purify them. Distill this water vntill the hed *
of the limbick doe wax red. Then change the receptory, and kepe
the white water verie close stopped. Then strengthen your fyer, *
and ther shall still a red oill of the collour of gould; the w*h*ich kepe
well to dissolue the red ferment and to multiply the red elixar 10
because it is our singular gould of nature not fyxed, and this
too humors ar drawen from our gould and siluer, w*h*ich is
our ☿, as Avicen said in his Epistell to his Sonn, saying: *
    It behoueth thee
to haue at the leste 60 pound of our gould and siluer to drawe
out his moisture. Lykwise, Guido saith: It behoueth thee to haue 15
moch water and moch oille. And soe thou shalte vnderstand that *
the ferment shall change with the vnp*er*fecte body of ♀. Thus,
℞ i^li of ♀ and i^li of ☽ made puer, and power them together for *
the white; and soe thou shalte doe with ♀ and ☉ for the red.

325r:
2   in ashes] in ~~a sh<ur>e fier first~~ ashes A
     In the right margin in Napier's hand] In balneo | or fimo A
3   with with a softe fyer first] first with a softe fire S2FS1S4S3, *m*. G
5   ther] then S2S1S4S3, *m*. G
     putrifi] <u>putrifi</u> *underlining in Napier's hand* A
6   fyx] <u>fyx</u> *underlining in Napier's hand* A
     vntill the] <u>vntill the</u> *underlining in Napier's hand* A
8   close] closly S2S1S4, well F, *m*. G
     strengthen] strength S2S1, *m*. G
13 said] did saye S2FS1S4, says S3, *m*. G
     Avicen said in his Epistell to his Sonn] *in red ink* A
     saying] Sonn saide he S2FS1S4, *abs*. S3, *m*. G
15 Guido saith] *in red ink* A
     It behoueth] <u>It behoue</u>th *underlining in red ink* A
     thee] *abs*. S2FS1, *m*. G
17 vnp*er*fecte] imperfect S2S1S4S3, *m*. G
18 ℞] *in red ink* A
     ☽] the moone S2FS1S4, *m*. G
     In the right margin in Napier's hand] & melted | together A
19 thou shalte] yow shall S2FS1S4S3, *m*. G

Then beate it out into thin plates, and cut it into small peces, and put the white into the white water, and the red into the red water seuerally by themselfe, and put them both in balneo, and they shalbe dissolued and shalbe like oille oliu<e> gren; and soe let it stand in a temp*er*at fyer vntill all be made like moulten pitche, and soe dry it by changinge from on cullour to another vntill it be white as the eyes of fishes in the form of p*er*ells. Then haue youe the first order of the elixar white. Multiply the earth with the milke of ☿ vntill it be thick like gum. Then it fasteneth ☿ and changeth all bodyes of the myn into syluer. And soe contynue the other vessell with the red vntill it be in the manner of fishes eyes of the collour of purple. Then multiply the red earth with the forsaid red oylle, as youe liste, and soe co*n*tynue it as youe did before vntill it be as it were gum, the w*h*ich changeth Venus and all other bodyes of the mine into p*er*fecte ☉, better then the mineralle or naturall &c. Finis.

20
*
*
* 25
*
30
*
35

---

24 fyer] heate or fier **S2S1**, heat of △ **S4S3**, *m*. **G**
26 fishes] 'f' *written on top of earlier* 'sh' **A**
27 the elixar] elixer **S2FS1S4S3**, *m*. **G**
28 of] or **S2S1**, *m*. **G**
vntill] till **S2S1S4S3**, *m*. **G**
35 &c Finis] &c Finis Capitula 33 **S2**, Finis Chapter 33th **S1**, *m*. **G**

Appendix 1: Letter from Humfrey Lock to Robert Dudley, 19 May 1572

National Archives, London, State Papers 70/123, fols. 148r–149r. (The transcription follows the editorial principles used for the edition.)

Right honorable and my very good Lorde, wheras by infformac*i*on 148r
ffrome my ffendes [*sic*] owte of England, I haue vnderstandyd that
yo*u*r Hono*u*r by untrewe reportes made of me haue conceavyd great
displeasure ageynste me, ffor the whiche, my good Lorde, I ame
right hartylly sorye. Yet consyderinge myne innocencye, I 5
receave ageyne so*m*me comfforte, hopinge that God of his
great m*e*rcye will bringe me ones ageyne into my countrye,
ye to aunswere myne accusers fface to fface, at what tyme
yo*u*r Hono*u*r shall well *p*erceave that I haue not comyttyd anye
ffault worthie of yo*u*r Hono*u*res so great displeasure. For God ys 10
my wytnes, I haue byn alwayes as well in Russia as in England
a trewe subiecte to my prynce and countrye, and haue hade
greate greffe whan I haue harde my prynce dishonoryd
and my countrie slaunderyd, not by Englyshemen but by Russes
thorowe and by the dealinge of Englyshemen. And ffor yo*u*r 15
Hono*u*res conceavyd displeasure, I moste hu*m*blye desyre yo*u*r good
Lordshipp, ffor Jesus Christes sake, that it may rest in dispence vntill
my returne into England, and yo*u*r Hono*u*res good ffavour to be
contenewyd towarde me as in tymes paste vntyll I haue made
made my purgac*i*on, and after as I shall be ffound worthie. 20
For I do not dought but the answeringe of my cause w*it*h the doing*es*
of my accusers shall redound as moche to there shame as there
ffalse reportes haue mynestryd vnto me discredyt ffor the tyme.
Yet, my good Lorde, notw*ith*standinge all my discredyte by myne
ennymyes and the Emp*er*ror*es* great goodnes vnto me bothe in 25
ffavour and lyvinge, yf I myght returne into my countrye

148r:
19–20   made made] *in manuscript*

in the good grace of the Quenes Maiestie, your good Lordshippes
ffavour, and my good Lordes of Burgley, I wold gladly forsake
all to enioye the lybertie of my countrie. For beffore God, I
speake it: my zeale ys suche to my prynce and countrye   30
that I wold chuse rather to lyve there in a pore estate,
than heare with great lyvinge, and I do meane, the almyghtie
God so permyttinge, shortlye to be in England, yf leave may
be obteynyd, the whiche deneyed, I shall moste humblye
beseche your good Lordshipp and my good Lorde of Burgley   148v
to be medyatores vnto the Quenes Maiestie ffor her gracious
lettre vnto the Emperrores Maiestie ffor my delyverye. For
althoughe the Quenes Maiesties requestes sent by her highnes
lettres and ambassadour ffor the ffre passinge of Englyshemen of   5
all sortes into Russia and ffourth ageyne at there pleasure
be ffullye grauntyd by the Emperrores Maiestie, yet I doght of
myselffe yf the Emperrour contenewe his buyldinge. And for
my accusers, my good Lorde, I can not name them all, but the
cheffe I suppose to be Mr Randall by his reportes and the   10
marchaunte Banester by his Lettres, whoo, yf they bothe had taryed
in England and the other that came beffore them, yt had
be moche better ffor those that causyd them to come into Russia
and ffor all other Englyshemen in the countrye, as it hathe
well appearyd synce there beinge heare, and is nowe   15
more apparaunt by the comynge of Mr Ienkynson, of whose
commynge the Emperrour lykyd so well that soo ffarre as I
can heare his Maiestie hath grauntyd hym all he requestyd
vnlesse it weare suche goodes as weare taken in the tyme
of his Maiesties displeasure ageynste the marchauntes. All-   20
thoughe the slaunderouse wordes of his ennymyes, the chefe
lordes of the Semskye, after his departinge owt of Russia
in A° 67, weare ffaire otherwyse, whoo synce that tyme
haue receavyd condinge punyshment ffor there treasones,
and he returninge ageyne into Russia contrarye to   25
mennes expectacion hath provyd hymselffe a good man,
clearyd all crymes, receavyd honour, the pryncys
great ffavour, wherby I suppose that yf Mr Ienkynson
had come in the place of Mr Randall without any fforecomers,
that all thies matters had than be dispatchyd and thowsaundes   30
of poundes savyd, whiche nowe standyth in dought to be

148v:
27   the pryncys] ~~and~~ the pryncys *in manuscript*

Appendix I: Letter from Humfrey Lock to Robert Dudley

recoveryd. Humblye besechinge your good Lordshipp to consydre
that I and such as I ame haue not brought the Emperroures
displeasure vppon the marchauntes but there owne garboyle, there
stryvinge, there hatinge one another, there entreprysynge
thinges vnlawffull, there discredytinge one another, ye and
there gredye sekinge to robbe one another, in suche a conffusyd
ordre that it was {not} lyke to contenewe, and all seamyd to the
Russes to be donne by the Quenes Maiesties auctorytie, whiche
was, as I thought, rather dishonour to her Maiestie and slaunder
to her countrye than otherwyse, so that very zeale constray-
ned me to ffynd falt with the abuses and somwhat to speke
therof not to Russes and straungers but to Englyshemen;
wherffore I was callyd a traytour and an ennymye to my
countrye bycause I cold not prayse there evyll doinges,
fflattre them, and say they dyd well. For they haue byne
in this country as the people of Iuda weare in the tyme of
Ieremie: they myght not heare of anye evyll to comme.
Suche as spak trewthe weare hatyd, and suche as could
fflattre them and bringe them lyes weare regardyd &
and beleavyd. Iuda eskapyd not the plage. Nether had
theis eskapyd yf the Quenes Maiestie of her great clemen-
cye had not spedely provydyd ffor them, and with her
Maiesties ambassage sent suche a man as God gave
great ffavour in the eyes of the prynce. Lett them nowe
therffore deale vpprightly. Yf they ffall ageyne yt wilbe
irecuperable. All thies thinges Mr Randall oversawe and
could not abyde to heare the trewth, but gave eare to
lyinge prophettes, wherffore he dyd no good but made all
thinges worse than they weare beffore his comminge into the
country. And ffor my accusacions, brevyty of tyme constraynyth me
to refferre them, yf God ˰{will}, vntyll my comminge into England, humblye
besechinge your good Lordshipp and my good Lord of Burgley in
the mene tyme in all trewth to stand, my good Lordes, and I shall
daylye next vnto the Quenes Maiestie pray for your honoures prospiryt<ie>,
helthe, and longe lyffe. In Russia the 19 of May A° 1572.
    Your Honoures humblye ffor ever duringe lyffe,
        Humffrid Locke

Appendix 2: Collation Patterns

The table below gives an overview of the agreements of the manuscripts in a few chapters selected for the collation: 1, 3 (301v: 1–302r: 41), 6–7, 11–13 (310v: 1–311v: 13), 17–18, 21 (316r: 1–316v: 46), 26 (320v: 1–22; 322r: 2–322v: 28), 28–32. All agreements were included, except for differences in spelling, use of abbreviations and sigils, and similar features. 'M' in the columns of G and S1 means that the reading is missing in the manuscripts, since Chapter 1 and the majority of Chapter 2 are absent in S1, and G lacks Chapter 33. Each instance of a grouping represents an agreement between two or more manuscripts in a specific reading. For example, the grouping AS3S4 occurs seven times against the other manuscripts. This should not be taken as an implicit indication that F, G, S1, and S2 agree on the other hand; the variation in the manuscripts is frequently more complicated, and the variants do not always fall into bipolar groupings. In fact, in two of the instances where A, S3, and S4 agree, none of the other manuscripts agrees with another: in one instance, F, G, S1, and S2 all have unique readings, and in the other instance F, G, and S2 have unique readings whereas S1 does not contain the reading at all since Chapter 1, where this instance of variation occurs, is missing in the manuscript. [1]

---

[1] For a more detailed discussion about this quantitative approach to collation, see Peter Grund, "'Misticall Wordes and Names Infinite': An Edition of Humfrey Lock's Treatise on Alchemy, with an Introduction, Explanatory Notes and Glossary" (Ph. D. diss., Uppsala University, 2004), 90–102.

Number of MSS	A Patterns	N	F Patterns	N	G Patterns	N	S1 Patterns	N	S2 Patterns	N	S3 Patterns	N	S4 Patterns	N
1 MS (unique)	A	96	F	126	G	537	S1	19	S2	25	S3	344	S4	24
							M	275						
2 MSS	AF	42	AF	42	AG	15	AS1	2	AS2	2	AS3	4	FS4	4
	AG	15	FG	19	FG	19	GS1	1	FS2	2	FS3	2	GS4	6
	AS1	2	FS2	2	GS1	2	S1S2	30	S1S2	30	GS3	30	S1S4	1
	AS2	2	FS3	2	GS3	6	S1S3	1	S2S4	3	S1S3	1	S2S4	3
	AS3	4	FS4	4	GS4	2	S1S4	1			S3S4	227	S3S4	227
3 MSS	AFG	85	AFG	85	AFG	85	AFS1	2	AFS2	15	AGS3	1	AS2S4	1
	AFS1	2	AFS1	2	AGS3	1	AS1S2	16	AGS2	4	AS2S3	1	AS3S4	7
	AFS2	15	AFS2	15	AGS2	4	FS1S2	14	AS1S2	16	AS3S4	7	FGS4	1
	AGS2	4	FGS2	4	FGS2	4	GS1S2	5	AS2S3	1	FGS3	1	FS3S4	7
	AGS3	1	FGS3	1	FGS3	1	S1S2S4	6	AS2S4	1	FS2S3	1	GS2S4	1
	AS1S2	16	FGS4	1	FGS4	1	S1S3S4	10	FGS2	4	FS3S4	7	GS3S4	30
	AS2S3	1	FS1S2	14	GS1S2	5			FS1S2	14	GS3S4	30	S1S2S4	6
	AS2S4	1	FS2S3	1	GS2S4	1			FS2S3	1	S1S3S4	10	S1S3S4	10
	AS3S4	7	FS3S4	7	GS3S4	30			GS1S2	5	S2S3S4	20	S2S3S4	20
									GS2S4	1				
									S1S2S4	6				
									S2S3S4	20				
4 MSS	AFGS1	4	AFGS1	4	AFGS1	4	AFGS1	4	AFGS2	59	AFGS3	3	AFGS4	3
	AFGS2	59	AFGS2	59	AFGS2	59	AFS1S2	42	AFS1S2	42	AFS1S3	2	AFS2S4	5
	AFGS3	3	AFGS3	3	AFGS3	3	AFS1S3	2	AFS2S3	1	AFS2S3	1	AFS3S4	5
	AFGS4	3	AFGS4	3	AFGS4	3	AGS1S2	6	AFS2S4	5	AFS3S4	5	AGS3S4	11
	AFS1S2	42	AFS1S2	42	AGS1S2	6	AS1S2S4	7	AGS1S2	6	AGS3S4	11	AS1S2S4	7
	AFS1S3	2	AFS1S3	2	AGS3S4	11	AS1S3S4	1	AS1S2S4	7	AS1S3S4	1	AS1S3S4	1
	AFS2S3	1	AFS2S3	1	FGS1S2	4	FGS1S2	4	AS2S3S4	1	AS2S3S4	1	AS2S3S4	1
	AFS2S4	5	AFS2S4	5	FGS3S4	7	FS1S2S4	1	FGS1S2	4	FGS3S4	7	FGS3S4	7
	AFS3S4	5	AFS3S4	5	GS1S2S4	1	GS1S2S4	1	FS1S2S4	1	FS2S3S4	4	FS1S2S4	1
	AGS1S2	6	FGS1S2	6	GS1S3S4	1	GS1S3S4	1	FS2S3S4	4	GS1S3S4	1	FS2S3S4	4
	AGS3S4	11	FGS3S4	4	GS2S3S4	5	S1S2S3S4	83	GS1S2S4	1	GS2S3S4	5	GS1S2S4	1
	AS1S2S4	7	FS1S2S4	1					GS2S3S4	5	S1S2S3S4	83	GS2S3S4	5
	AS1S3S4	1	FS2S3S4	4					S1S2S3S4	83			S1S2S3S4	83
	AS2S3S4	1												

Number of MSS	A		F		G		S1		S2		S3		S4	
	Patterns	N	Patterns	N	Patterns	N	Patterns	N	Patterns	N	Patterns	N	Patterns	N
5 MSS	AFGS1S2	137	AFGS1S2	137	AFGS1S2	137	AFGS1S2	137	AFGS1S2	137	AFGS2S3	2	AFGS1S4	1
	AFGS1S4	1	AFGS1S4	1	AFGS1S4	1	AFGS1S4	1	AFGS2S3	2	AFGS3S4	19	AFGS2S4	48
	AFGS2S3	2	AFGS2S3	2	AFGS2S3	2	AFS1S2S3	3	AFGS2S4	48	AFS1S2S3	3	AFGS3S4	19
	AFGS2S4	48	AFGS2S4	48	AFGS2S4	48	AFS1S2S4	47	AFS1S2S3	3	AFS1S3S4	1	AFS1S2S4	47
	AFGS3S4	19	AFGS3S4	19	AFGS3S4	19	AFS1S3S4	1	AFS1S2S4	47	AFS2S3S4	88	AFS1S3S4	1
	AFS1S2S3	3	AFS1S2S3	3	AGS1S2S3	10	AGS1S2S4	10	AFS2S3S4	88	AGS1S3S4	1	AFS2S3S4	88
	AFS1S2S4	47	AFS1S2S4	47	AGS1S3S4	1	AGS1S3S4	1	AGS1S2S4	10	AGS2S3S4	20	AGS1S2S4	10
	AFS1S3S4	1	AFS1S3S4	1	AGS2S3S4	20	AS1S2S3S4	19	AGS2S3S4	20	AS1S2S3S4	19	AGS2S3S4	20
	AFS2S3S4	88	AFS2S3S4	88	FGS1S2S4	11	FGS1S2S4	11	AS1S2S3S4	19	FGS1S3S4	1	AS1S2S3S4	19
	AGS1S2S4	10	FGS1S2S4	10	FGS1S3S4	1	FGS1S3S4	1	FGS1S2S4	11	FGS2S3S4	12	FGS1S2S4	11
	AGS1S3S4	1	FGS1S3S4	1	FGS2S3S4	12	FS1S2S3S4	12	FGS2S3S4	12	FS1S2S3S4	13	FGS1S3S4	1
	AGS2S3S4	20	FGS2S3S4	20	GS1S2S3S4	13	GS1S2S3S4	23	FS1S2S3S4	13	GS1S2S3S4	23	FGS2S3S4	12
	AS1S2S3S4	19	FS1S2S3S4	19					GS1S2S3S4	23			FS1S2S3S4	13
													GS1S2S3S4	23
6 MSS	AFGS1S2S3	8	AFGS1S2S3	8	AFGS1S2S3	8	AFGS1S2S3	8	AFGS1S2S3	8	AFGS1S2S3	8	AFGS1S2S4	10
	AFGS1S2S4	231	AFGS1S2S4	231	AFGS1S2S4	231	AFGS1S2S4	231	AFGS1S2S4	231	AFGS1S3S4	10	AFGS1S2S4	231
	AFGS1S3S4	10	AFGS1S3S4	10	AFGS1S3S4	10	AFGS1S3S4	10	AFGS2S3S4	9	AFGS2S3S4	9	AFGS2S3S4	9
	AFGS2S3S4	9	AFGS2S3S4	9	AFGS2S3S4	9	AFS1S2S3S4	335	AFS1S2S3S4	335	AFS1S2S3S4	335	AFS1S2S3S4	335
	AFS1S2S3S4	335	AFS1S2S3S4	335	AGS1S2S3S4	65	AGS1S2S3S4	65	AGS1S2S3S4	65	AGS1S2S3S4	65	AGS1S2S3S4	65
	AGS1S2S3S4	65	FGS1S2S3S4	47	FGS1S2S3S4	47	FGS1S2S3S4	47	FGS1S2S3S4	47	FGS1S2S3S4	47	FGS1S2S3S4	47
Total		1497		1497		1497		1497		1497		1497		1497

Table A. Quantitative patterns

Explanatory Notes

Ded. ll. 2–4 **Sir William Cycill . . . Tresurer of England**: For Cecil's titles, see p. 9.

Ded. ll. 18–25 **and showe himselfe . . . ar inclinde**: For a discussion of these lines, see pp. 3–4.

Ded. ll. 26–29 **Yet repent I not . . . I haue not seen**: For a discussion of these lines, see p. 10.

Ded. ll. 30–34 **wherby my seruice . . . vnto my arte**: "and" in l. 31 should probably be taken to mean 'if'; *OED* s.v. *and* conj.[1] C. 1.a. The sense would then be that, through this new knowledge that Lock has acquired, his duty [to Queen Elizabeth] will be restored if it is employed in the Queen's affairs. For "apployed," see note on l. 31; for "prince" in the meaning 'queen, female sovereign,' see *OED* s.v. *prince*.

Ded. l. 31 **apployed**: This verb, which is also used in S2, is not recorded in the *OED* or *MED*. It is most likely a blend of the verbs *apply* and *employ*. Although F has "applyed," the verb intended here may be *employ*, which seems to fit the context better (esp. in conjunction with the preposition *in*).

Ded. l. 37 **handes**: The reading in F and S2, "hand," is more suitable since the rhyming word in l. 35 is *land*.

Ded. ll. 70–74 **To perfecte . . . in this bock**: Cf. the "Russia note" in 316v: 27.

Ded. ll. 82–85 **that rune about . . . true and iust**: Cf. 148v: 32–149r: 13 of Lock's 1572 letter to Robert Dudley in Appendix 1. A presents this stanza in three lines: "that rune . . . as in them," "put . . . good," and "whose . . . iust." I have adjusted the lineation to fit the four-line stanza found elsewhere in the dedication. The two additional manuscripts, F and S2, both give the stanza in a four-line format.

Ded. ll. 87–89 **bringe to passe . . . globuse masse**: Cf. *ded*. ll. 118–19, 191–93, and 319r: 30 of the Treatise.

Ded. ll 94–105 **One saultes . . . losse thear light**: For similar lists of erroneous substances, see 299v: 9–10, 300v: 3–4, 323r: 8–10.

Ded. l. 109 **how bye theare worke thaye wine**: It is not exactly clear how to interpret this line. The main problem lies in the meaning of the verb "wine." Perhaps it should be taken in the intransitive sense of 'profit'; *OED* s.v. *win*¹ 6c and 7e. The line would then be highly ironical, implying that they do not actually profit at all, as shown in the previous stanzas. Another possibility, though a less likely one, is that "how" is actually a noun, meaning 'trouble, sorrow'; *OED* s.v. *how, howe* n¹ and *hoe* n³. The stanza would in that case need repunctuation, and a colon should follow "declame." The sense would be that they gain trouble/sorrow from their work. A third even more remote possibility, is that "wine" is a dialectal form of *won* 'dwell, reside'; *OED*; s.v. *won, wone*, and *win* v². The last two lines would then be fairly deprecating: what has been said in the previous stanzas is enough to show how they reside by their work (i.e., how they labor in their experimentation).

Ded. ll. 110–15 **Allthoughe I speake . . . crafte namid**: Cf. 321r: 30ll.

Ded. ll. 118–19 **the secretes to knowe . . . in one masse**: Cf. *ded*. ll. 87–98, 191–93, and 319r: 30 of the Treatise.

Ded. ll. 121–24 **Yf all men . . . on another starue**: Cf. 297v: 17–18, 301v: 20–33.

Ded. l. 125 **starue**: This is probably a misreading of *scorn*, found in F and S2, which rhymes with "forlorne" in l. 123.

Ded. ll. 142–205 **shall find him . . . his seed for nought**: These lines have been taken from F as they are missing in A. Their absence in A is explained by the fact that a leaf has gone missing between fols. 292 and 293. I have chosen F over S2, since the collation showed F to be more closely related to A than S2 is. See pp. 119 and 121.

Ded. ll. 170–74 **Although evill will . . . cast away**: Cf. 148v: 32–149r: 13 of Lock's 1572 letter to Robert Dudley in Appendix 1, and p. 7.

Ded. ll. 191–93: **assay to bring . . . globeous masse**: Cf. *ded*. ll. 87–98, 118–19, and 319r: 30 of the Treatise.

294r: 1 **Cap*itulu*m primu*m***: For a discussion of the title of the Treatise, see pp. 18–20.

294r: 2–31 **For becaus that . . . to be obserued in this crafte**: The Treatise begins with a number of anacoluthic clauses. The argument is that in other types of writing it is necessary to have an introduction to the text in the form of a prologue, preface, or epistle, but in this type of writing, which deals with "natural philosophy," "preseptes, admonitions, and warninges" should replace such an introduction.

294r: 13–14 **the wonderfull stone of philosofers called ye leess wordle**: Alchemical texts often refer to the philosophers' stone as *the less world* or *the microcosm*:

see e.g., Linden, *George Ripley's Compound*, 22; Reidy, *Thomas Norton's Ordinal*, 78; and Robert Steele, ed., *Three Prose Versions of the Secreta Secretorum*, EETS ES 74 (London: Kegan Paul, 1898), 87–88. Since alchemists thought that the philosophers' stone incorporated all the elements in perfect harmony, it was seen as a mirror image of the macrocosm: see Abraham, *A Dictionary*, s.v. *microcosm*. Cf. also 319r: 30ll. and 323r: 21ll.

294r: 32–296r: 26 **wherof the first commaundemente . . . in a smalle tyme etc.**: From 294r: 32 to the end of the first chapter, the Treatise is based on the alchemical text *Mirror of Lights*. See p. 34.

294r: 37 **be nothinge hindered**: Only A contains the reading *be*, which is possibly an attempt to clarify the otherwise strained syntax of the clause. For "nothing," see Glossary.

294r: 37–294v: 3 **But yf the workeman be [a] flebergeber . . . anythinge to good ende**: The precept in the Treatise has the same general content as the precept in the *Mirror of Lights*, but it is very different in formulation. Cf. *Mirror of Lights* (MS. R. 14. 45, fol. 70r): "ffor yif he tell it any othir he may lightly be holden a fals Menour [=a Franciscan friar] and dede therfor and neuer his werke be brought to an {ende}."

294v: 1 **flebergeber**: This is a variant spelling of present-day English *flibbertigibbet*, first recorded in the *OED* in 1549. This particular spelling with a final 'r' is not listed. The final 'ber' may have been influenced by the preceding 'ber' in 'fleber.'

294v: 6–8 **that in one he maye make fier . . . for the putrffacon of the stone**: Surprisingly, the Treatise seems here to combine features from the *Mirror of Lights* with features from that text's source, the *Semita recta*, unless they all existed in a copy of the *Mirror of Lights* that I have not found or that has not survived. Cf. *Mirror of Lights* (MS. R. 14. 45, fol. 70r): "that in some he haue fire and yn some clene materz"; *Semita recta* (Auguste Borgnet and Émile Borgnet, eds., *B. Alberti Magni, Ratisbonensis Episcopi, Ordinis Praedicatorum, Opera omnia* [Paris: L. Vives, 1898], 37.549) "in quibus fiant operationes ad sublimandum, ad solutiones et distillationes faciendum" 'in which operations for subliming are performed, and for making solutions and distillations.' See also note on 295v: 12–13.

294v: 8 **putrffacon**: i.e., *putrefaction*. An 'e' or an 'i' may have been skipped, but the single 'r' may also represent a syllabic spelling. For similar examples, see 302r: 3 "complmente," and 318r: 43 "infnitly." There are two other instances of the *-tion* suffix being spelled *-con* by Copyist 2 (see 294v: 26). The copyist may have omitted a tilde representing 'i' by mistake, or, more likely, they represent phonetic spellings, indicating that the *-tion* suffix was pronounced as it is today or in a similar way. Similar spellings have been recorded by Henry C. Wyld, *A History of Modern Colloquial English*, 3rd ed. (Oxford, Basil Blackwell, 1936), 293.

294v: 10–13 **For if it showlde be don . . . made of earth being wel glased**: The *Mirror of Lights* claims that the vessel must be either made of glass or glazed, not singling out vessels of earth specifically. Furthermore, the *Mirror of Lights* is more explicit about what happens with the medicine if it is dissolved in unglazed vessels. Cf. *Mirror of Lights* (MS. R. 14. 45, fol. 70r): "The fifthe co*m*maunde-me*n*t is þat all man*er* of vesseles þat medicy [=medicine] solued sholde be put yn þat þey be of glas or glased; ffor yif he put it yn vessell*es* of coper or of latyn, þe water wole wex grene, and yif it be yn yren, lede or ty*n*n, it wole be blak, and yif it be in erþe it wole p*er*issh þe vessell & mekill of þe medicyn be {loste}."

294v: 13–22 **The iiii precepte is . . . be in danger of thy life**: The manuscripts of the *Mirror of Lights* contain different versions of this precept. They agree with the Treatise precept in general content, but the versions display a great deal of variation in formulation. The precept in the Treatise is more closely related to MS. R. 14. 45 and MS. Harley 3542 than to e.g. MS. R. 14. 37 and MS. Kk. 6. 30, which present a much longer version.

294v: 17 **to posse**: This is probably a variant spelling or misreading for 'to pass,' which is found in all the other manuscripts.

294v: 25–30 **The 6th precepte is . . . but allso his worke**: The precept in the *Mirror of Lights* is first concerned with which time of the year is favorable for certain processes and when they are not to be attempted, and secondly with the importance of being patient in practicing alchemy. Cf. *Mirror of Lights* (MS. Kk. 6. 30, fol. 4v): "The 3ᵈ comaundme*n*te þat he kepe well his tymys of soluciou*n*s & dystylla-c*i*ons in su*m*er tyme. Sublimac*i*ons & calcinac*i*ons may be wroghte both in wynt-ter & in somer; & þat he be besye abou3te his werke tyll it be at an ende & no3ght werke o moneth & leve an͇{o}þ*er* & so lees his tyme & his gude in ydelnes."

294v: 30–35 **wherefore this worke, being once begone . . . at the will of the workeman**: This passage is not found in the *Mirror of Lights*. The insertion of the passage is probably an attempt to reconcile differing alchemical traditions, since the creation of the white sulfur is discussed in several other places in the Treatise. See 316v: 25, 321r: 6, and especially 317v: 16ll., 323v: 1–4.

294v: 34 **he maye be norished**: For the use of *he* to refer to inanimate objects in alchemical texts, see Grund, "Manuscripts as Sources."

294v: 35–295r: 9 **The 7 precepte is . . . they enter not into the bodies**: The number and order of the processes listed in this precept are different from the treatment in the *Mirror of Lights* (see the table below). Although the *Mirror of Lights* does not list projection, it all the same includes the discussion of what kind of powders may be used for projection.

Process	Treatise	*Mirror of Lights*
1	preparation	preparation
2	sublimation	sublimation
3	calcination	fixion
4	conjunction	calcination
5	putrefaction	solution
6	fixation	fermentation
7	imbibition	distillation
8	fermentation	coagulation
9	liquefaction	——
10	projection	——

295r: 8 {solued}: In A, the word "sv<olo>ved" has been smudged and "solued" has been added, probably as a clarification, without a cancellation of "sv<olo>ved."

295r: 9–11 **but remaine as ded thing*es* . . . wherby they should enter**: This passage is not found in the *Mirror of Lights*.

295r: 10 **made meltinge**: The construction *make melting* is not recorded in the *OED*. *MED* (s.v. *melting*) lists the construction with the meaning 'to effect the disintegration (of the finite world),' in a psalter. *MED* lists *melting* in this construction as a verbal noun (gerund). In the Treatise, *melting* seems rather to be an adjective with the meaning 'liquid.' See also 308v: 35 and 315v: 19. In 308v: 35, *melting* is coordinated with another adjective, *fluxible*, in the same kind of construction with *make*.

295r: 11–21 **Alsoe ther ar 4 manner of conditions . . . vsed in this crafte**: For a discussion of the overall difference in structure and presentation of the conditions between the *Mirror of Lights* and the Treatise, see pp. 34–36. In the AFG manuscripts, the material conditions do not seem to have been part of the original enumeration, although the manuscripts all state that there are four conditions. In A, "material" is a later addition, probably supplied in recognition of the fact that the text deals with the material conditions.

295r: 16 **holp{e}ne**: The word was originally "holpinge," but the 'g' has been canceled and the 'i' appears to have an 'e' written above it, indicating that the 'e' should replace it. The word *holpen* is used in all the manuscripts of the Treatise, and it is found in the marginal note in A that repeats some of this passage.

295r: 20–24 **Materealls noteth . . . vsed in this crafte**: The material conditions enumerated in the Treatise differ from those given in the *Mirror of Lights*. The Treatise is more inclusive in also giving vessels and furnaces. The reason for this difference may be that the Treatise differentiates between natural conditions, which comprise only substances, and material conditions, which include other items of use in alchemy in addition to substances. See the note on 295r: 11–21.

Cf. *Mirror of Lights* (MS. R. 14. 45, fol. 67r): "bodies of metallz, spe*rites*, saltis, stonys, floure*s*, skalys, poudre of glas, skales of iren, vinegre, swete water, ma*n*nys vreyne, & childes vreyn."

295r: 22 **powlders**: The *OED* lists instances of this spelling of *powder* from the fourteenth century to the seventeenth century (s.v. *powder*), deriving ultimately from French *pouldre* (or a similar spelling).

295r: 24–25 **andd all manner of thinges necessary vsed in this crafte, bodies or the 7 mettalls**: The reading "or" is found only in A, while the reading found in e.g. S2 (fol. 23v), "Bodies are the 7 mettales," is supported by the *Mirror of Lights*: "The bodyez of metalles arn þese" (MS. R. 14. 37, fol. 115r). At some point of the textual transmission of A (possibly by Copyist 3) 'bodies' seems to have been taken as one more item in the enumeration of material conditions, and 'are' has subsequently been changed to or misread as *or*.

295r: 29–30 **quicksilver Mercurie**: Only one manuscript of the *Mirror of Lights* includes quicksilver among the metals enumerated: the abridged version in MS. Ashmole 1423, p. 33.

295r: 31–34 **wherfore they onlie appe*r*tayne . . . betweene the sonne and the moone**: This passage is not found in the *Mirror of Lights*. It may derive from the alchemical tract *Thesaurus philosophorum* (see pp. 52–54). See 324v: 16–23.

295r: 34–35 **Wyth these bodies**: The phrase refers to *gold* and *silver*.

295r: 35–36 **forced, fixed, andd fermented**: *Fixed* is not included in the *Mirror of Lights*'s description. Cf. *Mirror of Lights* (MS. R. 14. 45, fol. 67r): "enforced & fermented."

295r: 36 **and so is made elixar, whitt or redd**: The *Mirror of Lights* does not mention explicitly the creation of an elixir; it states only that spirits and the perfect bodies once mixed will *color* other bodies. Cf. *Mirror of Lights* (MS. R. 14. 37, fol. 115v): "and w*ith* medly*n*g of hem they tyngyn bodyes and t*u*rnyn hem into whit and into reed."

295r: 36–295v: 3 **for Sol and Luna be turned . . . vnto p*er*fect Sol and Luna**: The *Mirror of Lights* claims that gold and silver are *called* souls (apparently as a result of their ability to turn a spirit into an elixir), not that they are *turned* into souls; for the terms *spirit* and *soul*, see p. 81. *Turned* may be a misreading at some point in the transmission for a word such as *termed*, *into* being added later to make sense of the construction.

This passage seems also to have been interpreted in a number of ways by the copyists of the Treatise. I have punctuated A so that the syntax is coherent. However, there is an indication in A that "[w]*y*th the volatylle spiritt" was taken to belong to the previous phrase "Sol and Luna be turned . . . ." A marginal note in A states "☉ & ☽ ar turned into | souls w*i*th the volatill | sprite." If the formulation in the marginal note is an indication of how the passage was perceived, it is unclear how the slightly detached "is made a stone" was interpreted syntactically. The *Mirror of Lights* points out that the spirit, when combined with gold and silver, is turned into a stone. This might also be implicit in A, although the formulation "wyth the volatylle spirit" may imply that the spirit by itself is sufficient. It is also worth noting that the reading in S2S3S4 is perhaps closer syntactically to the *Mirror of Lights* here, since these manuscripts use the construction "beinge turned."

295v: 3–9 **The elixar, being yoined . . . and lasteth for ever**: In the *Mirror of Lights* this passage is significantly different. First, the *Mirror of Lights* discusses imperfect bodies (to complement the earlier discussion about perfect bodies); it then introduces the four spirits, and describes alchemical gold. However, it does not include a discussion of the elixir joined with the quintessence. The description in the Treatise most likely derives from another source as a similar description of the creation of the elixir of life appears in several places in the Treatise: see 304r: 10–12, 312r: 19–27, 320r: 10–21, 323v: 19–21.

295v: 12 **and is a wilde water**: Both the Treatise and the English copies of the *Mirror of Lights* claim that mercury is a *wild* water, whereas the Latin version of the *Mirror of Lights*, Corpus Christi College, Oxford, MS. 175, and the *Semita recta* (the *Mirror of Lights*'s source) characterize mercury as a *viscous* water (*aqua viscosa*): see Grund, "'ffor to make Azure'," 35; Borgnet, *Opera omnia*, 552. Similarly, in the description of orpiment, which is not included in the Treatise, the *Mirror of Lights* refers to its "wildeshippes" (MS. R. 14. 45, fol. 68r), for Latin *viscositas* 'viscosity'; Corpus Christi College, Oxford, MS. 175 does not contain this passage. The origin of this reading is unclear. *Wild* in the sense of 'uncontrolled,' 'unrestricted,' 'unconfined' is contextually suitable; *OED* s.v. *wild* II 6.a. Perhaps it is a substitution made at some point of the transmission as the word, *viscous* or *viscosity*, was unfamiliar to the scribe. By way of comparison, it may be noted, however, that the Middle English version of the *Semita recta* edited by Halversen characterizes *mercury* as "a viscous water," and orpiment is said to have "ij viscositees in hym": Halversen, "The Consideration," 285, 288.

295v: 12–13 **runninge in the veynes of the earth**: The phrase "the veynes" is not found in the *Mirror of Lights*, but it is part of the description of mercury in the *Semita recta* "in visceribus terrae" 'in the innermost of the earth'; Borgnet, *Opera omnia*, 552. See also note on 294v: 6–8.

295v: 13–14 **a subtill substaunce of a whit coulor, coaguliid wyth his sulphur so imperceptably**: Cf. *Mirror of Lights* (MS. R. 14. 45, fol. 67v): "a sotill substaunce and white by at temperatif hete euenly ioyned togedre." The Treatise reading is clearly related to the reading of the *Mirror of Lights*, but, unlike the *Mirror of Lights*, the Treatise clarifies that mercury is joined (or coagulated) with *sulfur*. See also notes on 295v: 15–18 and 295v: 27–296r: 2.

295v: 15–18 **he resteth not but flyeth . . . neyther to rest in any place playne**: The *Mirror of Lights* does not make a distinction between the qualities of wateriness and those of dryness, while, by using the construction *although . . . yet*, the Treatise seems to stress that *wateriness* has a different function from that of *dryness*. Furthermore, in the Treatise, the dry quality is attributed to the internal sulfur of mercury, whereas a similar attribution is not found in the *Mirror of Lights*. Cf. *Mirror of Lights* (MS. R. 14. 37, fol. 116v): "And therfore it flees awey fro*m* ilk a pleyne place lyghtly and cleuez nat therto. For the watrynesse and dryenesse arn so temprid togidere in hym þat þey wole nat suffren hy*m* abyde in no pleyn place."

295v: 19–27 **and by this his swifte course . . . accordinge to the quality of the saide earth**: This section is not found in the *Mirror of Lights*.

295v: 27–296r: 2 **althoughet his coulor doe shewee . . . the stone of ph*ilosoph*ers ought to be taken**: The *Mirror of Lights* contains a much shorter description of mercury. For example, the *Mirror of Lights* excludes the passage on the internal sulfur that conveys the dryness upon mercury; cf. notes on 295v: 13–14 and 295v: 15–18. The *Mirror of Lights* states only that mercury and sulfur together constitute the matter of all metals. It is possible that the passage in the Treatise represents an attempt to reconcile different alchemical frameworks. Chapter 2, for example, places the emphasis on the role of mercury in the formation of metals. Perhaps, by stressing the importance of mercury in this passage, the Treatise anticipates the discussion in Chapter 2. Cf. 299v: 43–300r: 4.

295v: 30 **homogaine**: It is unclear whether the spelling intended is *homoganie* or *homogaine*. Neither word is found in the *OED* or *MED*. F reads "homogene," but adds a marginal comment declaring "Homogene w*ha*t it is, it is ☿ & called homgaine" (F, p. 10). Considering the close relationship between A and F, it is possible that a common ancestor of the two had the reading *homogaine* (or a similar spelling). The word may be a variant of *homogene* 'homogeneous' or *homogeny*, used in the sense of 'something homogeneous.' A marginal note in A stating "☿ is homogen" and the formulation of the comment in F suggest that *homogene* is intended. S2, on the other hand, reads "homogenie," suggesting that *homogeny* is intended at least in this copy.

296r: 4–6 **is of a harde substaunce . . . full of drye fewme**: The *Mirror of Lights* claims that sulfur is *burning* but does not mention "full of drye fewme."

296r: 8 **but the helpe of stronge waters**: A preposition such as *by* or *with*, found in the other manuscripts, needs to be supplied before "the helpe." The *Mirror of Lights* adds to the discussion that sulfur must be *boiled* with strong waters if oil is to be extracted.

296r: 10–13 **and also of one effect . . . his sulphur whit and redd**: This passage is not found in the *Mirror of Lights*. Instead, the *Mirror of Lights* manuscripts (except for MS. Kk. 6. 30) add 'quick' and 'dead' sulfur to the list of different kinds of sulfur, and discusses their characteristics. Cf. *Mirror of Lights* (MS. R. 14. 37, fol. 117r): "Qwyk is þat is drawyn out of þe erthe clene and nat moltyn, and þat is good for þe scabbe. And þe deed is þat is moltyn into cannis as þe apothecariez sellyn."

296r: 13 **The 3 spirit is arcenack**: The third spirit in the *Mirror of Lights* is orpiment, which is not discussed at all in the Treatise. MS. R. 14. 37, MS. Kk. 6. 30, and MS. Sloane 513, list arsenic as the fourth spirit, but neither the discussion of orpiment nor that of arsenic in the *Mirror of Lights* resembles the treatment in the Treatise. The passage may derive from another source. See the note on 296r: 22.

296r: 21–22 **for our matter . . . to be yoined wyth him**: Cf. 302r: 1–16.

296r: 22 **The ˏ{iiii} spirit is sal armonack**: The *Mirror of Lights* discusses different kinds of salts as a separate category, with a subsection on sal ammoniac. Although the discussion of sal ammoniac in the *Mirror of Lights* is extensive, it does not resemble the one found in the Treatise. The Treatise description may derive from another source. See the note on 296r: 13.

296v: 1–299v: 2 **The seconde chapiter sheweth the errours . . . as to their center and firste matter**: This part of the second chapter reuses the alchemical dialogue *Scoller and Master*. See pp. 36–39.

296v: 10 **of the worck of the stone of phi*los*op*h*ers ˏ{whether it} be founde**: The phrase "whether it" in A is a later addition, most likely in Forman's hand. It was probably made to compensate for the misreading of *if the works* as *of the works*; cf. the reading in G.

296v: 23–297r: 6 **Sonne, thow shalt vnderstande that of all stedfast kinde . . . in a thowsand years**: For this answer, see p. 38.

296v: 23–26 **Sonne, thow shalt vnderstande that of all stedfast kinde . . . be incresed ˏ{to} his lyke**: The syntax is slightly problematic in this passage. The source of the Treatise probably had a reading that combined elements found in the *Scoller and Master* manuscripts Corpus Christi College, Oxford, MS. 226, and MS. Sloane 3747: "ȝe sc*hal* vnderstand þat of al stidfast kyndes þer ys c*er*teyne t*er*me of grettnes, of growyng so þat in his spice may be encresyd his leke" (Corpus Christi College, Oxford, MS. 226, fol. 33v); "Ye shall vnderstonde that of all man*er* of stedfast kyndes there is a certayn t*er*me lymetid of gretnesse and growyng so that in his speciall kynde that is callid species it may be incresid and increse his liknesse" (MS. Sloane 3747, fols. 66r–66v). The passage may refer to the physical fact of pregnancy, suggesting that there is a certain time when bodies or creatures, solid or stable in form (?), can multiply; cf. *OED* s.v. *greatness*. Another possible interpretation (similar to the preceding one) is that there is a certain time in life when a body or creature is at its prime, and it is at this time that it grows and multiplies.

296v: 26–29 **because he is ethremogines . . . of trees barck, bodies, and leaues**: Cf. MS. Sloane 3747 (fol. 66v): "whiche thyngges byn callid ethremogenies whiche dyu*er*sen in sp*ecie* and not in gen*ere* in that they byn in p*artes* vnlike, as of best*es* flessh, blode, and bone, and in trees the barke, leif, and frute." MS. Ashmole 759, MS. Sloane 3747, and MS. Ferguson 91 make explicit that one example derives from the animal kingdom and one from the plant kingdom by referring to "of best*es*" and "in trees." The Treatise and most of the manuscripts of *Scoller and Master* leave out the explicit reference to *beasts*.

296v: 27 **ethremogines**: This word is not recorded in the *OED*. The *MED* records the form *ethromogenie* 'different in nature,' 'heterogeneous' (both instances from Chauliac's *Chirurgie*). The *MED* states that *ethromo-* is a form developed from the influence of *homo-* on *ethro-* (a by-form of *hetero-*); *MED* s.v. *ethromogenie*. This word is also found in four versions of *Scoller and Master*: MS. Ashmole 759, Corpus Christi College, Oxford, MS. 226, and both copies in MS. Sloane 3747. These four versions appear to use the word as a noun in the sense of 'heterogeneous things' (cf. MS. Sloane 3747 in the note on 296v: 26–29), and it should possibly be taken as a noun in the Treatise as well. Alternatively, the word should be taken as an adjective, where the final 'es' is a reduced form of 'ous.' See also 297r: 25.

296v: 30–31 **And those that be on in part . . . ar called homogenius**: The second *and* in the sentence makes the statement anacoluthic. This structure may derive from the source. Cf. MS. Ashmole 1490 (fol. 83v) "And they that be in p*arts* & species be all lyke and ar called homogenes the which diuerseth not in sp*ecies* nor in p*arts* but be all lyke as mettalle, of the w*hich* the leaste p*arte* is the hoole with himselfe."

296v: 34–297r: 3 **they turne [directly] . . . a naturall bodi of the fier**: There is no direct equivalent for this in any of the manuscripts of *Scoller and Master*, which exhibit great variation in content in this passage. Cf. e.g. MS. Ashmole 1490 (fol. 83v): "And therto it muste be brought .i. materia prima"; Corpus Christi College, Oxford, MS. 226 (fol. 33v): "þan they be t*ur*nyd & alt*er*atyd"; MS. Sloane 3747 (fol. 66v): "then they turne and encrese." Cf. 299v: 1.

297r: 4 **gaue his vnderstanding**: The reading *us* found in S2S4S3 for *his* in A is supported by the manuscripts of *Scoller and Master*. Cf. Corpus Christi College, Oxford, MS. 226 (fol. 33v): "geuyth us suche resonable discrecion"; MS. Ashmole 1490 (fol. 83v): "gaue vs such reason & wite."

297r: 11 **vitryall and commane sault, or sault armoniack**: None of the manuscripts of *Scoller and Master* include *sal ammoniac*; instead most of them list *alum rock* and *saltpeter*. *Scoller and Master* further specifies that these substances are useful in the sublimation of *quicksilver*.

297r: 15–23 **But here thow must vnderstand . . . in a shorter time then in the other stones**: This passage, which is not found in *Scoller and Master*, refers to a discussion later on in the Treatise. See Ch. 16, esp. 313r: 12–14. See also p. 61.

297r: 23–26 **And for our matter in the a*n*imall and vigitable stones . . . be enymies to thinges dissolvid**: The Treatise exhibits clear similarities with three manuscripts of *Scoller and Master* in particular: MS. Ashmole 759, MS. Sloane 3747, and MS. Ferguson 91. However, the Treatise connects the discussion with the previous description of the animal and vegetable stones, which is not found in *Scoller and Master*. See p. 61. Cf. MS. Sloane 3747 (fol. 67r): "for our matt*er* when it is dissoluid will no strange thyng of ethremogenies and also strange eyres byn contra*r*y to thyngges dissoluid."

297r: 25 **stranges**: The final 's' in "stranges" is probably written in anticipation of 's' in "thinges."

297r: 25 **etheromogines**: See note on 296v: 27.

297r: 27 **in putrificac*i*on**: This phrase is not found in *Scoller and Master*.

297v: 5 **radicaulle humydities**: The reading in the singular, "humiditye," which is found in the other manuscripts of the Treatise, is supported by *Scoller and Master*. For the concept of the *radical humidity*, see Michael McVaugh, "The 'Humidum Radicale' in Thirteenth-Century Medicine," *Traditio* 30 (1974): 259–83; cf. also *OED* s.v. *radical* A. 1a.

297v: 7 **phi*lozo*fie**: There is a letter, possibly an 'h,' preceding "phi" in the manuscript. Since it is unclear what this letter form represents, I have expanded the word as an instance of *philosophy*.

297v: 8 **For silver and gould is not commended**: 'for' probably does not have a causal meaning here (and in some other places), but rather functions as a simple connective close to 'and': see Ian A. Gordon, *The Movement of English Prose* (London: Longmans, 1966), 115. Cf. *Scoller and Master* (MS. Sloane 3747, fol. 67v): "and also gold and silu*er* byn not worth in this craft."

297v: 9–11 **as thay be silver and gould . . . by the rules of phi*lozo*fie**: This statement is not found in *Scoller and Master*. The argument seems to be that the names *gold* and *silver* are not used in alchemy to signify gold and silver. Rather, they are used as cover terms, that is to say, they can be employed to denote other substances in order to hide the identity of the real substances.

297v: 11 **{be} brought**: *Be* seems to have been added as an afterthought without a cancellation of the earlier *are*.

297v: 12–15 **For the phi*lozo*fers in times past . . . did hid it depely**: The sentence is anacoluthic. An *and* should be supplied after "science." Cf. *Scoller and Master*

(MS. Sloane 3747, fol. 67v): "For ph*ilosoph*res that were somtyme knewe the drede of God for the myschif that myght fall and therfore they hiddid it derkely."

297v: 15 **vnressononable**: i.e., *unreasonable*. The form is probably a result of dittography.

297v: 17–18 **and then had the ph*ilozofe*rs bene the cause of that distructione**: The Treatise may have left out the first part of the statement that is found in *Scoller and Master*. Cf. MS. Sloane 3747 (fols. 67v–68r): "and then shuld no man haue don for other and then hadde the ph*ilosoph*ers be cause of the distruccion of the worlde."

297v: 20 **workemen**: *Workmen* is not found in any of the manuscripts of *Scoller and Master*, which generally use *discrete man*. The reading "wicked men," which is found in all the other manuscripts of the Treatise, is supported by two manuscripts of *Scoller and Master*: Ashmole 1487 and Ashmole 1490.

297v: 22–30 **Therefore, to euery degre of working ... knit together be equallitie of elementes**: Cf. MS. Sloane 3747 (fol. 68r): "Therfore in eu*er*y degre of werkyng it is callid by a dyu*er*s name found by reason next to it in compleccion aft*er* the qualitees that may not be seyn ne felte but hauyng her werkyng in v*er*tues and also for his worthy kynde and m*er*velouse werkyng and for the p*ar*ticipacion that he holdith of thelementes." In *Scoller and Master*, the names are given to "it," not to "euery degre of working." What *it* actually refers to in *Scoller and Master* is uncertain. The sentence following this passage refers to the philosophers' stone, which seems a plausible referent. The exact meaning of the passage "the effecte ... sene" in the Treatise is unclear. *Scoller and Master* has a clearer reading, suggesting that the name of the substance that reason finds to be most like or closest to it (the stone?) in complexion according to the qualities (i.e., the four qualities hot, cold, moist, and dry) is used instead of the actual substance intended.

297v: 31 **names of names**: The reading "manner of names" found in all the other manuscripts is supported by *Scoller and Master*.

297v: 31–33 **hauing one only operatione ... he nedeth to his p*er*fectione**: Cf. *Scoller and Master* (MS. Sloane 3747, fol. 68r): "for he holdith in hym all thelement*es* and all that hym nedith to his p*er*feccion." Cf. 302r: 1–16.

297v: 31–32 **hauing one only operatione**: The meaning of *operation* in the context seems to be 'procedure of being formed or prepared' rather than 'function' (cf. 302r: 9–11). I have not found the former meaning recorded in the *MED* or *OED*.

298r: 2–3 **wherby the ph*iloso*fers doe vnderstand on the other**: Cf. *Scoller and Master* (MS. Sloane 3747, fol. 68r): "so that by on thyng they vnderstonde another thyng."

Explanatory Notes

298r: 3 **dyuised**: The reading "diuersed" found in all other manuscripts of the Treatise is supported by all the manuscripts of *Scoller and Master* except for MS. Ashmole 1490 (copied by Forman), which also has "devised" (fol. 81v).

298r: 4–12 **For the phi*losofe*rs haue in the moste darkeste manner . . . in the concauity and depenes of nature**: The passage in the Treatise differs significantly in content from the readings in the manuscripts of *Scoller and Master*. Cf. MS. Ashmole 759 (fol. 49r): "The phi*losophe*res in her te*r*mes spekyn analogice and bydden her children be ware of analogia. For in the most derkist man*er* all this matt*er* is hidde." The Treatise passage may have been influenced by 301v: 3–39.

298r: 15 **neither**: The S2S3S4 reading, "neuer," is supported by *Scoller and Master*.

298r: 17–21 **Sonne, the phi*losofe*rs in thes point*es* . . . ioyned to the tinturouse sowle**: This passage is not found in *Scoller and Master*. It is probably based on Ripley's *Concordantia* (Ripley, *Opera omnia*, 323). See also 305r: 31–35.

298r: 21–26 **Therefore, so<w>e in kind . . . whom he is fermented**: In *Scoller and Master*, this passage is part of the question and not the answer as in the Treatise.

298r: 21–22 **Therefore, so<w>e in kind that that is not therein**: The reading "sowe not," found in all the other copies of the Treatise, is supported by *Scoller and Master*. However, the Treatise may have omitted the middle part of the argument found in *Scoller and Master*, possibly owing to an eyeskip at some point in the transmission. Cf. Corpus Christi College, Oxford, MS. 226 (fol. 31v): "for suche as þu souwist such *schalt* þu repe & þerfore seche n*o*t in kynd þ*a*t ys n*o*t þerin."

298r: 21 **so<w>e**: The 'w' is highly tentative. The letter seems to have been changed a number of times, at one time being a 'v.' The outlines of a 'w' are visible, but it is uncertain if 'w' was intended to be the final reading.

298r: 22–26 **For as owre stone is lorde . . . whom he is fermented**: *Scoller and Master* does not discuss the philosophers' stone, but only gold. Neither does *Scoller and Master* claim that it is "highest in his simplissitye" after its fermentation. For "whom," see note on 298r: 25. Cf. Corpus Christi College, Oxford, MS. 226 (fol. 31v): "for truly S*ol* ys lord of stonys, noblest of bodyes, kyng & hed & best of hem alle."

298r: 25 **whom**: *When*, which is found in all the other manuscripts of the Treatise, is more suitable in the context.

298r: 26–31 **For as the questin is . . . or he that receiveth yt**: This passage seems to consist of a subordinate clause only, without a main clause. The reason for the syntax may lie in the reworking of this question and answer. As addressed in the note on 298r: 21–26, the Treatise includes a passage in the answer that is part of the question in *Scoller and Master*. In fact, in *Scoller and Master*, the answer starts with the passage that is left here as a subordinate clause in the Treatise. When this passage was restructured, the connection to the argument that is represented in *Scoller and Master* by the answer formula was left out. Cf. MS. Ashmole 759 (fols. 49r–49v): "Syr, as therof ye sey soth but yeit ye knowe not thentent of the ph*iloso*ph*e*rs whether is he a grett*er* kyng that in eu*e*ry kyngdom hath {his} reigne so that wit*h*out hym there is nothyng don, or ell*ys* is he that is kyng but of on realme and hath no thyng but of the grett*er* kynges yeifte, or ell*ys* is he a grett*er* leche that helith all seke bodyes, or ellys is he that resceyuyth the medecyn, all whiche power is in our golde."

298r: 27–28 **nothinge done withowt hime**: A form of *be* has probably been omitted after "nothinge." Cf. *Scoller and Master* (MS. Sloane 3747, fol. 68v): "wit*h*out hym their is no thyng don."

298r: 30–31 **that gyveth the mediciene**: Cf. *Scoller and Master* (MS. Sloane 3747, fol. 68r): "that helyth all seke bodies."

298r: 31–34 **For ouer golde & sylver ... in co*m*mon golde & co*m*mon sylver**: This passage is not found in *Scoller and Master*. See 308v: 1–4.

298r: 36–37 **because it houldeth in pouer both gould and silver**: *Scoller and Master* contains a simile about rubies, which is not found in the Treatise. Cf. MS. Sloane 3747 (fol. 68v): "for right as the rubie hath in hym theffecte of all p*re*ciouse stones right so our gold holdith in hym the v*er*tue of all ductible stones."

298r: 39 **tin<c>bir**: The reading is unclear in A. The unclear letter seems to be a smudged 'c.' The reading "tincture," which is found in all the other manuscripts, is supported by *Scoller and Master*.

298v: 5 **animall**: Instead of *animal*, *Scoller and Master* uses *amiable* 'friendly,' 'favorably disposed.' See also Geber's *Summa Perfectionis*, where almost exactly the same passage occurs: Newman, *The* Summa Perfectionis, 334, 670. Although it is quite possible that *animal* is a misreading of *amiable*, the Treatise may be using *animal* in the sense of 'vital,' 'maintaining or supporting life,' 'life-giving.' Although not recorded in the *MED* or *OED*, the sense is recorded for the Latin word *animalis*: see e.g. *OLD* s.v. *animalis*.

298v: 9–10 **before he hath receyued tincture**: This statement is not found in *Scoller and Master*. Cf. 298r: 17–21.

298v: 14–15 **Therfore, gould cannot giue that wh*ich* is not in him**: The reading found in S2 and S3 (and to some extent in S4), "nothinge maie goulde giue that is not in him," is closer to the reading found in most of the manuscripts of *Scoller and Master*. Cf. MS. Ashmole 1487 (fol. 108r): "Therfore he maye nothing gyue that he hathe not hymeselfe."

298v: 17–19 **nere their centure . . . soe soer fixed in them**: Cf. *Scoller and Master* (MS. Sloane 3747, fol. 69r): "nerer the centre of her sulph*ur* and m*er*cury and also her m*er*cury is not so sore fixed in hem." The Treatise is closely related to MS. Ashmole 759, MS. Sloane 3747, and MS. Ferguson 91 in its formulation. But the Treatise connects sulfur and mercury with *fixed* rather than *center*. The second instance of mercury has possibly been skipped at some point in the transmission owing to homeoteleuton.

298v: 19–20 **as it is in gould and siluer, w*hich* ar the p*er*fecte bodies**: The reading, "as in the perfect bodies," which is found in all the other manuscripts, is supported by *Scoller and Master*. The A reading may be an attempt at clarification.

298v: 21 **Farther**: i.e., *father*. Wyld notes this spelling of *father* and similar cases of unhistorical 'r' from the fifteenth century onwards. He argues that they constitute evidence that /r/ was no longer pronounced before consonants and could hence be inserted into words that did not historically contain an 'r': Wyld, *A History*, 298. "Farther" is, however, the only instance in the Treatise of unhistorical 'r,' and may thus be accidental.

298v: 26–27 **Sonn, nowe thou . . . of those I mean not**: Cf. *Scoller and Master* (MS. Sloane 3747, fol. 69v): "Nowe truly in that ye speke of thise metall*es* that byn corupt."

298v: 30–32 **for every thinge beinge dead . . . noe elemente may be w*ith* other**: Cf. *Scoller and Master* (MS. Sloane 3747, fol. 69v): "then is he corupt and he must so v*er*rely die that non element stond w*ith* another."

298v: 34 **matter**: The reading "manner" (found in GS2S3S4) is supported by *Scoller and Master*.

298v: 40–43 **For I haue proued by practice . . . then lead or other mettalls**: *Scoller and Master* presents this statement as part of the master's answer rather than the disciple's question.

298v: 41–42 **Yf thes bodies haue noe cause of solution into ☿**: The manuscripts of *Scoller and Master* contain a longer passage that describes why these metals should not be used in alchemical practice. The Treatise is probably based on a reading that was similar to the one found in MS. Ashmole 759, MS. Sloane 3747, and MS. Ferguson 91. There is possibly an omission in the Treatise, which may be the result of an eyeskip (*solution . . . solution*) at some point in the transmission. Cf. MS. Sloane 3747 (fol. 70r): "and thise yeve no cause of soluc*i*on in hemself but holdyn theymsilf ayenst the hete of the actif kynde and power and do kepe theymsilf all drie w*ith*oute soluc*i*on. For they ne shull eu*er* dissolue in our werk but with moche trauell and litell p*ro*fette."

299r: 3 **it had not nowe byn to seke**: i.e., 'not needing to be sought or looked for,' 'not hard to find'; *OED* s.v. *seek* v. III 19. b.

299r: 7–8 for by his porosity and hollownes, he houldeth abundance of ayer within him: *Scoller and Master* claims that the porosity and the abundance of air are additional similarities between tin and mercury. Cf. MS. Sloane 3747 (fol. 70r): "for the porositees that byn in hym, which is his holownesse, and for habundaunce of eyre that he holdith in hymsilf."

299r: 9 of ☿: This statement is not found in *Scoller and Master*, but seems to be possible to infer from the context.

299r: 9 qantiti: For a similar spelling, see "qallities" in 305v: 28.

299r: 12–29 Father, wherfore said youe . . . And the third is quicksiluer: Cf. 304v: 35–43.

299r: 18 the ☿ of kinde engendred in the earth: Cf. *Scoller and Master* (MS. Sloane 3747, fol. 70v): "the fyrst mercury is mercury of kynde whiche is the purist matter of kynde engendrid in therth." It is possible that a clause such as *which is the purest matter of kind* (found in *Scoller and Master*) has been omitted in the Treatise at some stage of the transmission owing to an eyeskip (*kind . . . kind*).

299r: 19–20 of the which mercuri is engendred: The Treatise differs from *Scoller and Master* in claiming that *mercury* is created from the *mercury of kind*. Most of the manuscripts of *Scoller and Master* claim that "of the which mercury all metall is engendrid" (MS. Sloane 3747, fol. 70v; my emphasis). On the other hand, some manuscripts, such as MS. Ashmole 1487 and MS. O. 2. 22, may point to the origin of the Treatise reading. Cf. MS. Ashmole 1487 (fol. 108v): "of the whiche quickesylver is ingendred and all manner of mettalles." Notable is that manuscript A adds "metallorum" 'of metals' after "sperma" (which is not found in any of the other manuscripts), possibly as a clarification.

299r: 20–22 Another ☿ ther is . . . of other gender cominnge betwene: The source of the Treatise probably combined features found in, for example, MS. Sloane 3747, and Corpus Christi College, Oxford, MS. 185. Cf. MS. Sloane 3747 (fol. 70v): "another mercury ther is wherwith other metall kyndely do dissolue theymsilf with hemsilf withouten enythyng of other gendre or kynde comyng betwene hem"; Corpus Christi College, Oxford, MS. 185 (fol. 133r): "The secunde mercury es quiksyluer þat wyth oþer metall by kynd dyssolue hemselfe wythowte any oþer body of oþer gender comyng betwen hem."

299r: 34–35 that of all things that groweth, things that may be felte ar beste: All the manuscripts of *Scoller and Master* also contain the reading *best*, except for MS. Ashmole 759 (fol. 51r), which reads "lest" 'least': "ye shall vnderstand that in all thynges that growen felyng thyngges byn the lest if they be departid into elementes." Whether the MS. Ashmole 759 reading is original or represents a later emendation of the text is uncertain, but the reading is clearly more in line with the argument in the father's subsequent answer to his son: when something has been returned into its constituent parts, the substance or parts that may be experienced by the senses are minute in comparison with aspects that cannot be experienced.

299r: 37 **feeling**: *Feeling* in the passive sense of 'perceptible to the senses' is recorded by the *MED* only once from Chauliac (s.v. *felen* [v1] 3c), and no instance is cited in the *OED*. Most of the copies of the *Scoller and Master* use this form. Cf. note on 299r: 34–35.

299r: 41–299v: 2 **For as all vegitable thing*es* ... as to their center and firste matter**: There is no direct parallel to this section in most manuscripts of *Scoller and Master*. MS. Ashmole 759 and MS. Ferguson 91, however, contain a similar passage: "But when myn*e*rall thyngges and vegitable that be of the kynde of omogenies byn dep*ar*tid into el*emen*tes by their kyndly deth then they t*ur*ne to erth or to wat*er* as to their centre and byn become yong ayen" (MS. Ashmole 759, fol. 51r).

299v: 2–300v: 30 **Sonne, thou shalte alsoe vnderstand ... and the co*m*bustibilyti of sulfure in the vnclean bodies &c**: This part of the second chapter represents an extract from *Perfectum magisterium*. See pp. 39–40.

299v: 8 **but by the poware of God*es* grace and gifte from aboue**: This statement is not found in the versions of *Perfectum magisterium* consulted. For other references in the Treatise to alchemy as a gift from God, see 294r: 26–27, 298r: 10–11, 301v: 22–23, 317r: 21–22, 319r: 12–13.

299v: 26 **yron**: "yron" most probably represents a misreading of *urine* at an early stage of the transmission of the Treatise. All the versions of *Perfectum magisterium* read *urine*.

299v: 27–28 **haue drawen of them the 4 element*es*, that is to say, they distilled of them first cleare water**: The Treatise is closer in formulation to the Middle English version MS. R. 14. 44 than to the Latin versions of *Perfectum magisterium*, which present longer, closely related versions. Cf. MS. R. 14. 44 (Part 4, fol. 1r): "& han drawe*n* of hem þe 4 elementys as thus þei dystylle of them cler watyr"; Arnold of Villanova, *Opera Arnaldi*, fol. 302r: "& ex illis primo extraxeru*n*t quattuor eleme*n*ta: vt cu*m* eis op*us* p*er*ficere*n*t separa*n*do *primo per* distillatione*m* ex predictis aqua*m* clara*m*" 'and from these they have first extracted the four elements so that they could finish their work with these by first separating from the mentioned substances a clear water through distillation.'

299v: 31–33 **vpon the which they poware the cleare water ... vntill the earth be made white**: The pouring of clear water mentioned in the Treatise may be a rendering of a reading such as *abluunt cum aqua* found in Zetzner, *Theatrum chemicum*, whereas the process of mixing is only mentioned in MS. R. 14. 44. Cf. MS. R. 14. 44 (Part 4, fols. 1r–v): "& in a whyle aft*yr* þei make þ*a*t erthe why3t wyt*h* þe watyr mengynge hem togedyr wyth sethyng & dystyllyng of hem so þat watyr oftyntymys tyl al þe erthe be mad qwhy3t"; Zetzner, *Theatrum chemicum*, 3: 128: "Postmodum cum aqua terram abluunt, & dealbant imbibendo ipsam, & decoquendo, & destillando aquam illam totiens, donec terra dealbetur" 'after that they wash the earth with the water and make it white by imbibing it and boiling and distilling the water until the earth becomes white.'

299v: 34–35 **and then they poware vpon the earth . . . vntill it hath dronke vp all his fier**: While the Treatise claims that the earth drinks the *fire*, the printed Latin versions stress *water*, *oil*, and *tincture*. MS. R. 14. 44 contains a reading closer to that of the Treatise, but adds water, oil, and earth. Cf. MS. R. 14. 44 (Part 4, fol. 1v): "thanne 3eue þei a3en to hym hys feer & hys oyle wrout & þat is to seye sethynge, sublymyng, & dystyllyng euere tyl þe forseyd erthe hath dronkyn al & þat is to seye watyr, oyle, feer, & erthe"; Zetzner, *Theatrum chemicum*, 3: 128: "deinde reddunt ei oleum suum ad ignem praeparatum, imbibendo, & destillando, quousque praedicta terra totum bibat, scilicet aquam, & oleum & tincturam" 'after that they give back to him his oil by a fire which has been prepared, by imbibing and distilling it until the mentioned earth has drunk it all, that is to say, the water, and oil, and tincture.'

299v: 36 **and then they haue calcioned the earth by a longe tyme**: This statement is not found in any of the versions of *Perfectum magisterium* consulted.

300r: 4–6 **blude, heare, eggs, vrin, sulfur, arcenike, saultes or any suchlike thinge**: The versions of *Perfectum magisterium* do not include "sulfur, arcenike, saultes." See also 296v: 12–14.

300r: 13–16 **Those they sublimed . . . the kind of the earth**: While the Treatise stresses that the result of sublimation is fire, *Perfectum magisterium* claims it to be air. The source text of the Treatise probably contained a passage that combined elements found in Zetzner, *Theatrum chemicum*, and MS. R. 14. 44. Cf. MS. R 14. 44 (Part 4, fol. 2r): "& þei sublyme hem so þat þei be of þe kende of þe eyre and þei resoluyn hem so þat they ben of þe kende of watyr and þei fyxyn hem so þat thei ben of þe kende of erthe and þei calcyne hem so þat þei ben of þe kende of feer"; Zetzner, *Theatrum chemicum*, 3: 129: "& istos sublimant sublimatione vulgari, non philosophica, & faciunt ipsos ascendere, ut sint naturae aereae; & postmodum fixant decoquendo, & calcinando, ut sint naturae terrae: postea solvunt, ut sint naturae aquae: postea destillant, ut sint naturae ignis" 'and those they sublime with an ordinary sublimation, not an alchemical one, and make them ascend, so that they become of an airy nature; and after that they fix them by boiling and calcining so that they become of the nature of earth; then they dissolve them so that they become of the nature of water; and after that they distill them so that they become of the nature of fire.'

300r: 17 **and calcioned them strongly**: This statement is not found in the versions of *Perfectum magisterium* consulted.

300r: 21–22 **for as I said before that mettalls ar not engendred out of their kinde**: The formulation "out of their kinde" is ambiguous. The determiner *their* may refer to the *spirits*, discussed in this section. However, this is unlikely since this connection has not been mentioned before (at least in the Treatise as it now stands). The only version of *Perfectum magisterium* that contains a similar formulation is MS. R. 14. 44 (Part 4, fol. 2r): "metallys beth not gendryd but of her owne kende." Perhaps *out* is a misreading at some stage for *but*. This assumption is supported by the formulation of the passage that the Treatise seems to refer to here: "and forasmoch as mettalls ar not engendred but of their sperme" (300r: 1–2). Another possibility is that *out of* should be interpreted as meaning 'outside the bonds or sphere of': *MED* s.v. *oute of* prep. 7. *Their* would then refer to "mettalls." See also 300r: 35, 300v: 46.

300r: 23 **outtaken of mercurie**: i.e., 'except mercury': see *OED* s.v. *outtaken* B.1.; *MED* s.v. *outtake(n* 3b.

300r: 26–27 **that ar sone absente and turned into a colle**: The source of the Treatise probably had the reading *are brent* (or a similar reading), which is found in MS. R. 14. 44 (Part 4, fol. 2r): "þat so sone arn brent & browt to nowt." This reading would subsequently have been misread as *absent*.

300r: 32 **but a receptakell and norishmente**: This phrase is not found in the versions of *Perfectum magisterium* consulted.

300r: 32–34 **soe lykewise in bringinge ... that is contrary to kinde**: Cf. *Perfectum magisterium* (MS. R. 14. 44, Part 4, fol. 2v): "on þe same ˄{wyse} comyn all oþer thynges of þe owne kende wythowtyn ony oþer myxtion."

300r: 35 **out of his kinde**: For a discussion of *out of*, see 300r: 21–22. The formulation *out of his kind* appears to be an elaboration on "strang thinge," stressing that the substance must be in accordance with the nature of the stone. Cf. *Perfectum magisterium* (MS. R. 14. 44, Part 4, fol. 2v): "þat our craft nedyth non mengyng of oþere materis but of þe owne kende." See also 300r: 32–34.

300r: 39–41 **Other ther be that mingle ... and all in vain**: The Treatise recommends a different set of procedures than the versions of *Perfectum magisterium* consulted. It is more closely related to the manuscript versions than to the printed texts. Also, unlike the Treatise, the *Perfectum magisterium* claims that salts can be prepared with spirits, not mentioning a mixing of bodies and spirits per se. Cf. MS. Ashmole 1384 (fol. 76v): "Et aliqui simul miscent cum corporibus, ut dictum est, calcinatis, preparatis et solutis uel cum spiritibus similiter preparatis et nichil inuenerunt" 'And some mix them together with bodies, as is said before, which have been calcined, prepared, and dissolved, or [they mix them] with spirits prepared in a similar fashion, and they found nothing.'

300v: 2–3 **but, beinge dissolued in his owne kinde, is made yᵉ ☿ of philosofers**: There is no direct correspondence to this statement in the versions of *Perfectum* *magisterium* consulted. It may have been added as a further explanation of the point. Cf. 300v: 21–22 and 301r: 28–30.

300v: 4–5 **neither in sprit*es* ioyned together, or by himselfe or by themselues, with the forsaid element*es***: Cf. *Perfectum magisterium* (MS. Ashmole 1384, fol. 76v): "Nec i*n* spiritib*us* p*er* se adi*n*uice*m* co*n*iunctis cu*m* ele*m*e*n*tis" 'neither in spirits by themselves joined together with the elements.' The formulation "or by himselfe or by themselues" may be an attempt at a clarification. The F manuscript of the Treatise, with which A shows a close relationship, contains only *himself,* suggesting that a possible ancestor of the two manuscripts had *himself,* or even *hemself* 'themselves.' *Themselves* may have been supplied because *himself* (referring to the stone?) is problematic in the context owing to the unclear referent. It may be noted that S2S3S4 only read "by themselfe." For *or . . . or* in the meaning 'either . . . or,' see *OED* s.v. *or* conj 3.a.

300v: 11–15 **lik as the sperme of man . . . w*hich* is heare to be taken for y*e* glase**: The Treatise introduces a simile between the human sperm and the sperm of metals, but the simile is never completed: the *like as* clause is not echoed by a similar clause containing the description of the metallic sperm. There may be an omission in the Treatise caused by homeoteleuton (cf. *matryce . . . matryce* in MS. R. 14. 44); perhaps the clause "w*hich* is heare to be taken for y*e* glase" is supposed to be the second part of the simile. Cf. *Perfectum magisterium* (MS. R. 14. 44, Part 4, fols. 2v–3r): "for ry3t as þe sp*er*me of man p*ro*fytyth nowt nor bry*n*gyth forth no frewt but yf yt be cast into þe matryce of a woma*n*, ry3t so m*er*cury neu*er* p*ro*fyth nowt but yf yt be cast into þe matryce of a woman þ*a*t it may be norschyd þerinne."

300v: 16 **Som ther be that mingell ⸫ wit*h* the p*er*fecte bodies**: Cf. *Perfectum magisterium* (Zetzner, *Theatrum chemicum,* 3: 129–30): "Quidam ipsum Mercuriu*m* cum corpore amalgamando miscuerunt" 'Some have mixed mercury itself with a body by amalgamation.'

300v: 21–24 **that yf no crud body . . . soe shall our medison be procreated**: This passage is not found in any of the copies of *Perfectum magisterium* consulted. It probably derives from Ripley's *Concordantia*; Ripley, *Opera omnia,* 326. See 305v: 18–19. *Perfectum magisterium* instead presents a discussion on the impossibility of joining a body and a spirit without the mediation of the soul (*anima*). The reading "yf" in A is probably a misreading for "of" (found in the other manuscripts of the Treatise). For "reuegitatid," see note on 305v: 18.

300v: 25 **Some haue mingled vnp*er*fect bodies**: The reading in S2S3S4, "haue mengyd vnp*er*fecte bodies with p*er*fect bodies," is supported by *Perfectum magisterium.*

300v: 28–30 **And the cause of their errour . . . in the vnclean bodies &c**: The Treatise presents an abbreviated and slightly different reason for the error than that found in the copies of *Perfectum magisterium* consulted, which do not mention the "vncleannes of ♄." I cite the part of the reason given in *Perfectum magisterium* that seems to correspond to the Treatise text. Cf. MS. R. 14. 44 (Part 4, fol. 3r): "and þe cawse of her erro*ur* is for þe erthe of sulph*ur*, bren*n*yng & fowl corrupcyon multyplying, ys ioynyd w*yth* þe inp*a*rf*y*3t body & cleuyth to hym in þe begynnyng of his growyng."

300v: 31–301r: 36 **Some I haue kowen that haue drawen a stronge water . . . that are not heare rehersed**: I have not been able to identify a direct source or sources for the rest of Chapter 2. The switch to the first person should probably not be taken as an indication that the following passages represent the author's own experience; rather it may have been transferred from a source text. See p. 55 for a discussion.

300v: 31 **kowen**: i.e., *knowen*. This is most likely simply a mistake where 'n' or a tilde has been left out.

300v: 38 **fier of the first degre**: To alchemists, *fire of the first degree* had a special technical meaning. Cf. Ripley, *Opera omnia*, 136: "Innaturalis est ignis occasionatus, ut calor in febribus, qui vocatur ignis humidus, artificialiter factus a Philosophis. Vocatur etiam ignis primi gradus, qui per similem co*n*temperantiam caloris vocatur Balneum, stufa, sterquilinium" 'Unnatural is the imperfect fire, like heat in fever, which is called humid fire, artificially made by the philosophers. It is also called fire of the first degree, which through mixing of heat in good proportions (?) is called Balneum, hot-air bath, dung pit.'

300v: 46 **out of the kind of his owne sparme**: See the note on 300r: 21–22.

301r: 13 **cruddy**: The meaning of the word "cruddy" seems to be 'crude,' 'unrefined.' The word *cruddy* is recorded in the *OED* only in the figurative meaning 'dirty,' 'unpleasant' (s.v. *cruddy*), all instances being from the twentieth century. The *MED* (s.v. *cruddi*) records one instance with the meaning 'granular,' which does not fit the context. It may also simply be a variant spelling of *crude* (cf. the reading of all the other manuscripts).

301v: 8–9 **the w*hich* chaos . . . worke being performed**: The syntax and meaning of this passage is unclear. Perhaps *found* in 301v: 8 should be interpreted as having the absolute meaning 'found suitable/good/useful' (though this meaning is not recorded in the *OED* or *MED*). The passage would then be read 'the philosophers found the chaos useful/suitable not only to bring about/accomplish nature's elemental processes, but also to hide in its caves the same processes while they are being brought about/accomplished.' Two slightly different meanings of chaos may thus be intended: first, chaos as the primordial matter; and second, chaos in a more metaphorical sense of 'disorder,' 'obscurity.' Many alchemists considered chaos, i.e., the primordial matter from which everything was thought to be created, to be the raw material of the philosophers' stone. Bringing order to the chaos by separating and recombining the elements was the most fundamental of alchemical processes: Abraham, *A Dictionary*, s.v. *chaos*.

301v: 14 **haue compiled**: The main clause has no overt subject. *They* (referring to the ancient philosophers) needs to be supplied.

301v: 37–38 **Ladie and Emprise of Phi***losofie*: Cf. 324r: 22.

301v: 40 **w<h>e**: The 'h' is a tentative reading, since the letter is smudged in the manuscript. There is a loop extending down, which resembles the descender of the 'h.' However, no other spelling of *we* as "whe" is found in the manuscript; neither is intrusive 'h' common in general in the manuscript. Cf. "whotnes" 324r: 11.

301v: 41–303v: 47 **out of on substance and masse . . . in the generatio*n* therof**: This section is taken from an anonymous collection of sayings by various alchemists, the *Dicta*. See pp. 41–42.

301v: 45 **argent viue**: The *Dicta* prefers the term *water* instead of *argent vive* 'mercury' (MS. Ashmole 759, fol. 69r). See 303v: 8–10.

302r: 1 **in himselfe**: The reading "in himselfe and w*i*th himselfe," found in FGS1S2, is supported by the *Dicta*.

302r: 4 **generation**: The *Dicta* uses *governance* instead of *generation* (MS. Ashmole 759, fol. 69r). See also, however, 302v: 31, where both the Treatise and the *Dicta* prefer *generation*.

302r: 4–6 **that thing hath in himselfe . . . by crafte drawen out of him**: Cf. *Dicta* (MS. Ashmole 759, fol. 69r): "that to this thyng belongyth is drawen out of hym and into hymsilf turned ayene."

302r: 6–7 **our stone is on substaunce and on matter, and of himselfe and w*i*th himselfe is made**: The *Dicta* (MS. Ashmole 759, fol. 69v) suggests that, while the phrase *of himself* should be taken with *matter*, *with himself* should be taken with *made*. The *Dicta* also reads *made one* instead of simply *made*.

302r: 10–11 **with one deede**: In the *Dicta*, the corresponding phrase constitutes a clarification of *disposition*, making clear that decoction is being discussed. Cf. MS. Ashmole 759 (fol. 69v): "s*cilicet* sedyng or decocc*i*on." Possibly "deede" is a misreading of an original (abbreviated?) *decoction*.

302r: 12 **the rotes of the mineralls**: The *Dicta* adds "and metall*es*" (MS. Ashmole 759, fol. 69v).

302r: 14 **moiste and dry**: This specification is not found in the *Dicta*.

302r: 16–21 **And by thes tokens he is knowen . . . endeth in the same that it beginneth**: The *Dicta* stresses that the work begins with red and ends with red (although its description actually starts with white). The Treatise, on the other hand, claims that red is the final color only. This leaves the Treatise statement that the opus begins and ends in the same way slightly cryptic, and it is easier to understand in the context of the *Dicta*. The Treatise may have adapted the source to fit other sections where black is the first color and red the final: see e.g. 322r: 27–32. Cf. MS. Ashmole 759 (fol. 69v): "for of the white comyth the redde and of the redde comyth blak and then white and then redde ayen, and thus the redde is the fundac*i*on of our werk, begynnyng and ende, and thus our werk is cercly, for it endith in the same where it began."

302r: 22–23 **Nor the sprite may not . . . a mean of the soulle**: For this imagery, see p. 81.

302r: 23 **of whom**: The reading "to whome," found in S1S2S3S4, is supported by the *Dicta*.

302r: 25–27 **but such a thinge as was hid . . . wax moulton in the fier**: The *Dicta* uses a slightly different formulation than the Treatise. Cf. MS. Ashmole 759 (fol. 70r): "but only that that was hidde before in the sprite and nowe is made opyn and as it shewed before it was molten."

302r: 28–30 **But in the begini*n*ge, the sprite of our mastery . . . which ph*i*los*o*fers calleth their gould**: The *Dicta* does not discuss the spirit, nor the quality of heat. Cf. MS. Ashmole 759 (fol. 70r): "this body which is fourme and p*e*rfecc*i*on and tyncture that all ph*i*los*o*ph*r*es sought is the gold of philosophy."

302r: 31–32 **semip*e*rfecte**: The *Dicta* simply uses *imperfect* (MS. Ashmole 759, fol. 70r). The Treatise reading may represent a misinterpretation of an original *imperfect*.

302r: 34–35 **the red man and the white wife**: For these symbols, see p. 80, and also 302v: 22–26.

302r: 37–38 **For without the ☉ and the ☽ . . . they be only the fermen*tes* of the ☉ and ☽**: The *Dicta* claims that gold alone is the ferment for both gold and silver, unlike the Treatise, which claims that both gold and silver are needed to ferment gold and silver.

302r: 39–41 **and ioineth the co*m*pound of element*es* . . . and in the earth fier**: The Treatise appears to attempt to join together two originally separate sayings in the *Dicta*. The *Dicta* provides no connection between the ferment and the subsequent discussion on the opposite properties of moist and dry, etc. After asserting that gold is the ferment of both silver and gold, the *Dicta* ends the saying by stressing that the red stone alone provides true tincture. (This last statement is absent in the Treatise: see note on 302v: 40–42.) The *Dicta* then begins a new saying concluding that the philosophers' stone is made of the four elements, and the four elements are nothing but moist and dry, etc. The Treatise seems to try to bridge the gap between the two separate statements by suggesting that gold and silver join the elements, the result being a fixed body (the stone?). In the Treatise, it is difficult to see where the relative clause ("which is noe other thinge . . . ") is intended to belong (to *fixed body? the compound of elements?*). The *Dicta* demonstrates that the properties were originally ascribed to the elements, which may suggest that this is also the intention in the Treatise.

302r: 42–44 **and ayer & fier is in them . . . but worketh inuisibly**: Cf. *Dicta* (MS. Ashmole 759, fol. 70v): "and other two that ar neither seien ne touchid as ayer and fyre."

302r: 44–46 **Therfore, yf thou canste turn the element*es* . . . that that is flyinge fixed**: Cf. *Dicta* (MS. Ashmole 759, fol. 70v): "Turne aboute theleme*ntes*, and thowe shalt fynde that thowe sechiste, whiche is no more but to make the moist to be drie and the fleyng to be fixe."

302v: 1 **fixed and flowinge**: The *Dicta* adds "body and sprite and soule" to the list after *fixed and flowing* (MS. Ashmole 759, fol. 70v).

302v: 3–4 **be all one and the same that ☿ and waters is; for water is the mother of all element*es***: The Treatise differs slightly from the *Dicta* in its formulation, but the basic idea seems to be the same. Cf. MS. Ashmole 759 (fol. 70v): "all thise thyngges byn but only on thyng and the same thyng that is m*er*cury, and wat*er* is moder of all el*emen*tes."

302v: 4 **grosnes and thicknes**: The *Dicta* reads *thinness* instead of *grossness* (MS. Ashmole 759, fol. 70v).

302v: 4 **sulfur**: The *Dicta* uses *earth* instead of *sulfur* (MS. Ashmole 759, fol. 70v). Cf. 301v: 45 for a similar example. The Treatise may have "updated" the *Dicta* formulation to fit its own terminology.

302v: 6 **then is it called earth**: Cf. *Dicta* (MS. Ashmole 759, fol. 70v): "then is wat*er* t*ur*nyd into erth." Perhaps the Treatise reading stems from a misreading of *turned* as *termed*.

302v: 6–8 **for the ph*ilosofe*rs calle . . . & stife or dry, earth**: The *Dicta* contains a slightly more elaborate version of this passage. For example, it adds that hot things are called fire. Despite its length, the *Dicta* formulation is closer to the shorter version of attributes found in all the manuscripts of the Treatise except A.

302v: 8–9 **And soe is our stone coaguled or curded, as male and female**: The Treatise formulation, which is fairly cryptic, is substantially different from the *Dicta*, which attributes a more active role to the stone. Cf. MS. Ashmole 759 (fol. 71r): "also our ston is a coaguler or a crudder, male and female and is likenyd to the crudde of mylke." See 302v: 26ll.

302v: 11–12 **and the curd that coaguled**: The group GS1S2S3S4 is closer in formulation to the *Dicta* than A (and F) since both the *Dicta* and the former group use the present tense. Cf. MS. Ashmole 759 (fol. 71r): "whiche is the crudder that congilith me*r*cury." Cf. e.g. 303v: 10.

302v: 14–15 **with his owne accorde he is coaguled into sulfur**: The *Dicta* states that "wit*h* his owen sulphur he is cruddid and congilid" (MS. Ashmole 759, fol. 71r), not that mercury is turned into sulfur.

302v: 15 **houlden**: The *Dicta* adds "fast from flight" (MS. Ashmole 759, fol. 71r), which clarifies the slightly cryptic formulation in the Treatise.

302v: 23 **coaguled by the name**: The reading "man," which is found in all the other manuscripts except F, is supported by the *Dicta*. See 302v: 26.

302v: 25 **sufferinge**: It is unclear what *suffering* means in the context. It should perhaps be interpreted as "being passive"; *OED* s.v. *suffer*. The *Dicta* reading suggests that the Treatise has left out the word *fire*: "and not suff*r*yng fyre" (MS. Ashmole 759, fol. 71v).

302v: 33–35 **pourgethe and entreth soe that the inner p*a*rte . . . which rednes is hidden**: The longer formulation in FS1S2S3S4 is closer to the *Dicta*. Cf. MS. Ashmole 759 (fol. 72r): "and that that is wit*h*yn is put outward and that that is wit*h*out is put inward so that the vtt*er* part clensith the inward p*a*rt, and the inward p*a*rt clensith the vtter p*a*rt till the redde that is naturally in the body hidde . . . ."

302v: 35 **burned**: The reading *turned*, found in S1S2S3S4, is supported by the *Dicta*.

302v: 39 **and afterward*es* turne that whitnes into rednes**: This statement is not found in the *Dicta*. It may have been added because the Treatise in general considers the red state of the elixir to be the ultimate goal. See 302r: 16–22, 302v: 40–41.

302v: 40–42 **because ther is noe stedfaste tincture . . . can noe more be changed**: The *Dicta* does not include this passage. It may instead derive from the discussion in 302r: 21–22 or from a passage left out at 302r: 39–41. See also the note on 302v: 39 for the importance of the red state of the stone.

302v: 44 **perfection**: Rather than stating that mercury is the perfection, the *Dicta* says that it is the "p*er*fecc*i*oner" (MS. Ashmole 759, fol. 72r), i.e., the thing that makes other things perfect.

302v: 45 **any other**: The *Dicta* is slightly more specific than the Treatise, stating "eny other sulph*ur* or eny other thyng" (MS. Ashmole 759, fol. 72r).

303r: 4 **not cold**: The *Dicta* does not mention the quality of mercury. Instead it stresses that "if . . . m*ercurie* were not the cause of it" (MS. Ashmole 759, fol. 72v), suggesting that if the procedure did not begin with mercury no pure gold could be the result. Cf. 303r: 9.

303r: 5 **the ascention of the sprite**: This statement is not found in the *Dicta*. Cf. 302r: 22–23.

303r: 11 **aery**: The *Dicta* does not claim mercury to be airy, but simply asserts that mercury is "eu*er* fleyng the fyre" (MS. Ashmole 759, fol. 72v). The Treatise reading may represent a misreading of an original *ever*, or *airy* may have been influenced by the characterization that follows, i.e., *flying the fire*.

303r: 14 **slainte**: i.e., *slain*. This form is not attested in the *OED* or *MED*. The form may signal double morphological marking, the 't' representing an unvoiced variant of the *-ed* inflectional morpheme added to the participial form *slain*. Oldireva Gustafsson, who has investigated past participle and past tense forms in a slightly later period (1680–1790), lists no similar cases: see Larisa Oldireva Gustafsson, *Preterite and Past Participle Forms in English 1680–1790: Standardisation Processes in Public and Private Writing*, Studia Anglistica Upsaliensia 120 (Uppsala: Acta Universitatis Upsaliensis, 2002). Cases of intrusive 't' have been recorded by Wyld in a number of words in the period, where the 't' may reflect a feature of pronunciation: see Wyld, *A History*, 309. There are no other possible instances of intrusive 't' in A. The *Dicta* uses "ou*er*come" (MS. Ashmole 759, fol. 72v).

303r: 15 **fixation**: The *Dicta* reads *perfection* instead of *fixation* (MS. Ashmole 759, fol. 72v).

303r: 18 **made of ☿ certain workes that abode the fier**: The *Dicta* remarks that as a result of the procedures it was mercury that abode the fire, not the products(?) that were made by the means of it. Cf. MS. Ashmole 759 (fol. 72v): "made by hym certeyn werk*es* that he abode vtt*er*ly the fyre."

303r: 25 **He is wild, and by his putrifaction he is made royalle and deare in his vertue**: The *Dicta* states that "he is vile by his putrefaci*on* and by his v*er*tue he is riall, noble, lyf, and dere" (MS. Ashmole 759, fol. 72v). Perhaps the Treatise reading "wild" has been influenced by the discussion in 295v: 12 (based on the *Mirror of Lights*), in which mercury is claimed to be "wild." See note on 295v: 12.

303r: 28 **matter of our stone**: The *Dicta* instead discusses *water of the stone* (MS. Ashmole 759, fol. 72v).

303r: 29–31 **and the other is refeirid . . . entention is all one thinge**: The exact meaning of the phrase "refeirid with" (also found in F) is unclear. Possibly, *refer* should be interpreted as having the sense 'reserve,' 'set aside' (recorded in the

OED s.v. *refer* 7c, in slightly different contexts). The sense of the passage would then be close to the reading found in S1S2S3S4 (and to some extent G) "reserued for," which is clearer in the context. It is interesting to note that Forman adds a marginal comment for this passage: "liquifaction | putrifaction | entention." He joins these three words together with a curly bracket and puts "on" 'one' on the other side of the bracket. Forman thus appears to have interpreted "entention" as an alchemical process instead of reading the phrase "entention is all one thinge" as 'the intention is all one thing.' However, how he made that interpretation fit with the syntax of the sentence is unclear. (Owing to the uncertainty of the syntax, I have left the punctuation open.) The *Dicta* presents a formulation similar, though not identical, to the one found in S1S2S3S4. Instead of promoting putrefaction, the *Dicta* prefers inceration, a process concerned with turning something into a wax-like condition. Cf. MS. Ashmole 759 (fol. 72v): "and that other p*art* is kept for his liquefacc*io*n and incerac*io*n, whiche is all on."

303r: 32–35 **putrifaction is ye reuolution . . . he encreaseth his quantity and goodnes**: The *Dicta* contains a much shorter version of this statement, simply saying that "to putrefie is to resolue that is congilid" (MS. Ashmole 759, fol. 72v). Cf. 305v: 33ll.

303r: 39 **fier and ☿**: The *Dicta* reads *air* instead of *mercury* (MS. Ashmole 759, fol. 73r).

303r: 39 **for the deed**: The *Dicta* supports the reading found in GS1S2S3S4, "for thee." A and F may have misinterpreted the reading *the* in their exemplar(s), assuming it to be the definite article instead of the personal pronoun. As a result, they would have tried to make sense of the construction by adding a noun ("deed") or by substituting *it* for *thee*.

303r: 43–44 **naturall**: "naturall" might be used as a noun in the meaning 'a natural thing,' although this meaning is almost exclusively attested for the plural *naturals* in the *OED*; s.v. *natural* n. 6.a. Alternatively, it may represent a misreading of *material*, which is recorded in S1S2S3S4. The *Dicta* has a substantially different reading stressing that "his [the fire's?] gou*er*na*u*nce is but wat*er* p*er*man*en*t" (MS. Ashmole 759, fol. 73r).

303r: 45–303v: 1 **Therfore, he is vnwise that studieth not in the gouerna*u*nce of the fier**: The *Dicta* does not directly instruct the reader to pay attention to the governance of the fire, but it indirectly hints at its importance. Cf. "whos [that of the water permanent] effecte standith in the gove*r*na*u*nce of the fyre" (MS. Ashmole 759, fol. 73r).

303v: 1–2 **and w*i*th him the body and the sprite be ioined together**: The *Dicta* also adds before this statement that "w*it*h hym the volatif is holdyn w*it*h the fixe" (MS. Ashmole 759, fol. 73r).

303v: 3 & co*n*fixation of elemente*s*: This phrase is not found in the *Dicta*.

303v: 3–12 **wherfore only water is sufficiente for vs . . . but in his mothers milke**: The Treatise is more condensed than the *Dicta*, and at the same time it contains elements not found in the *Dicta*, including the discussion about the mortified qualities. The *Dicta* focuses more on the nature of the water in question.

303v: 6 **the winge*s* of the winde**: The *wind* often symbolizes the vapor ascending from the dissolved body during sublimation: see Abraham, *A Dictionary*, s.v. *wind*. See also 302r: 22–23 and 310v: 39. It may also be an allusion to Psalms 18:10, 104:3.

303v: 8–9 **o*u*r water that wetteth not the hand**: This term is often used for *mercury* or *mercurial water*: see Abraham, *A Dictionary*, s.vv. *fire*, *sea* or *sea water*. See also 304v: 8.

303v: 13 **white & in poware red**: The *Dicta* adds "in dede" (i.e., 'in reality') after *white* (MS. Ashmole 759, fol. 73v). Cf. 302r: 31.

303v: 13–14 **is made by the addicion of the ☉**: The *Dicta* does not point out that the sulfur is made with the help of gold, but instead claims that "this sulphur is of the kynde of the sonne and the hert of the sonne" (MS. Ashmole 759, fol. 73v).

303v: 15 **priuity hid**: Cf. *Dicta* "prevy hidde gold" (MS. Ashmole 759, fol. 73v).

303v: 15 **dry water**: The *Dicta* also includes "stony wat*er*" after *dry water* (MS. Ashmole 759, fol. 73v).

303v: 18 **the stone**: The *Dicta* qualifies *stone* by saying "the white ston," probably in order to contrast it with the red stone, which it had previously used as a name for sulfur (cf. 303v: 16). The *Dicta* also includes "ayer" in its list before the white stone (cf. MS. Ashmole 759, fol. 73v).

303v: 20 **natures**: The *Dicta* instead uses *waters* (MS. Ashmole 759, fol. 73v).

303v: 21–24 **And this sulphur earth is p*er*fecte . . . of the w*hich* Hermes was firste finder**: This passage is not found in the *Dicta*, which instead has an unrelated discussion attributed to Raymond.

303v: 28–30 **And herin co*n*sider howe nature . . . muste our mastery be done**: The *Dicta* simply states "in this poynt take nature for thy merro*ur*" (MS. Ashmole 759, fol. 74r).

303v: 30 **whatsoever**: The meaning of *whatsoever* may here be *whoever*, although this sense seems to be highly restricted: see *OED* s.v. *whatsoever* 2b and *MED* s.v. *what-so-ever* 1b. It may also simply be a mistake.

303v: 34–35 **For the certainty of this worke is the coagula*ti*on of arg*ent* viue**: The *Dicta* does not mention the coagulation of mercury, but simply states "the certente of our science in the begynnyng of our mastrie is the begynnyng of ar-*gent* vi*ue*" (MS. Ashmole 759, fol. 74r).

303v: 37 **& beinge thus coaguled into his bodie**: Cf. *Dicta* (MS. Ashmole 759, fol. 74v): "if thou can cast him out of his body thou hast half the secrete, and if thou canneste turne hym into his body ayen."

303v: 38 **collored as white as snowe**: After this statement, the *Dicta* has a long section dealing with mercury and sulfur, which is absent in the Treatise.

303v: 38–39 **not any vaine water made of corosyues**: The *Dicta* does not mention corrosive water but instead emphasizes that "wat*er* of the skye" (MS. Ashmole 759, fol. 74v) is not the water intended.

303v: 40–41 **is had of our quick water of life, that is, ar***gent* **viue coaguled into sulfur**: The reading found in all the other manuscripts, "quick water or water of life," is closer to the reading in the *Dicta*. This section also ends the passage drawn from the *Dicta*. Cf. MS. Ashmole 759 (fol. 74v): "therfore quyk wat*er* or wat*er* of lyf is ar*gent* vi*ue*, whiche ar*gent* vi*ue* is congilid into his owen sulphur."

304r: 8–12 **by the reason wherof nature . . . the great elixar of life**: Cf. 323v: 1–22.

304r: 14 **entreth not medison**: The construction is probably a direct translation of the Latin "intrat non medicinas" 'is not brought into a medicine' found in Ripley's *Medulla*: *Opera omnia*, 140, 143. Cf. *OED* s.v. *enter* v. 14. See also 312v: 20–21, 313v: 3, 315r: 9–10.

304r: 15 **takinge**: The reading *lacking*, which is found in the other manuscripts, is more contextually suitable, since the point is that the mineral stone requires "a shorter manner of workinge" (304r: 17).

304r: 17 **whelle of phi***losofie*: In alchemical practice, the *wheel of philosophy* most often refers to the process of converting one element into another by means of its primary qualities or the process of unifying the elements. For example, since air and water share the quality of moistness, either of them could be turned into the other or they could be unified by means of this quality. The point of this process was to refine the base matter and to obtain the philosophers' stone: Abraham, *A Dictionary*, s.v. *opus circulatorium*; Roberts, *The Mirror*, 45–50. See also 305v: 33, 322r: 40.

304r: 18–30 **The elixar minerall . . . animall as minerall**: This passage is based on the *Medulla*, ascribed to the fifteenth-century English alchemist George Ripley: *Opera omnia*, 138–39. See p. 46.

304r: 20 **vicinall**: I have not been able to identify this word (also recorded in F). Most probably it simply represents a misreading of a form of *vitriol* (as found in the other manuscripts).

304r: 21 **fier elementalle**: The *elemental fire* is probably the ordinary fire which is otherwise referred to simply as *fire* in the Treatise. Cf. Ripley's (*Opera omnia*, 136) definition of *elemental fire*: "Ignis elementalis est, qui fixat, calcinat & comburit, & nutritur rebus combustibilibus" 'The elemental fire is the one that fixes, calcines, burns, and is nourished with combustible substances.'

304r: 26–27 **the sprite of the grene lyon or otherwise his blod**: The *blood of the green lion* is often used in alchemical texts to symbolize a dissolvent that may reduce all metals or a base material into the prime matter. Once this is accomplished, the lion acts as the medium or glue for the re-joining of the sulfur and mercury (the constituent parts of the prime matter) into a new perfected substance, the philosophers' stone: see Abraham, *A Dictionary*, s.vv. *green lion*, *blood*, *glue*. See also 318r: 27ll. and 324v: 2ll.

304r: 41–43 **that fermente, beinge broughte . . . the life of the stone**: Cf. 322v: 6–7.

304v: 8 **yet not wettinge that it totcheth**: Cf. 303v: 8–11.

304v: 9–23 **For ther is doble materia prima . . . lightly flyeth from the fyer**: This passage derives from the alchemical tract entitled *Notabilia Guydonis Montaynor*. See pp. 43–45.

304v: 10–14 **from water was . . . dep*a*rted the waters from water**: This passage paraphrases Genesis 1: 1–7.

304v: 10–11 **that is to say, water from water was before heaven and earth was made**: "from" may be a misreading of *for*, found in S1S2S3S4. Alternatively, *from* should perhaps be interpreted as a conjunction, meaning 'from the time when'; *OED* s.v. *from* C.; *MED* s.v. *from* conj.

304v: 35–43 **And althoughe quicksiluer of the mine . . . and water p*er*manente**: Cf. 299r: 12–24.

304v: 43–305r: 14 **the menstrue resoluble, hauinge . . . all imp*er*fecte bodies of the mine &c**: This passage resembles several other passages in the Treatise, most notably 320v: 15–26 and 320v: 44–321r: 3, and they may stem from the same source: George Ripley's *Medulla*. See p. 46.

304v: 45–305r: 2 **ioyned in the quintaescentia . . . together into a substance**: The syntax of this passage is unclear, and the punctuation that I have supplied is thus tentative. It is uncertain what the subject is of the finite verb phrase "ar putrified." Possibly the *of* before *sulfur* is a mistake and a new sentence should begin with *joined*, or an instance of *sulfur* has been omitted after *the fatness of sulfur* owing to homeoteleuton. The passage would then be read: "Ioyned in the quintaescentia by reason of the fatnes of sulphur, [sulphur] & ye sprite of the water ar putrified together into a substance."

305r: 19–306r: 17 **to declare that without ferment . . . is a secrete, and soe it shall remaine:** This section is an adaptation of the alchemical tract *Concordantia Raymundi et Guidonis*, attributed to the fifteeenth-century English alchemist George Ripley. See p. 45.

305r: 19–20 **noe true tincture, nor collour of gould nor siluer:** Cf. *Concordantia* (Ripley, *Opera omnia*, 323): "neque Sol neque Luna producatur" 'neither gold nor silver may be produced.'

305r: 24–25 **although that in naturall adrop be gould & siluer in poware:** Cf. *Concordantia* (Ripley, *Opera omnia*, 323): "sed Adrop est aurum & argentum potentia & non visu" 'but adrop is gold and silver in potentiality and not in visual aspects.'

305r: 25–29 **for that gould and y^t siluer . . . the assistance of gould and siluer vulgar:** This passage is not found in *Concordantia*.

305r: 29 **And Rhasis saith:** In the *Concordantia* (Ripley, *Opera omnia*, 323), the statement that adrop is gold and silver in potentiality (305r: 25) is ascribed to Rhasis. The statement that the Treatise attributes to Rhasis is instead said to be "secundum philosophos" 'according to the philosophers' in the *Concordantia*.

305r: 30–32 **meaninge therby y^t it muste be opened & made ayerouse to the entent yt may be fermented & muste be fixed:** There are substantial differences between the passage found in the Treatise and the corresponding passage in the *Concordantia*, both being quite obscure. The Treatise is closest to the two versions of the *Concordantia* found in MS. Ashmole 1479 and MS. Rawlinson B. 306: "It is opinyoind or chosen y^t our gould & siluer ar aerius which, to ye eand they may be fermented, must be fixed" (MS. Rawlinson B. 306, fol. 41v).

305r: 32 **For philosofers say:** Only one manuscript of the *Concordantia*, MS. Ashmole 1487, agrees with the Treatise in using *philosophers*; the other versions use *philosopher*, probably referring to Rhasis.

305r: 33–37 **meaninge therby that he houldeth . . . the stone is therby created:** This passage is not found in the *Concordantia*, which simply states "& ipsum Adrop appellatur aurum leprosum" 'and this adrop is called leprous gold' (Ripley, *Opera omnia*, 324).

305r: 39 **that is gotten adrope:** The S1S2S3S4 reading "that is gotten of naturall adroppe" is supported by the *Concordantia* (Ripley, *Opera omnia*, 324).

305r: 40 **the soulle of philosofers:** "soulle" is either a variant spelling or a misinterpretation of *Sol* 'gold' found in the *Concordantia* (Ripley, *Opera omnia*, 324). Cf. note on 315r: 35–37.

305r: 42 **the bodie is the fermente:** The *Concordantia*'s formulation is more explicit: "corpus est fermentum spiritus & spiritus fermentum corporis" 'the body is the ferment of the spirit and the spirit the ferment of the body' (Ripley, *Opera omnia*, 324).

305r: 42–45 the earth is the matter ... which is the stone: Cf. *Concordantia* (Ripley, *Opera omnia*, 324): "terra, in qua delitescit ignis, exiccat, imbibit & fixat aquam: et aer in quo delitescit aqua abluit, tingit perficitque terram & igne*m*" 'the earth, in which the fire is hidden, dries, imbibes, and fixes the water: and the air, in which the water is hidden, washes, tincts, and completes the earth and fire.'

305v: 1–2 without co*n*iunction: This phrase is not found in the *Concordantia*.

305v: 5 play they the foolles after ph*ilo*so*fe*rs: i.e., 'they act like fools according to the philosophers'; cf. *OED* s.v. *fool* n¹ 2b. The same wording is found in all the English versions of the *Concordantia* consulted.

305v: 6 For Raymond saith: The Treatise leaves out a passage from the *Concordantia* before the statement of Raymond's opinion that gives the reason why common gold and silver are not helpful in obtaining the ferment: the base material that should be employed is a thousand times better than common gold and silver after it has been purified: see Ripley, *Opera omnia*, 325.

305v: 11–13 But here note that ferment ... for the elixar and for the ferment: There is no direct equivalent of this statement in the *Concordantia*. The passage in the Treatise seems to replace a discussion in the *Concordantia* about mercury and the fermentation of the tincture with common gold: see Ripley, *Opera omnia*, 326.

305v: 18 reuegitated: "reuegitated" (which is also used in the *Concordantia* manuscripts Rawlinson 306, Ashmole 1424, and Ashmole 1479) seems to be used to translate *revivificari*, with the meaning 'to be revived,' 'to be restored into a natural condition or form'; cf. *OED* s.v. *revive* 11. This meaning of *revegetate* is not found in the *OED*, which does not record *revegetate* used transitively until the twentieth century and then only with the meaning 'to produce the growth of new vegetation.' Cf. also 300v: 23.

305v: 19–20 and p*a*rtly to be rubified: Cf. *Concordantia* (Ripley, *Opera omnia*, 326): "& sulphur auri partim rubificari" 'and the sulfur of gold partly to be rubified.' There is probably an omission of *sulfur* or *sulfur of gold* in the Treatise.

305v: 22–24 And soe of them wit*h* the ☿ ... the greate elixare both of white and red: This passage is not found in the *Concordantia* (Ripley, *Opera omnia*, 326–27). The *Concordantia* makes clear that in order to create the white and red elixirs the sulfurs of the bodies should be fermented with the two oils that have been reproduced. Cf. 304r: 12.

305v: 28 qallities: The 'a' is slightly extended, but there is no clear 'u.' For a similar spelling, see "qantiti" in 299r: 9.

305v: 30–31 **the true doctrine of phi*los*o*fe*rs**: Cf. *Concordantia* (Ripley, *Opera omnia*, 327): "Raymundi doctrinam" 'Raymond's doctrine.'

305v: 41 **gould and siluer vulgar**: The *Concordantia* (Ripley, *Opera omnia*, 328) does not claim that the gold and silver have to be vulgar, although that may be implied.

305v: 42–306r: 6 **And wheras the phi*los*o*fe*rs speaketh ... but altered into a spirituall quallity**: The Treatise may have misconstrued the middle part of the *Concordantia*'s argument. The *Concordantia* claims that the spiritual gold and silver hide in vegetable mercury and that these, and not the vegetable mercury itself, must be fermented with vulgar gold (although perhaps since vegetable mercury contains these spiritual substances the general tenor is the same in the two texts). Cf. *Concordantia* (Ripley, *Opera omnia*, 328): "atque hae authoritates, juxta se positae, specie quide*m* videntur discordare, cum tame*n* reipsa non discordent, quandoquidem denotant, quod Sol & Luna debeant alterari antequam ad fermentum adhibeantur, & quod Sol & Luna spiritualis Philosophorum, quae delitescu*n*t in nostro Mercurio vegetabili, sicuti tinctura lapidis albi & rubei, debe*n*t figi & ferme*n*tari secundum Raymundum cum auro vulgari, non quatenus est vulgare, sed ut alteratur" 'and these authorities, placed next to each other, seem on the face of it to disagree, although they do not in fact disagree, since they maintain that gold and silver should be altered before they are used for ferment, and that spiritual gold and silver of the philosophers, which hide in our vegetable mercury, as the tincture of the white and the red stone, must be fixed and fermented, according to Raymond, with vulgar gold, not as it is vulgar, but as it is altered.'

306r: 8–9 **Therfore, Raymond saithe with gould vulgar**: The same reading is found in all the English versions of the *Concordantia* consulted. Ripley, *Opera omnia*, 329, supplies "fermentamus" 'we should ferment.'

306r: 10 **And another phi*los*o*fe*r saith**: Cf. *Concordantia* (Ripley, *Opera omnia*, 329): "Et Guido dicit" 'and Guido says.'

306r: 11–12 **to signifie for fermente it must be taken for ☉ altered**: Cf. *Concordantia* (Ripley, *Opera omnia*, 329): "ad significandum quod pro fermento aurum alteratum debet accipi" 'to show that altered gold should be taken for ferment.'

306r: 13 ☿ **vegitable**: Cf. *Concordantia* (Ripley, *Opera omnia*, 329): "Spiritus nostri mercurii" 'the spirit of our mercury.'

306r: 40–41 **1590. Noviter scriptum *per* Simonem Forman. Iuni 19**: For a discussion of the dating of A, see p. 96.

307r: 1–309v: 32 **Hereafter followeth the declaration ... yeldeth delectable sauours. Finis**: Chapters 4–9 are based on the alchemical dialogue *Scoller and Master*. See p. 36.

307r: 1–9 **Hereafter followeth the declaration . . . for the better vnderstandinge of the same:** Although the discussion of the sublimations continues in the dialogue format in *Scoller and Master*, the Treatise simply gives the passage as expository text.

307r: 10 **For in the workinge of the stone ar many sublimations, wherof the firste is the sublimation of quicksiluer:** Unlike the Treatise, *Scoller and Master* makes a distinction between mercury and quicksilver by stating that there are many sublimations of mercury, but only one of quicksilver.

307r: 12–14 **from dry thing*es* not to him accordinge . . . let him be well sublymed:** MS. Ferguson 91 is the only manuscript of *Scoller and Master* to include a similar formulation about vitriol and salt. Cf. MS. Ferguson 91 (fols. 22v–23r): "y*o*u must first make mercur*ye* of argent vive w*hi*ch [is] done by sublimation of hem w*ith* (Vitriol) and co*m*mon (Salt) *p*repared" (the substances within parentheses are represented in the manuscript by sigils). Cf. Linden, *George Ripley's Compound*, 67.

307r: 17–23 **Also yf the sprite of quicksiluer . . . will wax blacke in his grindinge:** The Treatise reformulates the argument presented in *Scoller and Master*, and adds information that is absent in the manuscripts of *Scoller and Master*. Some of this information is echoed in 307r: 28–33. Cf. MS. Ferguson 91 (fol. 23r): "and in this maner y*o*u shall knowe whether it hath kept his actyve pouer or be incumbery*d*: when all they materialls be made reddi and putt into y*o*ur glasse to be sublimed, y*o*u must be whare in the riseng thereof that none of the argent vive arise up crude therew*ith*; for and the materialls be all incorporated, then the hard will not medle w*ith* the soft, and then it is incu*m*bred."

307r: 28–33 **wherfore quicksiluer may in noe wise . . . ioyninge him with the sprite of vitriall:** This passage is not found in *Scoller and Master*. See note on 307r: 17–23.

307r: 32–33 **throughe the p*a*rticipatinge and ioyninge him with the sprite of vitriall:** The construction and meaning of "participatinge" are unclear. The construction is probably *through the participating . . . with the spirit of vitriol*; cf. OED s.v. *participate* 2b. The sense seems to be 'by the sharing (of the nature or characteristics) of the spirit of vitriol,' since the point of the discussion is that quicksilver should obtain its working virtue and hence needs the qualities of heat and moisture.

307r: 33–35 **Therfore, beware that noe quicksiluer . . . will the matter wax blacke:** Cf. *Scoller and Master* (Corpus Christi College, Oxford, MS. 226, fol. 32v): "Therfor*e* be war þat no raw m*er*cure ryse not vp i*n* þe s*u*blimac*i*on fro þat þat ys hard, & if sume be q*ui*ck & su*m* be hard, þa*n* it wex blacke." See also 307r: 22–23.

307r: 36–37 **water or the stannes**: The A reading, "stannes" (or possibly "staunes"), is probably a mistake for or a variant spelling of *stones* (this being the only instance). The reading "other stones" found in S1S2S3S4 is supported by *Scoller and Master*. Cf. MS. Ashmole 759 (fol. 42v): "it will be white and fresshe on thy tonge as water or other stones." It is unclear what is actually meant by *other stones* in the context.

307r: 39 **sha<pt>**: The reading is unclear in A. The word intended is probably *sharp*, which is the reading in all the other manuscripts of the Treatise, a reading that is also supported by *Scoller and Master*.

307v: 4–42 **Take i**$^{li.}$ **of quicksilu**er **& grind . . . to clense an olde sore. Finis**: For a discussion on the relationship of MS. Ashmole 759 (a copy of *Scoller and Master*) and the Treatise in this chapter, see p. 39.

307v: 4–9 **Take i**$^{li.}$ **of quicksilu**er **& grind . . . by the heate of y**^{e} **fier**: This passage is not found in *Scoller and Master*. MS. Ashmole 759 begins the procedure with the introduction of vitriol and salt. Cf. 307v: 9–10.

307v: 12–13 **Then take your quicksiluer . . . grindinge them**: Cf. *Scoller and Master* (MS. Ashmole 759, fol. 42r): "then put theryn di coclyer [=half a spoon] of vynegre and se ye haue di li of ar$gent$ viue suerly knytte in a fustian [=cloth] and then wryng yn a part of your ar$gent$ viue thorough the fustian vppon your pouder and grynde all togeder."

307v: 15–16 **And here is to be noted . . . on pound of quicksiluer**: This passage is not found in *Scoller and Master* (MS. Ashmole 759).

307v: 17–22 **And when they ar throughly encorporated . . . a furnace of sublimation**: The Treatise and *Scoller and Master* (MS. Ashmole 759) relate similar processes but differ significantly in details. The Treatise discussion on the covers of the sublimation pots is not found in *Scoller and Master*. Cf. MS. Ashmole 759 (fol. 42r): "then put it into an olde pothard [=plate, vessel] and sette it vppon hotte eymers to dry all{wey} it steryng with a sclyce [=spatula] of tre till the moister be nygh out. Then grynd it vppon your molour [=grinding stone] ayen till no ar$gent$ viue appere. Then put {it} into a glasse to be sublymed yn and then sette it in a furneys with siftidde asshes or sotill sande."

307v: 31 **{be}**: The addition is made in a slightly browner ink than the text proper (see p. 97), and was probably made since the word *cold* was interpreted as an adjective. However, the original reading may have been the verb *cold* with the meaning 'to become cold': see *OED* s.v. *cold* v.

307v: 33–37 **Take out then the ☿ sublimed . . . like vnto roch allam**: *Scoller and Master* does not mention an iteration of the procedure, but rather prescribes that more sublimed mercury should be produced by starting the process again after the first mercury has been kept by itself. Cf. *Scoller and Master* (MS. Ashmole

759, fol. 43r): "kepe it by itsilf and then go {to} and make more and when ye haue sublymed a li or ii then put all yo*ur* sublymed m*er*cury togeder and sublyme it by itsilf ii or iii tymes and if it be wele wrought it shall be herde as aleym roche clere, clene, bitt*er*, and {sharp} on thy tonge."

307v: 38–42 **And the firste feces ... to clense an olde sore. Finis:** This passage is not found in *Scoller and Master* (MS. Ashmole 759), which instead expounds upon the virtues and deficiencies of vitriol.

308r: 1 **sublimaon:** i.e., *sublimation*. A tilde signaling 'ti' has probably been left out by Forman. See also 308r: 4, 5, and the marginal note in 307r: 11.

308r: 4–6 **Knowe that in this sublimaon ... center of their begininge:** This statement is not found in *Scoller and Master*.

308r: 7 **into ☿:** *Scoller and Master* uses *water* instead of *mercury*. See also the note on 301v: 45.

308r: 8–11 **For the soulle beinge dissolued ... maketh the bodye spirituall like vnto itselfe:** Cf. *Scoller and Master* (MS. Ashmole 759, fol. 43v): "for then the sowle is dissoluyd from the body into a sprite and therfore by sublimac*i*on either byn made sprites, fleyng togeder into aer. For then the wat*er* hath openyd the yate to the nounfleer t*ur*nyng hym into a sprite like to hymsilf." The "flyers gate" should probably be interpreted figuratively in the sense of 'the access to flight [i.e., the point of access or door of the flyer].' For *flyer* in the sense of 'volatile substance,' 'spirit,' see *OED* s.v. *flyer* 1.b. (*a*)., recorded in Ripley's *Compound of Alchymy*. Cf. also 302r: 22–23.

308r: 12–13 **takinge ther the inspiration of life of their owne moisture:** *Scoller and Master* adds "as a man doth of the ayer" (MS. Ferguson 91, fol. 23v). For "inspiration of life," see *OED* s.v. *inspiration*, and s.v. *breath* 5.a.

308r: 15–16 **wherfore the stainge of kind in this place is lefte in seruiage to flighte:** The argument of this passage is not absolutely clear. The manuscripts of *Scoller and Master* exhibit a wide variety of readings, which indicates that the copyists of *Scoller and Master* found this passage problematic. The Treatise reading is close to MS. Ferguson 91 (fol. 23v): "wherefor they flyeng togethe<r> ther kinde be left in saruage from flight, forasmuch as they be yoyned in sublimation." The same passage is found in *Rosarium abbreviatum* (Zetzner, *Theatrum chemicum*, 3: 678), which gives a slightly different version of the process: "ideo fugiens natura quamuis sit ei fuga essentialis, dimisit tamen fugae seruitutem" 'therefore the flying nature, although flying is essential to it, lost its servitude to flight.' The point of both *Scoller and Master* and *Rosarium abbreviatum* seems to be that the spiritual quality or the access to flight is lost, i.e., the substance is left in servitude, away from flight. Hence, they must be raised up into vapor, a procedure which is subsequently described.

The Treatise's "stainge" may be a misreading of *flying* at some point in the transmission, but it may also be reinforcing the sense of "seruiage."

308r: 19 **receyue life of every stedfaste kinde**: The reading found in S1S2S3S4, "receue life for th'inspiracion of th'ayre is the life of euery stedfast kinde," is supported by *Scoller and Master*. Cf. Corpus Christi College, Oxford, MS. 226 (fol. 32v): "for þe life of all stidfast kyndes commyth of inspiracion of þe eyre." The reading in A (and FG) may be the result of an eyeskip (*life . . . life*).

308r: 23–26 **This thirde sublimation showeth . . . into his bodie his water and ayer**: This passage is not found in *Scoller and Master*. Cf. 308r: 40–42.

308r: 30–31 **into an oyly substaunce, pure and clean, white, shinninge**: Cf. *Scoller and Master* (Corpus Christi College, Oxford, MS. 226, fol. 32v): ".i. to seie an ouly substans as white powder & clene."

308r: 32 **in the knittinge together of the elementes**: This phrase is not found in *Scoller and Master*. Cf. 297v: 29–30, 305r: 4–5, 306r: 31–32, 309v: 19–20.

308r: 34–35 **we calle that solution . . . the body like vnto himselfe**: This clarification is not found in *Scoller and Master*.

308r: 37–38 **then we say that . . . to abide with his body**: This statement is not found in *Scoller and Master*.

308r: 40–42 **congealle his water and ayer . . . together, the on with the other**: *Scoller and Master* simply states that earth congeals water. This may be the passage that the introduction of the chapter alludes to. See 308r: 23–26.

308r: 42–44 **soe that by this means . . . is by the other congeled**: The meaning of the passage in *Scoller and Master* seems to be slightly different from that of the Treatise passage. Cf. MS. Ferguson 91 (fol. 24r): "so that both dooengs is but one maner of workenge so that the one dissolveth noth, but if the other conjelle."

308v: 1–2 **by the reason wherof the heavy bodyes . . . to the center of their substanciall forme**: Cf. *Scoller and Master* (MS. Ashmole 759, fol. 44r): "For euery body is hevy and therfore it desirith ayen to his centre as erth."

308v: 2–9 **beinge made moch better . . . accordinge to the quallity of the fermente**: This passage is not found in *Scoller and Master*. Cf. 298r: 33ll.

308v: 18–19 **which drines we calle the bodie, and the moiste humiditie we call the sprite**: This clarification is not found in *Scoller and Master*.

308v: 19–23 **which sprite in the beginninge . . . be drawen from the heavy**: Cf. *Scoller and Master* (MS. Ashmole 1487, fol. 109v): "for the humidytye is holden onlye in the spirite, thoughe moiste and colde kindlye flye the fier. For then the bodye constraynethe hyme that he maye not flye from the heavye, ne maye not the lighte aboue without the heavye beneathe, ne the heavye benethe without the lighte aboue."

308v: 23 **compassion**: The corresponding word in *Scoller and Master* is *compaction* 'the state or condition of being compacted,' 'consolidation': see *OED* s.v. *compaction*[1]. Although perhaps a misinterpretation of *compaction*, *compassion* in the sense of 'fellow-feeling,' 'sympathy' may still be possible in the context, since the two properties or substances are attracted or drawn to each other because of their affinity.

308v: 24 **ixer is the begininge and endinge of all our worke**: The word "ixer" is also found in the version of *Scoller and Master* in MS. Ashmole 759 (fol. 44v), as "ex*ir*." Although the word is not recorded in the *OED* or the *MED*, it is recorded in Latin as a form or alternative word for *elixir*: see Ruska, *Turba Philosophorum*, 29; Barbara Obrist, ed., *Constantine of Pisa: The Book of the Secrets of Alchemy* (Leiden: Brill, 1990), 111, 212. However, it seems unlikely that *elixir* was the meaning intended originally considering the context. "ixer/exir" may instead be a misreading of a form of *either* (found in most manuscripts of *Scoller and Master*), referring to the contrary entities *moist* and *dry* or *heavy* and *light*. Cf. Corpus Christi College, Oxford, MS. 226 (fol. 33r): "eiþ*er* ys þe begy*n*nyng & þe ende." For the confusion of the letter forms 'y,' 'þ,' and 'x' in the fifteenth and sixteenth centuries, see Petti, *English Literary Hands*, 32. But although *elixir* was not originally intended, "ixer" may have been kept because of its connection with *elixir*: Forman even appears to have left enough space before "ixer" for 'El' to be added in A.

308v: 26–28 **for as by corruption . . . the generation of the second forme**: Cf. *Scoller and Master* (MS. Ferguson 91, fol. 24r): "for the corruption of the one is the generation of the other." Cf. 322r: 22.

308v: 30 **the ravens heed**: For the *raven's head*, see p. 82. Cf. also 320v: 22.

308v: 30–32 **that the bodies may be corrupted . . . can com forth noe encrese**: Cf. *Scoller and Master* (Corpus Christi College, Oxford, MS. 226, fol. 33r): "for þ*er* was neu*er* thy*n*ge borne ne growe neþ*er* had no soule but afft*er* putrifacc*ion*." See note on 308v: 32–33.

308v: 32–33 **Witnes our sauiour Christe . . . it bringeth forth noe fruite**: This paraphrase of the Bible (John 12: 24) is not found in *Scoller and Master* at this point. See note on 308v: 30–32. MS. Ashmole 759 (fol. 50r) does use the same quote, though much earlier and in Latin. See also Linden, *George Ripley's Compound*, 46.

308v: 34–36 **Soe lykwise in this worke . . . then turneth it to naughte**: The Treatise seems to present a slightly elaborated version of the *Scoller and Master* argument. The formulation "before ẜ fly" is not found in *Scoller and Master*; cf. 319v: 50. Cf. MS. Ferguson 91 (fol. 24r): "but if it rott it maye not melt nor solve, and if it solve not, it turneth to nought." For *melting*, see note on 295r: 10.

308v: 38–39 **for it is none other thing but the medlinge of earth**: The S1S2S3S4 reading, "medlinge of earth with water and water with earth," is supported by *Scoller and Master*. The reading in A (and FG) is probably due to an eyeskip. Forman seems to have felt that something was missing and has left a dash in the text.

308v: 39 **of ye fier**: This phrase is not found in *Scoller and Master*. Cf. 297r: 2–3.

309r: 1–3 **Therfore saith the ph*i*lo*so*fers . . . be congealed into glorius earth**: The manuscripts of *Scoller and Master* display a great deal of variation in content in this passage. The Treatise seems to be closest to MS. Ferguson 91 (fol. 24r): "therefor maye the ph*i*lo*sophe*rs that wi*th* this vapore there yoyne tinctures together till all be one body congeled together."

309r: 5–6 **For then it is called the dragon eating his owne wing*es***: For this symbol, see p. 82.

309r: 10–11 **Of the variable coullours . . . be pe*r*fecte coullours**: None of the manuscripts of *Scoller and Master* contain the exact reading found in the Treatise. The Treatise combines elements found in, for example, MS. Ferguson 91 and MS. Ashmole 1487. Cf. MS. Ferguson 91 (fol. 24v): "of w*h*ich colors tak<e> thou no hede, for the black and the white be the perfect colors"; MS. Ashmole 1487 (fol. 110r): "the w*h*ich collours ben variable; wherfore take no heed from blacke to white for they ben two kyndelye collours."

309v: 5–6 **forsoemoch that gould is brought . . . and worthines aboue siluer**: Cf. *Scoller and Master* (MS. Ashmole 759, fol. 45r): "forasmoche as Sol is aboue Lune in worthynesse."

309v: 6–8 **For this sublim{a}*tion* is not . . . yeldinge his vertue without number**: *Scoller and Master* emphasizes that "hit ys not only in s*u*blimac*i*on for to rere hym from the lowyst of þe vessel to the hyest but . . . " (Corpus Christi College, Oxford, MS. 226, fol. 33r; my emphasis). The phrase "make kind worke wi*th* kind" is not found in *Scoller and Master*. The Treatise reading may derive from a reading such as *make a simple kind a worthy kind*, which is closer to *Scoller and Master*. Cf. MS. Ashmole 1490 (fol. 83v) "to make a simple thinge a worthy thinge in kind." However, since the advice to let "kind worke wi*th* kinde" is found in other alchemical texts, e.g. Ripley's *Compound of Alchymy*, it may have been lifted from a source unrelated to *Scoller and Master*: see Linden, *George Ripley's* Compound, 28, 111. See also 309v: 12.

309v: 10–12 **therfor solution, congelation, & multiplica*t*ion . . . both into goodnes & quantity**: The source of the Treatise probably had a reading close to the one found in the version of *Scoller and Master* preserved in MS. Harley 6453. This is the only manuscript that includes the term *multiplication*. Cf. MS. Harley 6453 (fol. 29r): "therfore in soluc*i*on and congelac*i*on is all the worke of this cyence and all the multiplicac*i*on both in goodnes and quantitie."

309v: 16–17 **Therfore, Hermes saith as the worlde is made, soe is our stone:** This statement, which is found in all copies of *Scoller and Master* as well, probably derives ultimately from Hortulanus' eleventh-century Latin translation of and commentary to *Tabula Smaragdina*, attributed to Hermes Trismegistos: "[s]*ic mundus creatus est*, hoc est, sicut mundus creatus est, ita et lapis noster factus est" '*so the world has been created*, that is to say, as the world has been created, so is also our stone made': see Julius Ruska, ed., *Tabula Smaragdina: Ein Beitrag zur Geschichte der Hermetischen Literatur* (Heidelberg: Carl Winter, 1926), 185.

309v: 19–21 **wherfore thes quallities . . . is made the greate elixar of life:** The Treatise formulation seems to be a clarification or elaboration of the statement found in *Scoller and Master*. Cf. MS. Ashmole 1487 (fol. 110r): "between them together evenlye accordinge, peaceablye bounde, and p*er*petually ymade."

309v: 23 **noe fier by violence:** The reading "no violence of fire," which is found in all the other manuscripts of the Treatise, is supported by *Scoller and Master*.

309v: 30 **leaven to our paste:** Alchemists sometimes referred to the purified base material for the philosophers' stone as a paste or dough. This substance was believed to grow into the stone with the help of a ferment or leaven: see Abraham, *A Dictionary*, s.vv. *paste, ferment*.

310r: 39 **com ron:** i.e., 'come running.' According to Mustanoja, the construction *come* + the bare infinitive, "expressing simultaneous action and indicating the manner in which the action represented by [*come*] takes place," can be found throughout the Middle English period but becomes gradually less common and finally disappears: Tauno F. Mustanoja, *A Middle English Syntax, Part I: Parts of Speech*, Mémoires de la Société Néophilologique de Helsinki 23 (Helsinki: Société Néophilologique, 1960), 536–37. Mustanoja provides an example from Chaucer and one from Gower. No instance of *come* with a bare infinitive in this meaning is recorded in the *OED* after the fourteenth century: *OED* s.v. *come* v. B. 3.f. Visser lists three examples from the first half of the sixteenth century: Fredericus Th. Visser, *An Historical Syntax of the English Language, Part Three, First Half: Syntactical Units with Two Verbs* (Leiden: Brill, 1969), 1393 §1314. The construction in the Treatise may derive from an earlier source.

310r: 39 **enkitch:** Both A and F use this word, which is not recorded in *OED* or *MED*. It probably represents a misreading of *enrich* (found in S1S2S3S4), the 'r' being interpreted as a 'c' (which was later replaced by a 'k'). For the confusion of 'c' and 'r,' see Petti, *English Literary Hands*, 31. The meaning of *enrich* in the context seems to be to enhance the qualities of the amalgam or to increase the quantity of the amalgam.

310v: 6 **showerly:** Forman has added "surely" above this word, and "proceed" above "p*ro*sed" in 310v: 26. There are further instances in this chapter and the next where supralinear additions have been made: see 310v: 17, 18, 311r: 11. They are all added in a slightly browner ink and were probably supplied sometime after Forman initially copied the Treatise; for this feature, see p. 97. It is difficult to determine whether the additions "surely" and "proceed" are intended to replace the more idiosyncratic spellings of the words or whether they are intended as glosses. I have interpreted them as the latter in these two instances since there is no clear indication that they were intended as substitutions. In the case of 310v: 17, 18, and 311r: 11, however, I have interpreted the additions as improvements of the text, since they consist of letters or once of a word that seems to clarify the syntax.

310v: 16 **heate will**: The reading "water," which is found in all the other manuscripts of the Treatise, is more suitable in the context, since the passage deals with distillation. *Heat* might be influenced by the occurrence of *heat* earlier in the sentence. Alternatively, the clause should perhaps be interpreted as 'as long as any heat will distill it,' i.e., 'as long as the heat will last.'

310v: 26 **prosed**: See note on 310v: 6.

310v: 29 **as he is mercuried**: 'since he is turned into mercury'(?). The meaning of this clause is uncertain. The verb *to mercury* is attested in the *OED* only with the meaning 'to wash with mercury-water,' which does not fit the context in the Treatise. Perhaps it has the meaning of *mercurify* 'to change into the form of mercury' (*OED* s.v. *mercurify*), since raw mercury or quicksilver can be turned into mercury, one of the constituents of all metals. See also the discussion of *mercurized* in the note on 311v: 4.

310v: 37–41 **thou shalte moste showerly . . . as they be chaunged**: The "bird of humanity" may be the same as the *Bird of Hermes*, a symbol commonly employed for mercury: see Abraham, *A Dictionary*, s.v. *Bird of Hermes*. The term *nest* is often used to denote the vessel in which the alchemical opus is taking place: see Abraham, *A Dictionary*, s.v. *nest*. However, since the Treatise mentions the finding of the nest, it seems that *nest* should here be taken simply in the sense of 'dwelling-place,' 'location.' Cf. 303v: 4–12, 324r: 23–24.

311r: 18 **troblinge**: The verb *trouble* in the sense of 'to stir up' is perhaps intended: see *OED* s.v. *trouble* v. It may be a misinterpretation of *brobling* (or a similar spelling) 'bubbling,' which is found in S1S2S3S4 in the form "burblinge" (S2). For the verb *bruble*, see note on 313r: 1.

311r: 19 *(Marginal note)* **Then make | oill . . . yo**u***r* **medison**: This note is found in all the manuscripts of the Treatise except G.

311r: 32 *(Marginal note)* **But nevere . . . be made | oylle**: This marginal note is found in all the other manuscripts, except G, although they all read "before" instead of A's "lest."

311v: 4 **mercurized**: The word *mercurize* is not recorded in *OED* or *MED*. Neither have I found it in any Latin dictionaries. Although the exact meaning remains unclear, it may signify 'to join with mercury.' *MED* records the participle or adjective *mercurizate* with the meaning 'amalgamated with mercury' from Ripley's *Compound of Alchymy*. However, the meaning of the word in Ripley's text seems rather to be 'turned into mercury,' since the origin of metals is the subject of the passage: see Linden, *George Ripley's* Compound, 129. The meaning 'turned into mercury' is possible here too, though this seems less contextually suitable considering the subsequent statement that mercury should be "vapoured from the gould." See also 311v: 5, 312v: 34, 314r: 22, 315r: 16, 316r: 6, 317r: 33, 318r: 3, 5, 318v: 5.

311v: 10 **place of veri cleare yron**: The readings *plate* and *clean*, which are found in most of the manuscripts, are perhaps more contextually suitable, although the A reading is plausible. Cf. however, 312v: 42–43 and 313r: 33, where A uses *plate* and *clean*.

311v: 28 **bobles**: Although the word appears with a tilde above, it is doubtful whether it is a sign of an abbreviated 'r'; *bruble* in different forms occurs twelve times for *bubble*, but never with an 'o' spelling: see 313r: 1 for a discussion of the form *bruble*.

311v: 29–30 **and turne it into oylle**: This phrase is not found in any of the other manuscripts of the Treatise in this context. However, the phrase corresponds to some extent to the marginal note found in the passage corresponding to line 311v: 28 in all the other manuscripts (except G). Cf. S2 (fol. 33v): "After you haue taken owte your glas make oyle of it & then fermente." See also note on 311v: 36 "& oill." The reading in A may thus have originated in a marginal note that was subsequently merged with the main text. Alternatively, though perhaps less likely, the reading was originally part of the main text as in A, and then removed to the margin.

311v: 36 **& oill**: None of the other manuscripts of the Treatise contains this reading. However, all these manuscripts (except G) contain a marginal note in the passage corresponding to 311v: 36 in A, declaring "but first make it oyle" (S2, fol. 33v). See also note on 311v: 29–30.

311v: 36 **the chapters**: In A, a space is left between "the" and "chapters," possibly so that specific chapter numbers could be filled in.

311v: 37 **lix**: i.e., *like*. The 'x' is probably written in anticipation of 'x' in "wax."

312r: 1–27 **To fixe a malgam . . . and without &c. Finis**: For the source of this chapter, see pp. 57–58.

312r: 10 **3 dais**: Cf. MS. Sloane 3580B (fol. 53v): "3 daies & 3 nyghtes."

312r: 16 **vpon ☿ crud**: MS. Sloane 3580B (fol. 53v) adds "& vpon all bodyes."

312r: 23–24 **accordinge to the quallity of the medison**: MS. Sloane 3580B does not note this qualification.

312r: 25: **the said quintaescentia**: MS. Sloane 3580B (fol. 54r) adds "ye red elixir" before quintessence.

312r: 26–27 **it then healeth ... and without &c. Finis**: The reading in MS. Sloane 3580B (fol. 54r) is more elaborate: "to heale all lep*er*s, & to renewe youthe, & p*re*cyous for ye pestylence, & for all other sicknesses to ye houre of deathe appoynted by God, to who*m* be all honor & glorie; worlde w*i*toute ende. Amen." Cf. 324v: 14, and note on 324v: 23.

312v: 1–315v: 31 **The prologe to the mineral stone ... diuersitie in workinge &c. Finis capitulo 20mo**: Chapters 15–20 draw upon the alchemical tract *Medulla alkemiae*, attributed to the fifteenth-century English alchemist George Ripley. See p. 46.

312v: 1–31 **The prologe to the mineral stone ... chapters of this treatis. Finis**: This chapter introduces the discussion of the mineral stone found in Chapters 16 to 20, summarizing what will be dealt with in these chapters. The material for the chapter derives largely form George Ripley's *Medulla*, though there is no direct parallel chapter in the *Medulla*. See p. 47.

312v: 8 **as for the thing itself**: This statement is absent in F and G (and E1 and E2), which are otherwise related to A. The reading in A may be a later addition to make up for a passage felt to be missing, which should correspond to "as well" in 312v: 7. Cf. the reading in S1S2S3S4 where *as well* is complemented by "as also for the ordinary charges...."

312v: 23 **by the gates of putrifaction & fermenta*t*ion**: This may be a reference to George Ripley's *Compound of Alchymy*, which describes twelve different processes as twelve gates: *putrifaction* being the fifth gate, and *fermentation* the ninth gate. See Linden, *George Ripley's* Compound, 46, 69. See also 313v: 4, 317r: 17.

312v: 24 **the fier againste nature**: See note on 313v: 20–21.

312v: 35 **naturall adrop**: The *Medulla* (Ripley, *Opera omnia*, 141) does not mention adrop but simply states "cui conjungi debet" 'that with which it should be joined.'

312v: 39–313v: 16 **The calce, once mercurized ... for all gould alkamy of the firste order**: This passage is not found in any of the copies of the *Medulla* consulted, although the passage in 313r: 33 to 313v: 3 resembles the *Medulla* text. Both the Treatise and the *Medulla* suggest testing the substance on a hot plate of iron, and also point out that the mineral stone cannot be used as a medicine. I have not been able to identify a direct source for this passage. A similar recipe for making the fire against nature is found in MS. Ashmole 1479 (fol. 52v). It employs vitriol and vermilion together with saltpeter, and it states that the recipe is according to Raymond Lull. I have not been able to identify a similar recipe in Lull's writings.

313r: 1 **bruble**: The *OED* and *MED* only list one instance (in Middle English) of the metathesized form *bruble* for *burble* (the instance is a verb): see *OED* s.v. *burble* v., and n.; *MED* s.v. *burble* and *burblen*. A contains eleven instances of *bruble* in different forms, and one instance of *burble*, all in Forman's hand. (There are also two instances of *bubble*: see 311r: 20, 311v: 28.) F and G use the form *bruble* throughout, whereas S2 always contains *burble*. S1 and S4 commonly use *burble*, but there are a few instances of *bubble*. S3, on the other hand, exclusively prefers *bubble*. The variation between *burble* and the metathesized *bruble* may thus have existed in earlier stages of the transmission.

313r: 11 **water**: I have interpreted the 's' following "water" as a canceled 's,' but it may simply be a smudged 's.'

313r: 12–13 **the sprite of quintaescen*n*tia . . . shalbe minerales**: It is unclear what is meant by the statement that the spirit, or more likely vitriol and vermilion, "shalbe minerales."

313r: 42 **any more**: The reading "no," which is found in GS1S2S3S4, is more contextually suitable, since the loss of the sharpness of the water is discussed in the previous sentence.

313v: 11 **moch agreinge**: The reading found in all the other manuscripts, "more," is more suitable in the context than *much* in A, since this same statement is complemented by a *than*-clause.

313v: 19–25 **In the second p*a*rte of the min*er*all stone . . . which is not without the co*n*gelation of his water**: The Treatise here brings together three sentences that have been lifted from widely separated sections in the *Medulla*: see Ripley, *Opera omnia*, 143, 144, 145. None of the manuscripts consulted contains a similarly abbreviated version. For a detailed discussion of this passage, see pp. 47–49.

313v: 20–21 **with the fier againste nature compounded w*i*th the fier of nature**: Ripley discusses the *fire against nature* and *fire of nature* at the beginning of his *Medulla* (*Opera omnia*, 136–37). This discussion has not been included in the Treatise. I cite a passage here to illustrate the meaning of the two terms: "Ignis contra naturam dissolvit in potentia [MS. Ashmole 1490, fol. 68v: potentialiter], terit, interficit & destruit potentiam gubernativam formae lapidis: nam dissolvit lapidem in aquam nubis cum destructione formae (*al.* figurae) specificae. Vocatur autem ignis contra naturam, quia operatio illius est contra omnes naturales operationes, ut asserit Raymundus. Nam omnia, quae natura composuit, hic ignis destruit, & ducit ad corruptionem, nisi addatur illi ignis naturae. Quocirca, ut dicit Raymundus, hic jacent contrariae operationes. Nam sicut ignis contra naturam dissolvit spiritum corporis fixi in aquam nubis, & corpus spiritus volatilis constringit in terram congelatam & crystallinam; ita contrario modo ignis naturae congelat spiritum dissolutum corporis fixi in terram gloriosam . . . " 'The fire against nature dissolves with power (?) [MS. Ashmole 1490, fol. 68v: powerfully], grinds, kills, & destroys the governing power of the stone's form: it dissolves the stone into water of the cloud together with the destruction of its specific form (*alternatively* figure). And it is called fire against nature because its operation is against all operations of nature, as Raymond affirms. All that nature puts together this fire destroys and brings to corruption unless the fire of nature is added to it. Therefore, as Raymond says, here we have contrary operations. As the fire against nature dissolves the spirit of the fixed body into water of the cloud and restrains the body of the volatile spirit so that it becomes congealed and crystallized earth, so, by contrast, the fire of nature congeals the dissolved spirit of the fixed body into a glorious earth.'

313v: 22–25 **For so moch as it dissolueth . . . co*n*gelation of his water**: Cf. 315r: 24–26.

313v: 23–24 **lifteth in glorious earth**: There is a great deal of variation in A, and among the manuscripts of the Treatise in general, as regards the usage of the prepositions *in* and *into*. The variation occurs when they express "the motion or direction from a point outside to one within limits": *OED* s.v. *in* prep. Most of the variation occurs after the verb *put*. See e.g., 314r: 5, 315r: 19, 318r: 16, 324r: 4, and 324r: 31. The variation between *in* and *into* in this case may fall within this variation as well. Cf. 313v: 32.

313v: 27 **prepa*r*ard**: The second 'ar' has probably been added by mistake, since an abbreviation is already marked by a 'p' brevigraph, indicating 'par.'

313v: 28–29 **Then let the same pouder be put into an apte vessell for the purpose**: This statement is not found in any of the versions of the *Medulla* consulted.

313v: 46–314r: 3 **And this be calcioned into dry pouder . . . fixed & flowinge as wax**: Cf. *Medulla* (Ripley, *Opera omnia*, 146–47): "Cum hoc oleo (si fiat ex calcibus ☉. vel ☽ae) possumus incerare substantias caeterorum corporum elevatorum, quousque figantur & fluant" 'With this oil (if it is made of the calces of gold and silver), we may incerate the substances of the other elevated bodies, until they are fixed and flowing.' In the Treatise, it is the *oil* rather than the *substances of other bodies* that will be fixed and flowing. Cf. 322r: 16, which also draws upon this section of the *Medulla* but excludes this particular passage. The construction "this be calcioned" should probably be taken as a hortative subjunctive 'this should be calcined': see Matti Rissanen, "Syntax," in *The Cambridge History of the English Language*, vol. 3, *1476–1776*, ed. Roger Lass (Cambridge: Cambridge University Press, 1999), 228–29.

314r: 6–7 **may be made ye greate elixar**: This statement seems to replace the claim in the *Medulla* (Ripley, *Opera omnia*, 147) that "fit nigrum sicut pix liquida, &

haec est prima materia Mercurii & sulphuris" 'it will be black as molten pitch, and this is the prime matter of mercury and sulfur.' A long section in the *Medulla* is absent in the Treatise after this statement. For this passage, see 322r: 16ll.

314r: 8 **auugmented**: The spelling has most likely been caused by dittography, but it may also signal vowel length. (No other similar cases are found in the Treatise.)

314r: 9 **And when it is cast vpon a malgam**: Cf. *Medulla* (Ripley, *Opera omnia*, 151): "Et quando projicitur super Mercurium" 'And when it is projected upon mercury.' Cf. 322v: 15, where *mercury* is used.

314r: 13–16 **When we say by himselfe . . . it is said in the vegitall stone**: i.e., in Chapter 26. Needless to say, this statement is not found in the *Medulla*.

314r: 21–28 **Dissolue the bodye of gould . . . into the fier against nature**: This passage differs significantly from the versions of the *Medulla* consulted. Cf. *Medulla* (Ripley *Opera omnia*, 152): "Cum igne contra naturam fiat dissolutio corporis purissimi (cum sale cumbusto bene rectificato, aut cum arsenico fortificato ut mos est:) ex quibus purissime & sine arenosa grossitie in fundo remanente dissolvitur . . . " 'With fire against nature the dissolution of the very clean body should be made (with salt combust well rectified, or with fortified arsenic, as the custom is) from which it is dissolved very cleanly and without any sandy unclean mass remaining in the bottom.' The Treatise passage mentions "Lunarie," which is otherwise found only in Chapters 21 and 24: see note on 314r: 24. It is possible that the passage draws upon these later chapters or their source, which is unidentified. Cf. 316r: 6–18, 318r: 3–5, 318v: 5–12.

314r: 24 **Lunarie**: Alchemical writers often use the herb *Lunaria* or *Lunary* to symbolize the alchemical opus, although it sometimes symbolizes the prime matter or mercurial water: see Abraham, *A Dictionary*, s.v. *philosophical tree*; *Grosses universal Lexicon* (Halle, 1732–1754), s.v. *Lunaria*. It is unclear what the Treatise refers to here, although the most likely reference is the base material that is being refined into the prime matter, or mercury. Cf. 316r: 17, 318v: 10, 22.

314r: 29, 33 **after 7 dais [. . .] and set in balneo other 7 dais**: These two statements, specifying the time of the procedure, do not appear in the *Medulla*. Cf. 314v: 1.

314r: 37–40 **vntill all the sharpnes of the water . . . after the laste calcionation**: This passage is not found in the *Medulla*. Cf. 318r: 11–14.

314v: 10 **clos<e>d**: The syntax requires a verb form such as an imperative *close*, as found in all other manuscripts of the Treatise (except G). The form *closed* may be influenced by the *let . . . be* construction in 314v: 9.

314v: 13–15 **for then yt is a signe y^t . . . of the body dissolued**: The source of the Treatise was probably close to MS. Ashmole 1490 (fol. 71r): "signu*m* est qu*o*d

iam tinctura totaliter ˄{est} extracta ab igne contra naturam corporis dissolut<i>."
Ripley, *Opera omnia*, 154, presents a different version: "est indicium quod tunc tota tinctura corporis soluti, Solis, est extracta ab igne contra naturam" 'it is an indication that all the tincture of the body dissolved, gold, is then extracted from/by the fire against nature.' The reading in *Opera omnia* indicates that "of the body dissolued" should be taken with "tyncture." Cf. 314v: 2–3.

314v: 19 **7 daies**: Cf. *Medulla* (Ripley, *Opera omnia*, 154): "sex diebus" 'six days.'

314v: 20–21 **by distillinge**: This specification is not found in the *Medulla*.

314v: 29 **Nowe to com to the practice as we entend**: Cf. *Medulla* (Ripley, *Opera omnia*, 155): "Sed quia illud jam non pertinet ad nostru*m* propositum, procedamus ad partem practicae quam intendimus" 'But since this does not pertain to our aim, let us proceed to the part of the practice that we intend.' "illud" probably refers to the process of turning the oil into the great elixir. See 314r: 5–7.

314v: 30 **coulbuste salte**: I have not found the word "coulbuste" (which is also found in F) in the *OED* or the *MED*. It is probably a misreading for *combust* 'burnt,' 'acted on by fire,' which is found in GS1S2: see *OED* s.v. *combust*. *Combust* is also supported by the *Medulla*.

314v: 31 **partly**: All the other manuscripts of the Treatise, which contain the phrase *by itself purely*, are closer to the *Medulla*. This reading is probably a direct translation of "per se solummodo" as in Ripley, *Opera omnia*, 155, or "per se solu*m*" as in MS. Ashmole 1490, fol. 71r, both meaning 'by itself only.' See *OED* s.v. *purely* 2.a.

314v: 36 **matter**: All the versions of the *Medulla* consulted use "naturae" 'natures.'

314v: 37 **Then afterward let it coolle**: This statement is not found in the *Medulla*.

315r: 3–4 **And yf it shalbe found fyxed, yet let it haue another inceration**: Cf. *Medulla* (Ripley, *Opera omnia*, 156): "Etsi inveniatur fixum, habeat tame*n* adhuc aliam incerationem ad majorem securitatem" 'Even if it is found fixed, it should all the same have another inceration for greater certainty.' "And yf" probably derives from an original *etsi*, which was interpreted as *et si* 'and if' rather than 'even if.' Alternatively, *if* should be interpreted as having the meaning 'even if': see *OED* s.v. *if* 4.

315r: 6 **and after it is dryed with a softe fier**: The reading in S1S2S3S4, "and after it is dried wi*t*h a softe fire let it be calcined with a stronge fire, w*hi*ch thinge don increace it oftentimes & dry it wi*t*h a softe fire" (S2), is supported by the *Medulla*. The *Medulla* adds that the calcination should be carried out for eight hours. The reading "increase" in S1S2S3S4 is probably a misinterpretation of a form of *incere* (found in the *Medulla*).

315r: 7 **to the fier**: *To* is probably a direct translation of the Latin preposition *ad*, usually meaning 'to.' In the phrase "ad ignem" (*Medulla*, MS. Ashmole 1490, fol. 71r), however, it should probably be translated 'by the fire.' Five of the English copies of the *Medulla* contain the same reading as the Treatise.

315r: 9–10 **entreth not medison**: For this construction, see note on 304r: 14.

315r: 10–11 **but work<e>th in poware according to his elixeration**: This statement is not found in the *Medulla*. Cf. 312v: 20–27, 313v: 3–5.

315r: 15 **is in this manner**: A word, such as *it*, *which*, or *and*, needs to be supplied before *is*.

315r: 15–16 **that kibrik beinge dewly prepared, that is, mercurized & precipitated**: This statement is not found in the *Medulla*, which simply declares that the sulfur should be (very) pure. For a discussion of "mercurized," see 311v: 4.

315r: 35–37 **let the sole of the sonn vulger . . . Therfore, let yt ascend into heaven**: The sentence is anacoluthic: after the long relative clause the sentence begins again, although "yt" refers to the *soul of the sun*. The same anacoluthon is found in three of the English versions of the *Medulla*. Alternatively, though less likely, "let" should be interpreted as a main, transitive verb in the meaning 'allow the escape of': see *OED* s.v. *let* 7.a.

315r: 35–37 **let the sole of the sonn vulger . . . ioyned with the body**: In the Latin *Medulla* text, the relative pronoun in the phrase *without the which* refers to the *ointment*, not to the sun. It is possible that at some point the English text had *without the which soul*, which was subsequently misinterpreted as *without the which sol* 'sun.' See also 305r: 40.

315r: 37–38 **let yt ascend into heaven . . . from heaven into earth**: *Heaven* is often used in alchemical texts to refer to the top of the vessel, or alembic, and the bottom is similarly known as the earth: see Abraham, *A Dictionary*, s.v. *heaven*.

315r: 40–41 **a peace of whyte wax**: The reading "wh{i}te pece of wax," which is found in all other manuscripts of the Treatise (except G), is supported by the *Medulla*.

315r: 41–42 **norished with milke & meate**: After the philosophers' stone had been created, it was often seen as a child that needed to be fed with white and red mercurial waters to enhance its virtue and strength: see Abraham, *A Dictionary*, s.vv. *virgin's milk*, *green lion*, *philosophical child*. See also 317v: 27, 321r: 7–8.

315v: 1–5 **This medison fermented . . . adminstred vnto yt**: This sentence is anacoluthic. The *Medulla* has no *and* in the clause corresponding to 315v: 3 "and soe may it": Ripley, *Opera omnia*, 162.

Explanatory Notes 295

315v: 2–4 **putrified into sulfur, sep*a*ratinge . . . into a purple colloure**: The Treatise does not seem to render the *Medulla* argument in full, while it adds the phrase *putrified into sulfur*, which does not appear in the *Medulla*. Cf. *Medulla* (Ripley, *Opera omnia*, 162): "ratione separationis primae formae a meria [MS. Ashmole 1490, fol. 72r: a materia], ad eum finem, ut fiat magis formalis quam antea erat, quando primae qualitates remanebant non divisae, potest reduci ad colorem purpureum" 'by reason of the separation of the first form from the meria (?) [MS. Ashmole 1490: the matter] so that it should be more formal [see *OED* s.v. *formal* 1 (?)] than it was earlier when its first qualities remained unseparated, it can be turned into a purple color.'

315v: 5–6 **And thus in the minerall way . . . made the true elixar**: The meaning of the Treatise formulation is unclear. The source of the Treatise probably contained a reading close to MS. Ashmole 1490 (fol. 72r): "Sicq*u*e in op*e*re minerali ad album et ad rubeu*m* verissimu*m* elixer fiet" 'And so in the mineral opus for white and for red will the truest elixir be made.' See also W. Salmon, ed., *Medicina Practica, or Practical Physick* (London, 1692), 673.

315v: 7–9 **this elixar white or red . . . into a thine oylle**: Cf. *Medulla* (MS. Ashmole 1490, fol. 72r): "Sed hoc elixar album sive rubeu*m* in virt*u*te centu*m*platur et bonitate si cu*m* igne naturae confixata sibi sua quinta essentia in oleum tenue deducatur" 'But this elixir, white or red, is multiplied a hundred times in virtue and goodness if it is fixed with its quintessence with the help of the fire of nature and reduced to a thin oil.' An earlier Treatise manuscript may have had the reading *if his quintessence be fixed vnto it*. *If* was subsequently misinterpreted as *of*, which led to a restructuring of the syntax. See Salmon, *Medicina*, 674.

315v: 11–12 **w*h*ich turneth all other bodies into Sol or Lune according to the fermente**: This statement is not found in the *Medulla*. Cf. 314r: 7–8.

315v: 16–30 **This laste p*a*rte of the minerall stone . . . because of their diuersitie in workinge**: This chapter probably refers to a section in the *Medulla* that deals with a process that employs mercury only: see Ripley, *Opera omnia*, 157–60; Salmon, *Medicina*, 669–71. However, as made clear at the beginning of the chapter, this process is not discussed in detail "because of the tediousnes of his workinge."

315v: 19 **made meltynge**: For this construction, see note on 295r: 10.

316r: 1–2 **The preparation of the white oille for the animalle stone**: For this heading, see p. 97.

316r: 8–12 **For Moryen saith . . . the ashes that Morien speaketh of**: I have not been able to locate a source for this statement. It is not found in *Liber de Compositione Alchemiae quem edidit Morienus Romanus, Calid Regi Aegyptiorum*; Manget, *Bibliotheca*, 1: 509–19. Trinity College, Cambridge, MS. R. 14. 37 (fol.

4r) contains almost the same formulation as the Treatise: "and þanne it is callid asshen; wherof Morien*us* seyth that þe asshen þat arn in þe botme of þe vessell sette hem nat at naght which þat been in þe lowere place, and þey been þe dyademe of þe regne and þe asshen þat abydyn in þe botme & þe aqua vite vii tymes sublymid in his nat*ure* as it were eyr." It is unlikely, however, that this manuscript is the source of the Treatise; it is instead more likely that they share a common source.

316r: 17 **Lunarie**: See note on 314r: 24.

316r: 21 **gren lyon**: See note on 304r: 26–27.

316r: 31 **slainte**: i.e., *slain*. See note on 303r: 14.

316v: 5 *(Marginal note)* **This is the wife | set her not . . . a child without a | man**: In S1S2S3S4, this note is incorporated into the text and appears in a statement attributed to Morien: see note on 316v: 13–16. The F manuscript is more closely related to A in that it contains a marginal note, but the note appears after the line corresponding to 316v: 10 in A. A comparison with the source of the passage in 316v: 13–16 suggests that the material appearing in the note in A does not belong to the quotation from Morien. S1S2S3S4 thus seem to have embedded what was originally a marginal note into the text proper. See also note on 311v: 29–30.

316v: 13–16 **And of this coniunction speketh Morien . . . not don right in this worke**: This statement seems to derive ultimately from *Liber de Compositione Alchemiae quem edidit Morienus Romanus, Calid Regi Aegyptiorum*. Cf. Manget, *Bibliotheca*, 1: 516: "Hoc etiam te scire convenit, quod si corpus immundum perfecte non mundaveris, & illud non desiccaveris, neque ipsum bene dealbatum reddideris, & in illud animam non miseris, & ejus omnem faetorem non abstuleris, donec post suam mundificationem tinctura in illud decidat, nihil penitus hujus magisterii direxisti" 'It is necessary that you know that if you do not perfectly purify the unclean body, and if you do not dry it, and if you do not make it perfectly white, and if you do not infuse it with a soul, and if you do not remove all its feces from it, until tincture falls into it after the cleansing, you have not carried out anything successfully at all in this mastery.' See also the note on the marginal comment in 316v: 5.

316v: 15 *(Marginal note)* **The black collour . . . they be together**: This note appears in all the other manuscripts of the Treatise (except G). However, instead of A's reading "in the on as well as in the other," the other manuscripts read "in one as well as in both together" (S2).

316v: 27 **caste heaven vpon earth**: Cf. note on 315r: 37–38.

316v: 27 *(Marginal note)* **This teacheth the | imbibition . . . for to doe it my | selfe**: This note (the "Russia note") appears in all the manuscripts (except G). For the importance of this note for the dating of the Treatise and the identification of the author, see pp. 2 and 9.

Explanatory Notes 297

316v: 39 **the darknes of his ecclips**: The *eclipse* is often used to signify the putrefaction of the raw material when it is dissolved into its constituent parts: see Abraham, *A Dictionary*, s.vv. *eclipse, nigredo*.

316v: 43–45 **wherof Moryen speaketh, saying . . . ferment to be prepared**: I have not been able to identify the source of this statement. It is not found in *Liber de Compositione Alchemiae quem edidit Morienus Romanus, Calid Regi Aegyptiorum*: see Manget, *Bibliotheca*, 1: 509–19.

317r: 14 **the chapters of fermentation**: i.e., Chapters 12 and 13.

317r: 14 **year after**: This reading is uncertain. I have interpreted the reading as "year after" (written together in the manuscript) in the sense of 'years after': see *OED* s.v. *year* 1a.β. It is possible that this slightly awkward reading is a result of a misinterpretation of 'h' as 'y,' two letters which are very similar in many sixteenth-century secretary hands: Samuel A. Tannenbaum, *The Handwriting of the Renaissance* (New York: Columbia University Press, 1930), 47, 86. Alternatively, 'y' should perhaps be interpreted the same way as 'y' in y^e (i.e., the old thorn), thus giving the reading *thereafter*. However, this is less likely since 'y' in this role is only found in the fixed y^e and y^t and once as y^m in the manuscript, and *thereafter* is also awkward in the context.

317r: 17 **George Reply in English miter called his XII Gates**: This is a reference to George Ripley's (1415?–1490?) treatise in verse entitled *The Twelve Gates* or more commonly *The Compound of Alchymy*: see Linden, *George Ripley's Compound*. For a discussion, see p. 59.

317r: 21 **hath ordeyned to the vnderstanding therof**: i.e., 'has selected to receive the understanding of it'; cf. *OED* s.v. *ordain* II.

317r: 26 **Of ☽ and ♀**: For this heading, see p. 97.

317r: 27–28 **Forasmoch as this 22 chapter . . . them into too partes**: This statement is an indication that Chapters 21 and 22 perhaps constituted a unit in a source text.

317r: 29–32 **with another body . . . whit and yellowe**: i.e., copper. Cf. 317r: 43, 318v: 2–3, and 324v: 26ll.

317v: 18 **Raymon saith**: I have not been able to identify a source for this statement.

317v: 27 **the milke of his mother**: See note on 315r: 41–42.

317v: 32–33 **and the yonge man wayned . . . com to perfecte age**: Cf. 310v: 37–41.

318r: 1–2 **Of the dissolution of ☉ . . . & red oille**: For this heading, see p. 97.

318r: 13 **sprite of of the water**: The double *of* is presumably due to dittography.

318r: 19 **showe with a yellowe collour**: I have not found the construction *show with* in the *OED* or the *MED*. Perhaps it is a misreading early on in the transmission of the Treatise for *shine with*. See also 323r: 42.

318r: 27 **the blod of the lyon**: See note on 304r: 26–27.

318r: 30 **slainte**: i.e., *slain*. See note on 303r: 14.

318r: 43 **infnitly**: The writing in A is very cramped at this point. The spelling "infnitly" may be the result of minim confusion, or it may reflect syllabic /n/. See note on 294v: 8.

318v: 2–3 **the grene bodie of feminine kind**: i.e., copper. See also note on 317r: 29–32.

318v: 10 **Lunarie**: See note on 314r: 24.

318v: 22 **Lunarie**: See note on 314r: 24.

318v: 25 **the nexte chapter before**: i.e., Chapter 23.

318v: 29 **Vince et viue**: i.e., Latin 'be victorious and live.'

318v: 33–34 **includ them in on chapter into too p*ar*tes**: The verb *include* probably carries the meaning of 'divide' as well, hence triggering the use of *into*; cf. *OED* s.v. *into* 1b. and 7. Alternatively, *into* may be the result of anticipating the subsequent *two* ("too" in A).

318v: 43 **appropriat**: "appropriat" probably has the meaning 'got hold of,' 'extracted,' when it is taken with "out of the said animall stone." This particular meaning is not recorded in the *OED*, which does give a similar meaning: 'to make the private property of any one': see *OED* s.v. *appropriate* v.

318v: 48 **loware astronomy**: i.e., *alchemy*. Alchemy, being the study of the metals in the earth, was often perceived as an earthly counterpart of heavenly astronomy, which was the study of the celestial bodies: see Abraham, *A Dictionary*, s.v. *astronomy [earthly]*.

319r: 9–10 **wherfore Moryen saithe . . . co*n*serninge wordly thinges**: I have not been able to identify the source of this statement. It is not found in *Liber de Compositione Alchemiae quem edidit Morienus Romanus, Calid Regi Aegyptiorum*: see Manget, *Bibliotheca*, 1: 509–19.

319r: 14–17 **Noye . . . Moyses . . . Sallomo*n* . . . Machabeus**: The biblical figures Moses, Noah, and Solomon are well attested as alleged alchemical authorities: see Raphael Patai, *The Jewish Alchemists: A History and Source Book* (Princeton: Princeton University Press, 1994), 18–40. However, I have not been able to find any other mention of Judas Maccabeus possessing the secret of the philosophers' stone. For the biblical passages alluded to in this section, see Genesis 6: 11–22, Exodus 36: 8–38, 1 Kings 6, and 1 Maccabees 4: 36–61.

319r: 22–23 **a holly woman, the syster of Moyses**: This is probably a reference to Mary the Jewess (or Maria Hebraea), allegedly an alchemist of the third century AD. Early on in the history of alchemy she was equated with Moses' sister Miriam. For a discussion, see Patai, *The Jewish Alchemists*, 60–80.

319r: 27–28 **in penultima etate mundi**: i.e., 'in the penultimate age of the world.' Since the age referred to is supposed to follow the "laste age" (319r: 25), the designation of the age as the *penultimate age* is odd. The 'n' in "penultima" may be a misreading at some point for 'r,' suggesting that the word was originally *perultima* 'the very last.' The reading in S1S2S3S4, *perultimus*, lends support to this assumption, although the Latin construction in these manuscripts is ungrammatical in the context: "in perultimo etas mundi."

319r: 31 **the leess wordle**: See note on 294r: 14.

319r: 35 **aptation**: This word is not found in the *OED* or *MED*, but it is recorded in some Latin dictionaries: see e.g. *DML* s.v. *aptatio*. The meaning of the word in the Treatise seems to be 'constitution,' 'build,' which is consistent with the Latin meaning recorded. See also 319r: 41.

319r: 35–37 **ther is nothing . . . may be compared thervnto**: This is the only instance of a zero relative pronoun in subject position in the Treatise. According to Rissanen, the construction is common in sixteenth-century texts: Rissanen, "Syntax," 298.

319r: 43–319v: 5 **it is of Morien the ph*ilos*ofer . . . vz., with the ferment of ☉ and ☽**: The passage probably stems from the alchemical tract *Notabiliora philosophorum*. See p. 51.

319v: 14 **vz., 40 daies to his blacknes**: In all the other manuscripts (except G where it does not occur at all), this statement appears as the first phrase in the marginal note that occurs in A at 319v: 14. See the following note.

319v: 14 *(Marginal note)* **40 dais in his blacknis . . . siccasion or | dryinge**: This marginal note is found in all the other manuscripts of the Treatise (except G). In FS1S2S3S4, however, the note begins with the statement that is incorporated into the text in A (319v: 14), i.e., "vz., 40 daies to his blacknes." Furthermore, all the other manuscripts prescribe forty days for the going out of the blackness, not three hundred and forty as in A. More or less the same prescription about time is found in one of the alchemical poems, "The standing of the glass," that follows the Treatise in three of the manuscripts: "The glasse with the medison must stand in the fyer | 40 dais till it be black in sighte | 40 daies in the blacknes to stand he will desier | and then 40 daies more vntill it be white | and 30 dais in his dryinge, yf thou liste to doo righte" (A fol. 325v). See pp. 15–17 and 102.

319v: 24 **tella foliata**: "tella" is probably a misreading for *terra*, which is found in S1S2S3S4. The reading seems to have existed in a common ancestor of AG and probably F, which has a related reading, "stella" ('star'). The phrase *terra foliata* 'the foliated earth' appears in many alchemical texts signifying "the sublimated earth, the purified body of the Stone which awaits re-animation by the return of its previously separated soul": Abraham, *A Dictionary*, s.v. *white foliated earth*.

320v: 3–322v: 28 **Nowe of the vegitall stone . . . the chapter followinge &c. Finis**: For a discussion of the structure of this chapter and its relationship to Ripley's *Medulla*, see pp. 49–50.

320v: 7–8 **Raymon in his Op*er*tory called black, blacker then blacke**: This is probably a reference to Raymond Lull's *Apertorium*: see Pereira, *The Alchemical Corpus*, 64. The *Medulla*, which is the direct source of the Treatise, does not state from which work by Raymond the phrase is taken. Before this statement, the *Medulla* includes a long passage that is absent in the Treatise. It discusses how some alchemists have tried to produce the fire of nature with the help of ordinary wine, which is an erroneous procedure, according to the *Medulla*.

320v: 8–9 **the menstrue resoluble engendred of the body dissolued**: This statement is not found in the *Medulla*. It may derive from the passage in 320v: 16–18.

320v: 10 **or the water therof**: The S1S2S3S4 reading, "tarter" 'wine stone' for *water*, is supported by the *Medulla* (Ripley, *Opera omnia*, 165).

320v: 12–13 **of mettaline kinde only and is a simple substance**: The Treatise seems to connect two unrelated statements in the *Medulla*. In the *Medulla* (Ripley, *Opera omnia*, 165), the phrase "is a simple substance" describes vegetable mercury, whereas the Treatise omits an intermediate passage and connects it with the metallic water.

320v: 14–15 **by vertue wherof they doe multiply their similitud*es***: The Treatise appears to have omitted the second half of the *Medulla*'s argument. MS. Rawlinson B. 306 (fol. 63r) seems to give the clearest reading of the *Medulla* copies: "by vertue wherof they may multiplie their simyllytuds in these m*er*curies, w*hi*ch are w*i*thout [i.e., such mercuries?] and be medysonable to mans bodie to expell from them many deseases and to restore to formor youth and to kepe the bodie in health vntell the tyme apoynted of God." *They* seems to refer to *parts* (see 320v: 13), but it is unclear what is actually meant by the phrase *their similitudes* (even in the *Medulla*).

320v: 18–321r: 9 **For when we haue attayned . . . the fyer and overcom it**: The Treatise replaces the *Medulla* discussion with a long passage that is not found in the *Medulla* copies consulted. I have not been able to identify a source text for this passage. For a similar passage, see 304v: 43–305r: 14.

320v: 22 **the ravens hed**: See p. 82.

320v: 23 **which say**: The reading "we," which is found in GS1S2S4, is more syntactically suitable in the context: the A (and F) reading, *which*, leaves the sentence without a main clause.

321r: 8 **milke and meate**: See note on 315r: 41–42.

321r: 26 *(Marginal note)* **M^r sb aqua me**r: This elliptical note is found in all the manuscripts (except G). I have kept the original abbreviations in my transcription since it is not absolutely clear what they stand for. Possibly, a Latin phrase such as "mercurius sublimatus aqua mercurii" 'sublimated mercury [is?] water of mercury' is intended.

321r: 27 **Kotolory**: This is probably a variant spelling or misinterpretation of *Catalonia*, found in the other manuscripts of the Treatise and supported by the *Medulla*.

321r: 31–36 **and is a puer viniger as Antazagurras said . . . made open vnto me**: This passage derives from the alchemical tract *Notabiliora philosophorum*. See p. 51.

321r: 37–321v: 2 **take puer argall . . . sufficient for your worke**: This passage is very close to the beginning of Chapter 28 of the Treatise. Most likely, this passage is based either on the source text of Chapter 28, *Thesaurum philosophorum*, or on Chapter 28 itself. Cf. 324r: 3–16.

321v: 3–5 **For Avicen saith . . . accomplishment of the elixar**: This statement ultimately derives from (Pseudo-?)Avicenna's *Declaratio lapidis physici Avicennae filio suo Aboali*: see Zetzner, *Theatrum chemicum*, 4: 876. A longer portion of this text is quoted in 325r: 13–15. The quote here may have been taken over from the source of 325r: 13–15, i.e., the *Thesaurus philosophorum*. Cf. note on 321r: 37–321v: 2.

321v: 7 **wilbe lacke**: The phrase most probably represents a misreading of *will be black*, as found in all the other manuscripts of the Treatise. However, the phrase in A, though syntactically awkward, may have been influenced by the meaning of *lack*, implying the lack of a specific color.

321v: 27–28 **wherin dwelleth our beaste againste nature**: This statement is found in the *Medulla* in its treatment of the mineral stone: see e.g., MS. Ashmole 1490, fol. 69v. It appears in the section of *Medulla* that has been cut out of the discussion in Chapter 17 of the Treatise. See note on 313v: 19–25.

321v: 32–33 **in the chapter of dissolution**: i.e., Chapter 10.

321v: 42 **within too howars**: Cf. *Medulla* (Ripley, *Opera omnia*, 171): "infra horam" 'within the hour.'

322r: 2–3 **it shall not cease to worke, without outward heate added vnto yt**: This is probably a direct translation of the Latin "non cessabit operari sine igne extrinseco illi administrato" 'it will not cease working although no external fire/heat has been added': Ripley, *Opera omnia*, 171. For this kind of absolute participial clause in English, see Rissanen, "Syntax," 322. Cf. 322r: 1–2.

322r: 4–322v: 18 **then may scarcly suffice . . . of ♀ dissolued by himselfe**: The passage based on *Medulla*'s chapter on the vegetable stone ends in 322r: 4, but is resumed in 322v: 18. At this point, the *Medulla* (MS. Ashmole 1490, fol. 73r) declares "et deinceps ordine sicut in op*ere* aquae compositae vsque ad eius totale co*m*pleme*ntum* p*er* o*mn*ia procedatur" 'and after that proceed in everything in due order as in the work on the composite water until its complete accomplishment.' The Treatise omits this instruction but inserts the procedure from the chapter on the mineral stone here. Some parts of the procedure have been related before (in Chapter 17, 313v: 33–314r: 13), but they are treated more fully in this chapter. For a discussion of this passage, see pp. 49–50.

322r: 15–16 **soe moch the more it shalbe thine and subtill**: For a passage from the *Medulla* that is absent after this statement, see 313v: 46–314r: 5.

322r: 17–18 **in fier of the first degre being moiste**: The *Medulla* text in MS. Ashmole 1490 (fol. 70r) indicates that the phrase *being moist* may refer to *fire*: "in igne prime [sic] gradus humido nigrescit" 'in moist fire of the first degree it turns black.' Cf. Martinus Rulandus, *Lexicon alchemiae sive dictionarium alchemisticum* (Frankfurt, 1612), s.v. *ignis*; see also note on 300v: 38. However, it may also refer to the following "it" (i.e., the oil). Cf. 322r: 20–21.

322r: 22 **the corruption of the first form is the genera*t*ion of the second**: Cf. *Medulla* (Ripley, *Opera omnia*, 147): "corruptio unius est generatio alterius" 'corruption of the one is generation of the other.' Cf. 298v: 33–34, 308v: 26–28.

322r: 23 **a bodie naturall**: All the versions of the *Medulla* consulted state that the body created is a *neutral* body: see Ripley, *Opera omnia*, 147.

322r: 26–27 **should hould of the nature of quintaessentiae**: "should hould of" is probably a direct translation of "teneat de" found in the source: see Ripley, *Opera omnia*, 148. Salmon, *Medicina*, 662, translates "must partake of," which may be the sense intended. I have not been able to find the construction *tenere de* in Latin dictionaries.

322r: 31 **say**: The reading *then know*, found in S2S3S4 (and to some extent in FS1) is supported by the *Medulla* (Ripley, *Opera omnia*, 148): "tum cognosces" 'then you will know.'

322r: 32 **lokinge**: The A reading is supported by three of the *Medulla* copies, while *looked*, which is found in all the other manuscripts of the Treatise, is supported by the remaining eight copies of the *Medulla* consulted.

322r: 32–39 **beinge first discended . . . of the beste p*er*fection**: It is not wholly clear what the referent of *being first descended* is. The problem may have come into the Treatise from its source, since all the *Medulla* manuscripts are similarly ambiguous. An exception is MS. Ashmole 1507 (fol. 114r), which supplies "the same" within square brackets, indicating that it is an addition. See p. 50. Ripley, *Opera omnia*, 148, and Salmon, *Medicina*, 663, give *the sun* as the referent, which seems possible in the context.

322r: 33 **laterall**: The word "laterall" (or "lattal" in all the other manuscripts of the Treatise) may be a misinterpretation of *natural*. Most of the copies of the *Medulla* use *natural*, except for two copies that also use *lateral*: MS. Ashmole 1479 and MS. Rawlinson B. 306. Consequently, this reading may already have existed in the source of the Treatise. For the reading in S1, where a correction to *natural* appears, see pp. 120–21.

322r: 34 **parte**: The reading *part* in A probably represents a misinterpretation of the word *pale* at some point in the textual transmission, as found in the *Medulla* (MS. Ashmole 1490, fol. 70r): "in Occidentem pallidam" 'to the pale west.' The reading *pale* is found in GS1S2S3S4 either in the spelling *pale* or *pail*, but it is uncertain whether the word intended in the manuscripts is *pale* 'pallid,' 'faint,' or *pale* 'enclosed area,' 'enclosure.' The *OED* records examples with both spellings for both words; *OED* s.v. *pale* n. F reads "peele," probably from *peel* '[palisaded] enclosure': see *OED* s.v. *peel* n²; *MED* s.v. *pel* n(1). There thus seems to have existed some variation in the tradition as to whether the reading should be interpreted as an adjective or as a noun.

322r: 35 **alterate**: The reading "alterable" 'capable of being altered or changed,' which is found in all the other copies of the Treatise, is supported by the *Medulla*.

322r: 40 **the whell of phi*losofie***: See 304r: 17.

322r: 40–41 **the oyll is simplified of the body altered**: i.e., 'the oil of the altered body is simplified': see Ripley, *Opera omnia*, 149.

322r: 42–43 **what other thinge is ther then on w*hi*ch nature hath made to be engendred**: i.e., 'what other thing is there than one which nature has caused to be created.' The construction "hath made to be engendred" is a literal rendering of the Latin construction *fecit generari*. For this use of *make* in causative constructions, see Visser, *An Historical Syntax of the English Language* 3. 2, 2261 §2068.

322v: 8–9 **a naturall vnctuosity of the fermentall body**: The longer reading "a naturall vnctuositie for the sowle together with the ounctuosity of the fermentall body," found in all the other manuscripts of the Treatise, is supported by the *Medulla*. The omission in A is probably owing to an eyeskip (*unctuosity . . . unctuosity*).

322v: 10–11 **who yet beinge absente may in noe wise be formed**: The argument of the *Medulla* text is different. Cf. Ripley (*Opera omnia*, 150): "qui ea absente nullo modo formari potest" 'which [the stone] if it [the soul or unctuosity] is absent cannot in any way be formed.' The phrase 'yet being absent' seems to be a direct translation of the Latin ablative absolute 'ea absente'. See note on 322r: 2–3. Yet may be a misinterpretation at some point in the textual transmission of *yt* 'it'; for the spelling *yt* for *yet*, see *OED* s.v. *yet*. If this is the case, the misinterpretation must have existed early on in the textual tradition, since all witnesses contain the reading *yet*. It may even have existed in the source: MS. Ashmole 1479 and MS. Rawlinson B. 306 contain the same reading as the Treatise.

322v: 18–25 **And when the elixar shalbe . . . whell of phi*losofie***: The discussion from the *Medulla*'s chapter on the vegetable stone resumes in this passage after the interpolation of the discussion of the mineral stone. See note on 322r: 4–322v: 18.

322v: 19–20 **being first rectifyed into thine oylle**: This phrase refers to the menstrue rather than the elixir: see Ripley, *Opera omnia*, 171–72; Salmon, *Medicina*, 681.

322v: 24–25 **the whell of phi*losofie***: See 304r: 17.

322v: 25–28 **that ther is nothinge in the wordle . . . in the chapter followinge &c Finis**: This passage is not found in the *Medulla*. See 319r: 2–9.

323r: 1 **Of the animall stone**: For this heading, see p. 97.

323r: 4–8 **We saie rich and poore . . . co*m*pacte together in element*es***: This passage is not found in the *Medulla*. Cf., however, 323r: 15–16.

323r: 11–13 **wherfore is is said that . . . that God hath gyuen him**: This statement is not found in the *Medulla*. For similar statements about alchemy as a gift from God, see note on 299v: 8. The A reading "is is" is probably a misreading of "it is" as found in, e.g. S2.

323r: 14–15 **the mother and root*es***: The FS1S2 reading, "the rootes and mothers," is supported by the *Medulla*.

323r: 15–18 **Yet the element*es* of the moste eccellent creatur . . . of the sprit*es* and bodies of mettalls**: The Treatise differs to some extent from the *Medulla* versions consulted. Cf. Ripley (*Opera omnia*, 174): "Veruntamen elementa supradictorum non intrant in opera Elixirium, nisi per virtutem & commixtione*m* cum elementis spirituum ac corporum metallicorum" 'But the elements of the things mentioned above do not enter into the works of the elixirs, except by the help (?) of the spirits and metallic bodies and the mixture with their elements.' Instead of attributing the elements to "the moste eccellent creatur in whom they ar soe evenly co*m*pounded," the *Medulla* discusses the elements of the abovementioned substances, presumably wine, eggs, hair, and blood. See 323r: 9–10. Perhaps this formulation was imported from 319r: 40ll., which similarly discusses the microcosm. See also note on 323r: 19, 323r: 27–28.

Explanatory Notes 305

323r: 19 **all those that ar animalls**: The *Medulla* discusses "haec, quae sunt naturalia" 'those that are natural': Ripley, *Opera omnia*, 174. The Treatise reading may represent a misreading of an original *naturals* at some point in the transmission. Alternatively, the reading *creature* in 323r: 15 and *animals* here may be interconnected. At some point, *creature* may have been interpreted in the sense of 'something living,' 'a creature' rather than in the general sense of 'something created,' which seems to be intended in the context. See 323r: 27–28, 319r: 40. As a result, the substance discussed was seen as the best of animals (or creatures). If *creature* should indeed be interpreted in the sense of 'a living creature,' *animals* may represent a further elaboration on this theme. The motivation behind discussing the substance as a living creature and situating it among other living creatures remains unclear. Cf. Reidy, *Thomas Norton's* Ordinal, 55 ll. 1715–18.

323r: 20 **which phi*loso*fers haue taken for the nearest matter of our stone**: Cf. *Medulla* (Ripley, *Opera omnia*, 174): "quod Philosophi acceperunt tanquam propinquissimum mundo minori seu microcosmo" 'which the philosophers have taken as the closest thing to the less world or the microcosm.'

323r: 22 **the lesse wordle**: See the note on 294r: 13–14.

323r: 22–24 **wherin ar the figures, courses ... sprit*es*, and naturall vertues**: The same reading as that of the Treatise is found in the *Medulla* copy MS. Ashmole 1480 (Part 5, fol. 15r), except for the fact that MS. Ashmole 1480 reads "complexions and humors" instead of "complection of humors." The other *Medulla* copies display a great deal of variation in formulation, but the content is similar.

323r: 24–26 **the w*hich* bodie after Aristotle ... vntill they repugn not**: The Treatise leaves out a part in the *Medulla* in which this part of the work is likened to a noble bird. The subsequent passage, which is attributed to Aristotle, appears in the *Medulla* as well, though the *Medulla* adds, after *divided into four parts*, "quia unaquaeq*ue* pars, ut ille inquit, habet unam naturam particularem" 'since every part, as he says, has its own particular nature': see Ripley, *Opera omnia*, 175. The passage derives ultimately from pseudo-Aristotle's *Secreta secretorum*: see Steele, *Three Prose Versions*, 88; M. A. Manzalaoui, ed., Secretum Secretorum: *Nine English Versions*, EETS OS 276 (Oxford: Oxford University Press, 1977), 174.

Compassed is probably a misreading early on in the transmission for *composed*: cf. Ripley, *Opera omnia*, 175, "sunt iterum conjungendae" 'should be joined again.' However, this may already have existed in the source, as *compassed* occurs in two copies of the *Medulla*: MS. Ashmole 1479 and MS. Rawlinson B. 306.

323r: 27 **vz., sulfur and argent viue, alias water & fier**: The only copy of the *Medulla* that contains a formulation similar to that of the Treatise is MS. Ashmole 1507 (fol. 119r), though the terms are reversed: "vizt water and fire, [al*ias* sulphure and ☿ry]." The square brackets are found in the manuscript, and seem to indicate that material has been added that was not part of the original text. See notes on 323r: 27–28, 323r: 29–31; see also p. 50. Most of the other manuscripts simply give *fire* and *water*.

323r: 27–28 wherby is made a naturall bodie of an excellent form: The only copy of the *Medulla* that contains this reading is MS. Ashmole 1507 (fol. 119r), where the reading appears within brackets. See note on 323r: 27.

323r: 29–31 but must be brought to passe . . . secrete of this worke: Only one of the copies of the *Medulla* contains this passage, MS. Ashmole 1507 (fol. 119r). However, it adds the passage within brackets. See note on 323r: 27.

323r: 31–39 Yet to ye experte workman . . . as is said before; w*hich* **oylls**: The Treatise appears to move away from the *Medulla* here, but picks up the *Medulla* discussion again at 323r: 39 after *which oils*. Instead, the Treatise here refers to another chapter in the Treatise, most likely Chapter 25: see 319v: 8–14. When the *Medulla* discussion resumes, the Treatise appears to take the oils that were prepared in Chapter 25 as the base substances for the procedures beginning in 323r: 39. The *Medulla*, by contrast, continues to discuss the substance that it has described earlier in the procedure, i.e., the less world.

323r: 42 showe: The S1S2S3S4 reading, *shine*, is found in ten of the eleven copies of the *Medulla*, but MS. Ashmole 1487 reads *show*.

323v: 3–4 and as the verie sulfure of nature, p*rocreat* **and made of arg***ent* **viue**: The only copy of the *Medulla* that is close to the Treatise is MS. Ashmole 1507 (fol. 119v), although it also adds "some quantity of Lune (in manner of amalga*m*) being put to it," which is also found in the other copies of the *Medulla*. The reading *is*, which is found in FS1S2S3S4, instead of *as* in AG is supported by the *Medulla*.

323v: 4–15 Then let the said sulfure . . . like a burninge cole: This passage is not found in the *Medulla*. It may refer to Chapter 25. See 319v: 36–320r: 10.

323v: 7 said for in the chapter of the animall stone: i.e., Chapter 25; cf. 319v: 37ll. The reading "for" is most likely a misreading of "afore" and/or reflects a misconception of the structure of the clause. It is less likely that "for" stands for *fore* 'before,' 'earlier,' since this usage seems to have been rare after the fourteenth century: see *OED* s.v. *fore* and *afore*.

323v: 16 Then let the second sulphur named the white and red sulfur: The phrase "the second sulphur," which is the reading in all the copies of the Treatise, is probably a misconception of a phrase such as *the two sulfurs*. Digits, such as *2*, are often found in early Modern English manuscripts representing both the cardinal *two* and the ordinal *second*; cf. e.g. 296r: 2. Cf. *Medulla* (Ripley, *Opera omnia*, 176): "haec duo sulphura, videlicet, album et rubeum" 'these two sulfurs, that is, the white and the red.' See also 323v: 5–6.

323v: 23–34 **And here in this place ... in thes fewe chapters folowinge &c**: This introduction is not found in the *Medulla*.

324r: 2 *(Recipe in Napier's hand)*: A recipe in Richard Napier's hand appears between the heading and the text of Chapter 28. It is unclear at several points, mainly because of its overlap with the heading of Chapter 28. Versions of the same recipe appear in Napier's hand in fol. 352v in A. See pp. 97–98.

324r: 3–325r: 35 **Take therfore the argull ... mineralle or naturall &c. Finis**: Chapters 28 through 33 are based on the alchemical tract *Thesaurus philosophorum*. See pp. 52–54.

324r: 3 **Take therfore the argull of wyne, grind it to smalle duste**: The description in *Thesaurus philosophorum* is longer and more detailed. Cf. MS. Ashmole 1459 (p. 187): "R̊ argoll of most strong wine, wh*i*ch cleaveth to the vessells side, and after it is dryed at the sunne, bake it in an hott oven as you doe great loaves of bread, and then it will be tart on the tongue, and of gray collour; powder it smale."

324r: 4–5 **and vnto on pound ... the pouder of argull**: The longer reading in the other manuscripts of the Treatise, "powre vpon it aqua vite; vnto one pownde of powder put a pinte of aquavite" (S2), is closer to the reading in *Thesaurus philosophorum*.

324r: 5–6 **Then stop the vesselle ... stand soe 24 howars**: *Thesaurus philosophorum* (MS. Ashmole 1459, p. 187) adds that this must be done in a "cold moyste place."

324r: 10 **loweste partes**: Cf. *Thesaurus philosophorum* (MS. Ashmole 1459, p. 188): "in the entrayles of the earth."

324r: 10 **pace**: The meaning of the A reading, which is also found in F, is unclear. Perhaps it is an alternative spelling of *pass* in the sense of 'bring to its end,' 'accomplish': see *OED* s.v. *pass* 21. Alternatively, it may be a misreading of "place," found in the other manuscripts, which is closer to *Thesaurus philosophorum*'s *dispose*.

324r: 11 **whotnes**: Similar spellings with 'wh' for historical 'h' before 'o' are given by Wyld, *A History*, 307. Eric J. Dobson argues that such spellings are evidence for the development of a /w/ glide in this and similar contexts: *English Pronunciation 1500–1700*, 2 vols. (Oxford: Clarendon Press, 1968), 2: 997. This is the only spelling of this kind in the Treatise. Cf. note on 301v: 40.

324r: 11 **daye**: Cf. *Thesaurus philosophorum* (MS. Ashmole 1459, p. 188): "till they agree together." The reading *day* is found only in A. It is possible that the reading "agree" was misread at some point as *ayre* (found in the other manuscripts of the Treatise): 'y' and 'g' are very similar in some handwritings of the sixteenth century; cf. Petti, *English Literary Hands*, 31–32.

324r: 15 **and doe this 15 times**: *Thesaurus philosophorum* presents a longer passage that outlines the process in more detail, but it does not state how many times the process should be carried out. Instead, the Treatise is closer to e.g., Ripley, *Opera* *omnia*, and MS. Ashmole 1407, which also assert that the process should be carried out fifteen times, adding "and yf you will a 100 tymes" (MS. Ashmole 1407, Part 4, p. 24).

324r: 16 **then thou haste the fyer of nature**: Cf. *Thesaurus philosophorum* (MS. Ashmole 1459, p. 188): "For now kindleth our fire, that is called naturall fire, influent or flowing from sunne or {and} starrs."

324r: 17 **from fyer, vz., water from fyery earth**: This statement is not found in *Thesaurus philosophorum*.

324r: 20 **our eagle, our ☿, our sal armoniak**: *Thesaurus philosophorum* adds a number of other terms that are not included in the Treatise. Cf. MS. Ashmole 1459 (p. 188): "our eagle, our mercurie, our aqua vitae gotten out of earth and water, w*h*ich is our sal arm*on*iac alchymised, and sal anabrone, sep*a*rate allready, filtred and distilled, calcined & dissolued, our heaven, our quintessence."

324r: 23–24 **fewe doe find this bird ... Kepe it well &c**: Cf. *Thesaurus philosophorum* (MS. Ashmole 1459, p. 188): "Wherefore few men can catch this heavenly byrd, and therfore they wander as fooles, and find noe fruite." For the symbol of the bird, see note on 310v: 37–41.

324r: 26–29 **Nowe, farthermore, let vs ... great elixar of life &c. Finis**: In *Thesaurus philosophorum*, the material corresponding to Chapter 29 of the Treatise is the heading for chapter two. For a discussion of this feature, see p. 53. Although the heading in *Thesaurus Philosophorum* is very similar in wording to the Treatise chapter, the content differs substantially. *Thesaurus philosophorum* stresses that the following chapter is about producing, first, water *and oil* of mercury, which are the tinctures of gold and silver; secondly, white and red oil, which are *ferments for metals and precious stones*; and thirdly, potable gold and elixir vitae.

324r: 31 **℞ ergo in no*m*ine domine**: i.e., '℞ ergo in nomine domini' ('Therefore take in the name of the Lord'). This formula is not found in *Thesaurus Philosophorum*. For references to God in the Treatise, see note on 299v: 8.

324r: 33 **without the heate of the fyer**: *Thesaurus philosophorum* (MS. Ashmole 1459, p. 189) adds "for our water is fire, burning the bodyes more sorer then common fire." Part of this passage may have been omitted in the Treatise owing to an eyeskip (*fire ... fire*).

324r: 33–36 **is our ☿ mettalin ... nor soulle hath life**: Cf. *Thesaurus philosophorum* (MS. Ashmole 1459, p. 189): "☿ mettaline, animate w*i*th a vegetable spirit of wine minerall, in the places of his generac*i*on vegetable, mixt with the substance of wine inanimate; for it is animated by influence of the sunne, without the w*h*ich noe bodye hath life in him." The syntax in S1S2S3S4 is clearer than in A, since these manuscripts add "bycause liuinge thinges," though the meaning is still difficult to interpret. The argument may be that the substance is called *vegetable* because of its (place of) origin and because it is influenced by the sun.

324r: 36–38 **Soe it is to saie . . . is made of 3 natures**: In *Thesaurus philosophorum*, these statements are reversed: according to Raymond, there are 3 natures, specified as water, air, and fire.

324r: 38–39 **and is the water wherof mettells are procreated**: The reading in S1S2S3S4, "and is the water of mettales for bycause of one only nature all mettals are procreated," is closer to *Thesaurus philosophorum*.

324r: 39–40 **which is the menstrue of Raymond, which is resoluble**: Cf. *Thesaurus philosophorum* (MS. Rawlinson D. 1217, fol. 3r): "it is ye menstrue resoluble of which Raymond spake in his Alphabet in the letter D."

324r: 40 **by the which too extremityes be ioyned together**: *Thesaurus philosophorum* gives a further specification of what the menstrue resoluble actually is, a passage which is not included in the Treatise. Cf. MS. Ashmole 1459 (p. 189): "The menstrue resoluble is a vapour potentiall existent in whatsoever bodye mettalline, ioyning together two extreames." Cf. 321r: 15–18 and 321v: 20–22.

324v: 3–5 **for because by quietnes . . . the heate of the sonn**: This passage differs significantly from the corresponding passage in *Thesaurus philosophorum*, both being fairly opaque. Cf. MS. Ashmole 1459 (p. 189): "by reason it is eiuen [MS. Rawlinson D. 1217, fol. 3r: "{euen}"] in his owne tartary [=territory?] & place; for there was hee drawne from the matrix of the earth by the attractiue virtue of the sunne & moone."

324v: 10 **circulatinge**: The sentence is anacoluthic. *Thesaurus philosophorum* uses an imperative *circulate* instead.

324v: 14 **vnto the tyme apointed**: This statement is not found in *Thesaurus philosophorum*. See note on 324v: 23.

324v: 16–32 **And here to com to . . . and shall com to perfecte elixar &c**: Chapters 31 and 32 together make up a single chapter, the third, in *Thesaurus philosophorum*.

324v: 23 **vnto perfecte ☉ and ☽**: *Thesaurus philosophorum* adds a passage about the medicinal virtues of the great work: "and allsoe by that worke all manner of sicknesses is perfectly cured & mans body preserued from all corruption vntill the hower prefixed of God" (MS. Ashmole 1459, p. 190). See note on 324v: 14.

324v: 25 **did Alkamus the philosofer invent**: For *Alkamus*, see pp. 53–54. *Thesaurus philosophorum* does not attribute the less work to Alkamus. Instead, it simply states that "the lesse worke which only serveth to the scyence of alchymie, which alchymist philosophers found to transmute every bodye mettalline" (MS. Ashmole 1459, p. 190).

324v: 27 **transmuptation**: i.e., *transmutation*. Wyld records intrusive 'p' only between two consonants; Wyld, *A History*, 309. Since there are no other instances of this spelling in the Treatise and no other instances of intrusive 'p,' it may simply be an accidental spelling.

324v: 28–29 **and therfore to neither body . . . of the ☉ and ☽**: The argument in the Treatise seems to be similar in sense to that of *Thesaurus philosophorum*, but the two texts differ in formulation. Cf. MS. Ashmole 1459 (p. 190): "and therefore alterable into a bodye newtrall." The longer reading in the Treatise may suggest that since Venus, i.e., copper, is corruptible it cannot be turned into gold and silver, implying that it can be turned into a substance (a neutral body?) that can be used as the base matter for the elixir. See also note on 324v: 29–32. Alternatively, *neither* may be a misreading of *neutral* at some point.

324v: 29–32 **the w***hich* **after it doth putrifie . . . and shall com to p***er***fecte elixar &c**: In *Thesaurus philosophorum*, it is the milk in which Venus is putrified that is called alchemical mercury or the aqua vitae of metals. What is more, Venus is not called the small ferment of gold and silver; instead, with the addition of ferment from gold and silver, Venus can be turned into a perfect elixir. Cf. MS. Ashmole 1459 (p. 190): "w*h*ich, after his perfect putryfaction in his owne mothers milke, that is to say, in our alchymists mercury and water of life metaline, by a little ferment of Sol and Luna shall become elixir white or redd."

324v: 36–38 **The salte is made . . . great heate of the fyer**: The reference is to Chapter 28. See p. 53.

324v: 42–325r: 1 **and the same pound vpon 10**^{li} **and that ten pound vpon 100**^{li}: This statement is not found in *Thesaurus philosophorum*.

325r: 1–2 **when all is dissolued by putrifaction and made thine**: Cf. *Thesaurus philosophorum* (MS. Ashmole 1459, p. 190): "then must you puryfie that milke, when you haue dissolued as much as you list, till it be very thinne." The other copy, MS. Rawlinson D. 1217 (fol. 3v), is closer to the Treatise in that it reads "putrefie" instead of *purify*. However, *purify* is written in the margin and "putrefie" underlined, which indicates that the reading *purify* should be substituted.

325r: 3 **with with a softe fyer**: The double *with* is presumably the result of dittography.

325r: 3 **caste away the first water**: Cf. *Thesaurus philosophorum* (MS. Ashmole 1459, p. 190): "that a litle fainte water may come, w*h*ich cast away, or keepe it to double calx ♀."

325r: 4–5 **and change your receptory . . . com a thick water**: Cf. *Thesaurus philosophorum* (MS. Ashmole 1459, p. 190): "then with more heate a silver water will come somewhat thicke."

325r: 6–8: **Distill this water . . . verie close stopped**: This passage is not found in *Thesaurus philosophorum*.

325r: 8–9 **Then strengthen your fyer . . . the collour of gould**: Cf. *Thesaurus philosophorum* (MS. Ashmole 1459, p. 190): "And at the last will come a red oyle of golden collour."

325r: 13–15 **as Avicen said . . . drawe out his moisture**: This quote is also found in the *Thesaurus philosophorum*. It ultimately derives from (Pseudo-?)Avicenna's *Declaratio lapidis physici Avicennae filio suo Aboali*: see Zetzner, *Theatrum chemicum*, 4: 876. See also 321v: 3–5.

325r: 16 **Lykwise, Guido saith: It behoueth thee to haue moch water and moch oille**: *Thesaurus philosophorum*, which also contains this quote, adds in Latin "vt sit tanta multitudo tincturae quanta et olei" 'so that the quantity of tincture is as large as that of oil' (MS. Ashmole 1459, p. 191).

325r: 18–19 ℞ i^{li} of ♀ . . . **them together for the white**: *Thesaurus philosophorum* presents a slightly different version of this recipe, and the proportion of the two substances differ. Cf. MS. Ashmole 1459 (p. 191): "Therefore to the white elixer melt together lib. i of ♀ as it cometh from the mine with ʒ 1 of fine ☽."

325r: 21–22 **and put the white . . . red water seuerally by themselfe**: There is no corresponding statement in *Thesaurus philosophorum*, which simply states "into i lib. weight of this water" (see MS. Ashmole 1459, p. 190). The source of the Treatise may have contained a reading similar to the one found in Ripley, *Opera omnia*, 334–35: "pone Lunam & Venerem in aqua spissa, & Solem cum Venere in oleo rubro" 'put Luna and Venus in the thick water, and Sol and Venus in the red oil.'

325r: 24 **soe let it stand in a temperat fyer**: Cf. *Thesaurus philosophorum* (MS. Ashmole 1459, p. 191): "worke these two together betweene two glasses circulatorily in temperate fire."

325r: 25 **by changinge from on cullour to another**: Cf. *Thesaurus philosophorum* (MS. Ashmole 1459, p. 191): "and pass it soe from collour to collour."

325r: 29–31 **And soe contynue . . . the collour of purple**: *Thesaurus philosophorum* presents a longer, slightly more explicit process. Cf. MS. Ashmole 1459 (p. 191): "But if it be of gold, at the begining for the redd worke then shall you continue it still in drye fire of greater heate till the said white earth like to fishes eyes become wanne and be made like purple redd."

325r: 34 **Venus**: *Thesaurus philosophorum* (MS. Ashmole 1459, p. 191) uses the symbol for *mercury*, '☿,' which may have been misinterpreted in the Treatise for the similar-looking ♀ 'Venus' or 'copper.' Cf. e.g. 317r: 8–10, 317v: 36–38, 320r: 35–36, 324v: 22–23.

Glossary

This selective glossary focuses on alchemical terms, but it also includes some rare, archaic, or obsolete words or meanings of words. Some terms have been included since their spelling may make them opaque. For the definitions, I have relied on the *OED* and to a lesser extent the *MED*. For alchemical terminology, I have primarily used Abraham (*The Dictionary*), Rulandus (*Lexicon*), and *Grosses universal Lexicon*. I have also had occasion to consult Latin dictionaries for previously unrecorded words in English, such as the *OLD* and *DML*. In addition to the glossary of lexical items, I have included a brief list of proper nouns, which covers names of alchemical authorities that are referred to in the Treatise.

I have structured the glossary in the following way. First, I give an inventory of forms or spellings of the word in question. Words that have widely different variant spellings have been given separate entries, but cross-references are provided to a main entry where the definition is supplied. The spellings or forms are followed by a word-class definition of the item (see below). I have given the definition of the word within single quotes. The definition is followed by a reference to the place in the text where the item occurs; *ded.* refers to the *Dedication*. An * has been added after a reference if the word is discussed in the EN. If there are more than three instances of the word in the text, I cite three examples, usually the first three instances of the word, and the citations are followed by '*etc.*,' indicating that there are additional occurrences in the text. Different meanings of the same word have been separated with the help of a semicolon. If there are several different forms of a word, especially in the case of verbs, the definition is given in the base form. So there is not always a one-to-one correspondence between the forms and the definitions. For some terms with specific alchemical significance such as *mercury* and *tincture*, it has proved very difficult to determine what the exact meaning of a word is in every single instance. In these cases, I have given several related meanings without separating them with semicolons, and I have not tried to define the instances more narrowly.

Abbreviations:
adj.=adjective; *adv.*=adverb; *comp.*=comparative (adjective); *conj.*=conjunction; *constr.*=constructed with; *n.*=noun/noun phrase; *num.*=numeral; *pl.*=plural; *ppl.*=participle; *prep.*=preposition; *pron.*=pronoun; *pt.*=past tense; *refl.*=reflexive; *sup.*=superlative (adjective); *v.*=verb; *v. n.*=verbal noun

accion *n.* 'operation,' 'effect' 322r: 19
accomplishe *v.* 'complete' 323r: 18
acte *n.* 'a state of accomplishment or reality,' 'real existence' 304v: 17, 19, 305r: 28 *etc.*
additamente *n.* 'addition' 302r: 13
adrop, adrope *n.* 'mercury [one of the components of all metals]'(?) 305r: 25, 26, 33 *etc.*
aduste *adj.* 'brown' 317r: 31
aerouse, ayerouse, ayrouse *adj.* 'of the nature of air,' 'full of air' 298r: 20, 305r: 31, 33
aery *adj.* 'airy,' 'air-like,' 'full of air' 303r: 11, 305r: 8
aerynes *n.* 'air,' 'the quality of air' 322v: 8
after *prep.* 'according to' 299v: 25, 300r: 3, 302r: 36 *etc.*
alkamistick, alkamistik *adj.* 'alchemical' 324v: 30, 34
alkamy *n. in the phrase* "golde/gould alkamy": 'alchemical gold,' 'gold produced through alchemical procedures' 295v: 7, 312v: 20, 21 *etc.*
allam, allom, allames, alloms, alomes *n.* 'alum,' 'a whitish transparent mineral salt' *ded.* l. 94, 296v: 12, 297r:7 *etc.* See "roch"
alle *n.* 'ale' 309v: 31
amalgamed *ppl.* 'mixed [by amalgamation]' 301r: 4
and *conj.* 'if' *ded.* l. 31*
animal, animall, animalle *adj.* 'related to the animal stone' 304r: 30, 310v: 35; 'vital,' 'life-giving' 298v: 5*; *in the phrase* "animall stone": 'one of the three kinds of the philosophers' stone,' 'the most potent stone said to be able to transmute base metals into silver and gold and to heal diseases' 297r: 23, 304r: 3, 6 *etc.*
animalls *n. pl.* 'living creatures' 323r: 19*
apployed *ppl.* 'employed'(?) *ded.* l. 31*
appropriat *ppl.* 'extracted'(?) 318v: 43*
apropriate *adj.* 'proper,' 'peculiar' 305r: 23
aptation *n.* 'constitution' 319r: 35*, 41
aqua *n.* 'water'; *in the phrase* "aqua permanent": 'mercury [one of the components of all metals]' 299r: 23; *in the phrase* "aqua viscosa": 'viscous water,' 'mercury' 299r: 19. See "water"
aquauite, aquavite *n.* '[unrectified] alcohol' 321r: 44, 321v: 19, 324r: 4 *etc.*; 'mercury [one of the components of all metals]' 324v: 31. See "water"
arcenack, arcenick, arcenike, arseneke, arsenick, arsnicke *n.* 'arsenic,' 'one of the substances known as spirits' 296r: 13, 296v: 13, 297r: 29 *etc.*
ardent, ardente *adj.* see "water"
argall, argalle, argull *n.* 'argol,' 'the tartar deposited from wines' 321r: 37 (x2), 324r: 3 *etc.*; 'a cover term for the base material for the philosophers' stone'(?) *ded.* l. 114
argent viue *n.* 'mercury' 299v: 10, 301v: 45, 302v: 2 *etc.*

arseneke, arsenick, arsnicke *n.* see 'arcenack'
assay *n.* 'a test'; *in the phrase* "put them in assay": 'try them,' 'test them' 300v: 25
assay *v.* 'try,' 'attempt' *ded.* l. 191
astronomy *n. in the phrase* "loware astronomy": 'alchemy' 318v: 48*
attractiue *adj.* see "vertue"
augmente, augmented, auugmented *v.* 'increase,' 'multiply' 314r: 8, 11, 315r: 1 *etc.*
aurum *n.* 'gold'; *in the phrase* "aurum potabile": 'drinkable gold,' 'the elixir,' 'a powerful medicine' 312r: 26, 318v: 38, 320r: 20. See "potable"
ayer *n.* 'air,' 'one of the four elements' 299r: 8, 299v: 29, 300r: 15 *etc.*; 'air,' 'gas' 308r: 8, 10, 12 *etc.*

balination *n.* 'bathing,' 'exposing a substance to heat in a balneum' 314v: 8. See "balneo"
balneo, ballneo *n.* 'a water bath kept at less than boiling temperature' 301r: 23, 310v: 9, 14 *etc.*
barme *n.* 'ferment,' 'yeast' 309v: 31
be *prep.* 'by' 297v: 29
bodi, bodie, body, bodye, bodies, bodyes *n.* 'a distinct form or kind of matter,' 'a substance,' 'a metal' 295r: 9, 21, 24 *etc.*; 'the body of a substance as opposed to the soul or the spirit,' 'the raw or base material' 302r: 23 (x2), 302r: 29 *etc.* See "sole," "sprit," "still"
breninge *ppl.* 'burning' 323v: 3
bruble, burble *v.* 'bubble,' 'boil' 313r: 1*, 316v: 17, 318v: 8
brubles *n. pl.* 'bubbles' 316v: 22, 318r: 21, 318v: 19 *etc.*
brublinge, brublinges *v. n.* '[the process of] bubbling' 317v: 13, 318r: 43, 319v: 17 *etc.*
bruse *v.* 'beat small,' 'grind down' 314v: 37
burble *v.* see "bruble"

calaminaryes *n.* 'an ore of zinc' 297r: 31
calce *n.* 'calx,' 'a powder produced by thoroughly burning or roasting a mineral or metal' 310v: 20, 311r: 4, 24 *etc.*
calcin, calcine, calcioned, calcyne, calsioned, calsioneth *v.* 'reduce to a powder by roasting or burning' 299v: 36, 300r: 15, 17 *etc.*; 'burn in the fire to a calx' 318r: 8
calcinacion, calcination, calcynacon, calsinacion *n.* 'the process of reducing by fire to a calx or powder' 294v: 7, 26, 295r: 2 *etc.*
calcioninge *v. n.* 'calcination,' 'reducing by fire to a calx or powder' 300r: 40, 313r: 27, 320v: 38
cankared *ppl.* 'ulcerated,' 'gangrened' 295v: 7
carle *n.* 'man,' 'villain' *ded.* l. 154
carte *n. in the phrase* "carte load": 'the specific quantity which it is customary to load at one time,' 'a large quantity' 299r: 36

centure *n.* 'center' 298v: 17

chase *n.* 'pursuit (of an enemy)' *ded.* l. 66

chefest *adj. (sup.)* 'principal,' 'greatest' 324v: 11

circulate, circulatinge *v.* 'go through continuous distillation' 320r: 13; 'subject a substance to continuous distillation' 324v: 10

circulate *ppl.* 'circulated,' 'put through continuous distillation' 317v: 28, 318r: 31

circulation *n.* 'the process of refining a substance through continuous distillation' 305v: 23, 311r: 12, 324v: 22

circulatori, circulatorie, circulatory *adj.* 'used for the process of continuous distillation' 311r: 28, 311v: 17, 315v: 10 *etc.*

circumduced *ppl.* 'caused to move or turn around' 322v: 24

clernes *n.* 'a clear or transparent substance' 305v: 12, 306r: 19

cleuinge *ppl.* 'cleaving,' 'sticking' 297r: 20

coagulacion, coagulation *n.* 'the action or process of coagulating,' 'solidification' 303v: 3, 35

coagulate *v.* 'carry out the process of coagulation' 303r: 31; 'become solid,' 'solidify' 317v: 12

coaguled, coaguleth, coaguliid *v.* 'convert into a solid mass,' 'make solid,' 'curdle' 302v: 7, 8, 13 *etc.*; 'join through coagulation' 295v: 14, 302r: 29(?); 'bring about coagulation' 302v: 26, 303r: 21; 'become solid,' 'solidify' 302r: 27, 302v: 12

coescentiall *adj.* 'united or inseparable in essence or being' 305v: 1, 320v: 13

cold, could, coulde *n.* 'one of the four principal qualities of the elements,' 'coldness' 295v: 28, 298v: 12, 299r: 9 *etc.*

collorike *adj.* 'choleric,' 'of a hot or fiery nature' 303v: 39

colorably *adv.* 'feignedly,' 'in a way as to conceal the real purpose' 322v: 2

com *v. in the phrase* "com sharp(e)": 'become biting [in flavor]' 310v: 11, 320r: 31, 321v: 36

combred *ppl.* 'hindered,' 'obstructed'(?) 307r: 23

combustibilyti *n.* 'inflammable quality' 300v: 29

comixion, commixcion, commixtion *n.* 'the action or process of mixing together' 295v: 25, 302v: 44, 303r: 37 *etc.*

commendations *n. pl.* 'expressions of approval,' 'recommendations' 294r: 9

commended *ppl.* 'recommended' 297v: 8, 298r: 13

commixed *ppl.* 'mixed together' 307r: 20, 322v: 9

compacte *ppl.* 'compacted,' 'firmly joined together' 314v: 41, 323r: 7

compassed *ppl.* 'composed,' 'joined together' 323r: 25*

compassinge *ppl.* 'travelling in a circuitous course' 322r: 38

compassion *n.* 'fellow-feeling,' 'sympathy'(?) 308v: 23*

compleccione, complection, complections *n.* 'the combination of the four primary qualities (*cold, dry, hot, moist*)' 295v: 28, 297v: 25, 299r: 7 *etc.*; 'combination [of humors etc.]' 319r: 32, 323r: 23; 'complection,' 'the color of the skin' 298v: 25

complemente, complmente *n.* 'completion' 302r: 3, 304r: 41, 319v: 6
composed *ppl.* 'produced' 318v: 39; 'compassed,' 'enclosed'(?) 319r: 30
composision *n.* 'production [by combination of various parts]' 324v: 33
compounded *ppl.* 'put together,' 'united' 313v: 20, 323r: 7, 16; 'composite' 313v: 30
comprehended *ppl.* 'enclosed' 298r: 11; 'included,' 'comprised' 307r: 7
concauity *n.* 'hollow,' 'cavity' 298r: 11
concreate, concrete *adj.* 'united,' 'connected' 305v: 1, 320v: 13
condicions, conditions *n. pl.* 'prerequisites,' 'categories'(?) 295r: 14, 15
confer *v.* 'compare' 310v: 36
confixat *ppl.* 'firmly fixed together' 303r: 33
confixation *n.* 'the action of fixing firmly together' 303v: 3
congealed, congealeth, congealing, congealinge, congeall, congealle, congeled, congelinge, congell *v.* 'convert into a solid mass,' 'make solid' 299r: 32, 299v: 33, 35 *etc.*; 'become solid,' 'solidify' 312v: 18, 313r: 28, 317r: 6 *etc.*
congealing, congealinge, congeling *v. n.* 'the action of turning into a solid substance,' 'solidification' 299v: 12, 300r: 40, 313r: 32 *etc.*
congelation, congelatione *n.* 'the action of turning into a solid substance,' 'solidification' 297r: 28, 308v: 11, 309v: 10 *etc.*
coninge *n.* 'knowledge,' 'learning' 302v: 18
coningly *adv.* 'skilfully,' 'expertly' 322v: 24
conioyned *ppl.* 'joined together' 301r: 3, 316v: 31, 317r: 1 *etc.*
coniunction, coniunctions *n.* 'the action or process of joining together,' 'union' 295r: 2, 299r: 30, 302v: 27 *etc.*
consequently *adv.* 'subsequently' 305r: 41
continue *v.* 'maintain,' 'retain' 313r: 32
contrarietis *n. pl.* 'opposing or contrary principles or substances' 305r: 4
contrition, contrytion *n.* 'an action similar to grinding or pounding caused by fire' 303r: 27, 304r: 10, 317r: 10
conuenably *adv.* 'suitably,' 'properly' 309v: 21
convenient *adj.* 'suitable,' 'appropriate' 313r: 30
conveniently *adv.* 'suitably,' 'appropriately' 315r: 34
cordiall *adj.* 'restorative,' 'reviving' 312v: 21
corrosyue, corosyues *n.* 'a substance that corrodes by chemical action,' 'an acid' 296v: 14, 303v: 39, 307v: 41
corrupte, corrupteth *v.* 'decay,' 'disintegrate' 298v: 30, 31
corruption, corruptione *n.* 'destruction,' 'decomposition' 297r: 28, 298v: 33, 308v: 26 *etc.*
coulbuste *adj.* 'combust'(?) 314v: 30*
craft *n.* 'art,' 'science' 294r: 31, 294v: 23, 36 *etc.*; 'alchemy' 299r: 25; 'skill' 302r: 5, 304v: 31
craftely *adv.* 'skilfully' 303r: 16; 'acutely' 300r: 27
creatur, creature *n.* 'a created being [animate or inanimate]' 319r: 40, 323r: 5, 15*
crocfer *n.* 'crocus ferri,' 'oxide of iron' 297v: 1

crowne *n.* 'the top of a vessel or receptacle' 311v: 18, 320r: 29
crud *adj.* 'raw,' 'unrefined' 300v: 21, 301r: 23, 25 *etc.*; *in phrases with* mercury: '[the metal] mercury,' 'quicksilver' 301r: 19, 22, 303v: 41 *etc.*
crud *n.* 'a raw or unrefined substance' 307r: 22
cruddy *adj.* 'crude'(?) 301r: 13*
curd, curde *n.* 'curd,' 'a substance similar to a curd in appearance,' 'the male principle of the philosophers' stone,' 'sulfur [one of the components of all metals]' 302v: 9, 11, 12 *etc.*
curded *ppl.* 'curdled,' 'congealed' 302v: 8

darkly *adv.* 'in an obscure or mysterious manner' 299v: 24
decked *ppl.* 'covered,' 'clothed' 317r: 36
decocted *ppl.* 'brought to perfection by heat' 305v: 7
decoction *n.* 'the action of boiling,' 'the action of perfecting by heat' 296r: 4, 301v: 46, 303v: 31 *etc.*
dedused *ppl.* 'reduced [to a different form]' 315v: 8
deed, deede *n.* 'action,' 'procedure' 302r: 11, 303r: 29; *in the phrase* "in deed": 'in reality' 302r: 31
deed *adj.* 'dead' 295v: 1
depart, departed *v.* 'divide into parts,' 'separate' 304v: 14, 323v: 5
departing *v. n.* 'separation' 299r: 39
depenes, depnes *n.* 'deep place,' 'cavity' 298r: 12, 301v: 8
destitute *adj.* (*constr.* "of") 'deprived of,' 'devoid of' 301v: 31
directe *ppl.* 'directed,' 'addressed,' 'dedicated' *ded.* l. 79
dispocicion, disposision, dispotition *n.* '[relative] position,' 'situation' 297v: 10, 315r: 40; 'procedure' 302r: 10
dispose, disposed *v.* 'give' 306r: 38; 'deposit' 322r: 41
dissend *v.* 'distil by descent [where the liquid descends down separated from the gross material]' 315r: 34
dissention *n.* 'distillation [where the liquid descends down separated from the gross material]' 308r: 27
dissolu, dissolue, dissolued, dissolueth, dissoluinge, dissolvid *v.* 'dissolve,' 'reduce into a liquid condition,' 'separate,' 'disintegrate' 296v: 20, 297r: 24, 26 *etc.*; 'become liquified,' 'melt,' 'break up' 310r: 23, 25, 311r: 8 *etc.*; 'perform the process of dissolution' 296r: 24
dissoluinge *v. n.* 'dissolution,' 'disintegration,' 'the process of reducing into a liquid condition' 299v: 12, 320v: 37, 38
dissolution *n.* 'dissolution,' 'disintegration,' 'the process of reducing into a liquid condition' 310r: 1, 310v: 26, 313v: 24 *etc.*; 'a solution' 311r: 27
dissolutiue *adj.* see "menstrue"
disspite *n.* 'anger,' 'indignation' 300v: 41
distill, distillled, distilleth, distillinge, distillud *v.* 'distil,' 'purify,' 'separate' 299v: 28, 300v: 17, 300v: 33 *etc.*; 'become vaporized and then condensed

into liquid,' 'undergo distillation,' 'condense from the still' 310v: 17, 314r: 23, 316r: 17 *etc.*; 'carry out a distillation' 299v: 35
distillacion, distillation, distillations *n.* 'distillation,' 'vaporization' 294v: 26, 305r: 9, 310v: 1 *etc.*
distillinge *v. n.* 'distillation,' 'vaporization' 313r: 26, 313v: 43, 320v: 38
dragon *n.* 309r: 5. See p. 82
drames *n. pl.* 'medicinal potions' *ded.* l. 150
drawe, drawen, draweth, drawing, drawinge, drewe *v.* 'extract' 299v: 27, 300r: 29, 300v: 31 *etc.*
dresse *v.* 'prepare' 300r: 38
drie, dry, drye *n.* 'one of the four principal qualities of the elements,' 'dryness' 298v: 12, 299r: 9, 302r: 40 *etc.*; 'a dry substance' 307r: 22
drines, drynes *n.* 'dryness,' 'dry quality or substance' 295v: 17, 29, 310v: 31 *etc.*; 'one of the four principal qualities of the elements,' 'dryness' 304v: 25, 308v: 18, 319v: 22 *etc.*
drith *n.* 'dryness,' 'one of the four principal qualities of the elements' 305r: 6, 306r: 33
dry *adj. in phrases with* fire *or* furnace: 'containing ashes, coals or sand' 296v: 21, 301r: 24, 302r: 14 *etc.* See "balneo" and "drie"
duske *v.* 'obscure,' 'dim' 309v: 24

eake *adv.* 'eke,' 'also' *ded.* l. 95
earth, earthe, erth, yerth *n.* 'earth,' 'one of the four elements' 300r: 16, 302r: 40, 41 *etc.*; 'an earthlike substance,' 'a solid substance,' 'a body' 295v: 20, 23, 25 *etc.*; 'that which is in the bottom of the vessel' 316v: 27. See "heaven"
ecclips *n.* 'eclipse' 316v: 39*
effect, effecte, effectes *n.* 'purport,' 'tenor' 294r: 7; 'effect,' 'efficacy' 296r: 10; 'result,' 'consequence' 297v: 24; 'action,' 'influence' 323r: 22
eftsonne *adv.* 'again,' 'from time to time' *ded.* l. 128
eleament, elemente, elementes, ellementt *n.* 'one of the substances (*air, earth, fire, water*) of which all material bodies are compounded' 295v: 24, 30, 297v: 30 *etc.*
electuaryes *n. pl.* 'medicinal conserves or pastes' 309v: 31
elementalle *adj.* 'common,' 'ordinary' 304r: 21*
elementally *adv.* 'by way of an element,' 'in an elemental manner or sense' 303r: 34
elementated *ppl.* 'compounded of elements,' 'joined in elements' 322v: 24
elixar, elixare, elyxar, elixars, elyxars *n.* 'a transmuting agent,' 'the philosophers' stone' 295r: 36, 295v: 3, 4 *etc.*; *in phrases with* white: 'an elixir that transmutes a substance into silver' 295r: 36, 298v: 1, 305r: 13 *etc.*; *in phrases with* red: 'an elixir that transmutes a substance into gold' 295r: 36, 298v: 1, 305v: 32 *etc.*; *in the phrase* "elixar of lif": 'an elixir that heals bodily diseases and prolongs life' 295v: 4, 304r: 4, 304r: 12 *etc.*
elixerat, elixeratid *v.* 'turn into an elixir' 313v: 20, 315r: 14

elixerate *ppl.* 'turned into an elixir' 304r: 19, 312v: 24, 313v: 3 *etc.*
elixeration *n.* 'the process of turning into an elixir' 306r: 23, 26, 318v: 32
elyuated *ppl.* 'raised,' 'sublimed' 300v: 23, 305v: 19
embibe, imbibe, imbibed, ymbibed *v.* 'absorb,' 'soak up' 305r: 43; 'soak,' 'saturate with moisture' 314r: 1, 318v: 20, 319v: 44 *etc.*
encere *v.* 'turn into the consistency of wax' 315r: 5
enclyninge *ppl.* 'having a tendency towards' 317r: 39, 318r: 21
endeuour *v.* (*refl.*) 'exert oneself,' 'make an effort' 319r: 39
enduwed *ppl.* (*constr.* "with") 'invested with,' 'possessed of' 304r: 7, 323r: 5
engendreth, engendred, ingendred *v.* 'produce,' 'generate,' 'create' 295v: 20, 32, 298r: 15 *etc.*
enkitch *v.* 'enrich'(?) 310r: 39*
entend *v.* (*refl.*) 'devote himself,' 'endeavor' 301v: 29
entermedle *v.* (*refl.*, *constr.* "with") 'mix oneself up with,' 'have to do with' 294v: 14
entreat, entreate *v.* (*constr.* "of") 'treat of,' 'discuss' 301v: 3, 303v: 48, 304r: 13 *etc.*
entreth *v. in the phrase* "entreth not medison": 'is not brought into [the form of] medicine' 304r: 14*, 312v: 20, 313v: 3 *etc.*
epistelles, epistles *n. pl.* 'prefaces,' 'letters of dedication' 294r: 6, 15, 17
equallitie *n.* 'equal distribution' 297v: 29
erthlynes *n.* 'earthiness,' 'the quality of earth' 322v: 7
es vst *n.* 'burnt copper,' 'cupric oxide' 297v: 1
espoused *ppl.* 'united' 309v: 22
estem *v.* 'consider,' 'think' 310v: 34
etheromogines, ethremogines *n. pl.* 'substances that are heterogeneous in nature' 296v: 27*, 297r: 25
evacuations *n. pl.* 'substances that have been emptied out of a vessel' 314v: 15
examinacion, examination *n.* 'a test,' 'a trial' 295v: 8, 296r: 16, 297r: 33 *etc.*
exhale *v.* 'raise,' 'exalt' 321v: 23
extremeties, extremities, extremitis, extremityes *n. pl.* 'extremes,' 'two things at either extreme end of a scale' 305r: 4, 320v: 23, 321r: 17 *etc.*
eyen *n. pl.* 'eyes' 301r: 15

faintnes *n.* 'faint quality,' 'faint substance'(?) 317v: 5, 6
faste *adv.* 'firmly,' 'tightly' 307v: 19, 27
fasteneth, fastened *v.* 'make firm or solid,' 'fix' 300r: 15, 300v: 10, 324v: 1 *etc.*
feces *n. pl.* 'dregs,' 'sediment' 307v: 34, 35, 38 *etc.*
feeling *ppl.* 'perceptible to the senses' 299r: 37*
ferment, fermente, fermentes *n.* 'ferment or yeast that ferments a substance into perfection,' 'a substance that gives form to or multiplies the philosophers' stone or elixir,' 'the philosophers' stone' 301r: 9, 11, 302r: 30 *etc.*
ferment, fermente, fermented, fermentinge, firmented *v.* 'subject to fermentation,' 'cause fermentation in' 295r: 36, 296r: 18, 298r: 20 *etc.*

fermentall *adj.* 'pertaining to a ferment,' 'of the nature of a ferment' 304r: 10, 314r: 8, 322v: 9 *etc.*

fermentation *n.* 'the process of acting as a ferment or yeast so as to cause transmutation,' 'the process of giving form to or multiplying the elixir or philosophers' stone' 295r: 4, 311r: 1, 36 *etc.*

fier, fyer *n.* 'fire,' 'one of the four elements' 299v: 29, 35, 300r: 13 *etc.*; *in the phrase* "fire of the first degre": see EN 300v: 38; *in the phrase* "fier againste nature": 'a dissolvent,' 'a substance that reduces the base material into its prime matter' 304r: 14, 19, 312v: 24 *etc.*; *in the phrases* "naturall fire" and "fier of nature": 'a substance that joins the base material and the released spirit,' 'a medium' 310v: 25, 313v: 21*, 39 *etc.* See "menstrue" and "sole"

fiery *adj.* 'hot as fire,' 'red hot' 313r: 33, 315r: 3

fild *v. pt.* 'filed,' 'polished' *ded.* l. 9

firmented *ppl.* see "ferment" *v.*

fix, fixe, fixed, fixeth, fixing, fyx, fyxed *v.* 'make firm or solid,' 'turn into a semi-solid or viscous condition' 295r: 6, 8, 35 *etc.*

fixacion, fixation *n.* 'the process of turning a liquid substance into a solid or semi-solid, viscous substance' 295r: 3, 297r: 21, 22 *etc.*

fixatorie *adj.* 'used for the purpose of fixing' 314v: 33

flebergeber *n.* 'flibbertigibbet,' 'a very talkative person' 294v: 1*

flegma *n.* 'water,' 'watery substance,' 'feces'(?) 300v: 33, 36, 316r: 43 *etc.*

fluxible, fluxuble *adj.* 'fluid,' 'flowing' 302v: 7, 308v: 35, 317v: 36 *etc.*

forced *ppl.* 'strengthened,' 'reinforced' 295r: 35

form, forme *n. in the phrase* "substanciall forme": 'the principle of existence in a substance and the source of all that is essential to that substance' 304r: 18, 308v: 2

formall *adj.* 'that which concerns the order,' 'that which pertains to the procedure' 295r: 12

froward *adj.* 'adverse,' 'unfavorable' *ded.* l. 20

gaue *v. pt.* (*constr.* "over") 'abandoned' 301r: 15

ghoste *n.* 'soul' *ded.* l. 46

globe *n.* 'a round mass' 312v: 37

globeous, globouse, globus, globuse *adj.* 'having the shape of a sphere' *ded.* l. 89, l. 193, 304v: 6 *etc.*

gome *n.* see "gum"

gooths *n. pl.* 'goats' *ded.* l. 96

grose, gross, grosse *adj.* 'coarse,' 'unrefined' 306r: 20, 314v: 32; 'preliminary,' 'basic' 304r: 6(?), 320v: 36, 322r: 34 *etc.*

grosnes *n.* 'a coarse or unrefined substance,' 'feces' 302v: 4, 314v: 23

gum, gummes *n.* 'a gum-like substance' 302v: 18, 321r: 28 (x2) *etc.*

handlinge *n.* 'management with the hand,' 'manipulation' 324v: 25
hard *ppl.* 'heard' 303r: 17, 321r: 34
hear, heare *n.* 'hair' *ded.* l. 95, 299v: 11, 26 *etc.* See "here"
heat, heate, heete *n.* 'one of the four principal qualities of the elements,' 'heat' 299r: 10, 302r: 29, 304v: 25 *etc.* See "hote"
heaven *n.* 'the top of the alembic' 305r: 7, 315r: 37*, 38; 'that which is in the top of the alembic' 316v: 27. See "limbeke"
heed *n.* 'head' 308v: 30, 311v: 20, 320r: 28. See "still" and "ravens"
her *adv.* 'here' 307r: 23
here *n.* 'hair' 323r: 9. See "hear"
herehence *adv.* 'from this point forward' 322r: 35
hole *adj.* 'whole' 300v: 41, 321r: 42
homogaine *adj.* 'homogeneous,' 'consisting of parts all of the same kind' 295v: 30*
homogene *adj.* 'homogeneous,' 'consisting of parts all of the same kind' 296r: 21
homogenius *adj.* 'homogeneous,' 'consisting of parts all of the same kind' 296v: 31, 303r: 22
hote *n.* 'one of the four principal qualities of the elements,' 'heat' 298v: 12, 307r: 27, 32 *etc.* See "heat"
houlden *ppl.* 'restrained,' 'held back' 302v: 15*
humiditie, humidity, humydities *n.* 'fluid,' 'moisture' 308v: 16, 18, 309r: 9 *etc.*; *in the phrase* "radicaulle humydities": 'natural and necessary fluids' 297v: 5*
humor, humors *n.* 'fluid,' 'watery substance' 304r: 28, 325r: 12; 'the four chief fluids of the body' 319r: 33, 323r: 23
husbandrie *n.* 'farming,' 'agriculture' *ded.* l. 203
hydinges *n. pl.* 'secrets,' 'means of concealment'(?) 299r: 4

imbibe, imbibed *v.* see "embibe"
imbibition, imbybition, imbibitions *n.* 'the process of soaking or saturating' 295r: 4, 315r: 28, 316v: 35 *etc.*; 'a solution' 319v: 29
imbybinge *v. n.* 'imbibition,' 'soaking or saturating' 313r: 26
imperfect, imperfecte *adj.* 'base,' 'impure' 300r: 39, 305r: 13, 308v: 7 *etc.*
imperiall *adj.* 'majestic,' 'exalted' 322r: 38
incerate *v.* 'turn into the consistency of wax' 314v: 38
inceration *n.* 'the process of turning into the consistency of wax' 315r: 1, 4
inclinable *adj.* 'favorably disposed' 322r: 24
indifferent *adv.* 'indifferently,' 'moderately,' 'fairly' 311r: 30
infinite *adj.* 'indefinite in nature,' 'indeterminate' 301v: 16
ingendred *ppl.* see "engendreth"
ingrossed *ppl.* 'made dense' 297r: 19
inspissation *n.* 'thickening' 298v: 4
ixer *n.* 'elixir'(?) 308v: 24*

kibrik *n.* 'sulfur' 315r: 15
kind, kinde *n.* 'nature' 297r: 4, 5, 298r: 21 *etc.*; 'nature,' 'essential character' 296v: 21, 33 (x2) *etc.*; 'nature,' 'type,' 'genus' 295v: 1, 297r: 8, 32 *etc.*
kindly *adv.* 'naturally,' 'as a matter of course' 299r: 21, 40, 308v: 20
knit, knite, knitinge *v.* 'join,' 'tie' 297v: 29, 305r: 4, 309v: 20
knittinge *v. n.* 'joining,' 'combining' 306r: 31, 308r: 32
Kotolory *n.* 'Catalonia' 321r: 27*
kymia *n.* 'an alchemical vessel [used primarily for distillation]' 314r: 6, 314v: 26, 315r: 19 *etc.*

laies *n. pl.* 'songs' *ded.* l. 164
largly *adv.* 'in full,' 'fully' 306r: 28, 317r: 13, 323v: 33
laterall *adj.* 'situated on the side,' 'toward the side' 322r: 33*, 36
leaue, leave, leaveth *v.* (*constr.* *-ing*-form) 'cease,' 'stop' 294v: 28, 310v: 17, 320v: 48; (*constr.* "of" i.e., *off*) 'discontinue' 320v: 39
leaven, leavene *n.* 'ferment' 309v: 27, 30. See "ferment"
leed *n.* '[the metal] lead' 298v: 22, 299r: 2
lente *adj.* 'gentle,' 'slow' 313r: 21
leprous *adj.* 'base,' 'impure' 305r: 36
lese *v.* 'lose' 294v: 29
letter *n.* *in the phrase* "after the letter": 'literally' 299v: 26
letteth *v.* 'hinders,' 'prevents' 297r: 27
lifte *ppl.* (*constr.* "vp") 'encouraged' 301v: 30
lightly *adv.* 'easily,' 'readily' 299v: 9
limbeke, limbick, limbicke, lymbicke *n.* 'alembic,' 'a vessel used for distillation' 310v: 6, 14, 311r: 7 *etc.*
liquefaction, liquifaction *n.* 'the process of reducing into a liquid condition' 295r: 4, 303r: 30, 317r: 32 *etc.*
liquifie, liquified, liquifyed *v.* 'reduce into a liquid condition' 296v: 34, 303v: 11; 'become liquid,' 'dissolve' 324r: 13
liquour *n.* 'a liquid,' 'an alcohol'(?) 314r: 23, 316r: 17, 318v: 22 *etc.*
liste, lyste *v.* 'desire,' 'like' *ded.* l. 141, 325r: 32
lix *prep.* 'like' 311v: 37*
loosed *ppl.* 'dissolved' 299r: 22
lothsumnes *n.* 'loathsomeness,' 'a feeling of loathing or disgust' *ded.* l. 42
loweste *n.* 'the lowest part' 309v: 7
luckely *adv.* 'prosperously' 322r: 32
Luna, Lune *n.* 'silver' 295r: 37, 295v: 3, 305r: 41 *etc.*
Lunarie *n.* 'base material,' 'prime matter,' 'mercury' (?) 314r: 24*, 316r: 17, 318v: 10 *etc.*
lute *n.* 'clay or cement [used to close up an orifice]' 307v: 21, 310v: 8 (x2)
lute, luted *v.* 'cover with clay,' 'close up with the help of clay' 307v: 19, 310v: 6, 13 *etc.*

lyon, lyone *n.* 'a dissolvent' 304r: 26*, 316r: 21, 318r: 27 *etc.*
lyvely *adj.* 'vital,' 'necessary to life' 322v: 27

maintenance *n.* 'upholding or keeping' 301v: 34
malgam, malgame, malgams *n.* 'amalgam,' 'a mixture of mercury and a metal' 310r: 3, 5, 10 *etc.*
marcury, mercuri, mercurie, mercury, mercurye, marcuries *n.* '[the metal] mercury,' 'one of the substances known as spirits,' 'mercury [one of the components of all metals]' 295r: 35, 295v: 11, 296r: 12 *etc.*
masterie, mastery, mastrye *n.* 'special skill,' 'mastery of a subject' 301v: 12, 302v: 18; 'the alchemical opus' 302r: 9, 13, 20 *etc.*
materia *n. in the phrase* "materia prima": 'prime matter,' 'the original substance from which everything derives' 304v: 9 (x2), 20
matrix *n.* 'the womb,' 'a place in which something is produced' 300v: 13, 303v: 29
mean *n.* 'a medium' 295r: 34, 298v: 6, 302r: 23 *etc.*
mean, mene *adj.* 'moderate,' 'temperate' 312r: 9, 312v: 45, 313r: 10 *etc.*; 'intermediate' 309r: 12, 315r: 36, 40
measell *n.* 'a leper' 298v: 25
measure *v.* 'regulate,' 'control' 303r: 42
meate *n.* 'food,' 'red mercurial water'(?) 315r: 42*, 321r: 8
medcin, mediciene, medicin, medicine, medison, medisone, medisons *n.* 'the elixir,' 'the philosophers' stone' 294v: 9, 11, 295r: 5 *etc.*
medled *ppl.* 'mixed' 302v: 28
medlinge *v. n.* 'mixing' 308v: 39
meltinge, meltynge *ppl.* 'melted,' 'fluid,' 'liquid' 295r: 10*, 308v: 35, 315v: 19
menerall *adj.* see "mineral"
menstruall *adj.* 'of the nature of a menstrue' 305r: 39. See "menstrue"
menstrue, menstrues *n.* 'a dissolvent' 304r: 22, 320v: 28, 321r: 15 *etc.*; *in the phrase* "menstrue resoluble": 'a substance that joins contrary natures, such as mercury and sulfur,' 'a medium' 304v: 42, 305r: 3, 320v: 8 *etc.*; *in the phrase* "dissolutiue menstrue" or "resolatiue menstrue": 'a dissolvent,' 'a substance that reduces the material into its prime matter' 320v: 14, 27, 320v: 31 *etc.* See "fier"
mercuri, mercurie, mercury, mercurye *n.* see "marcury"
mercuriall *adj.* 'of mercury,' 'consisting of mercury' 310v: 28, 311r: 5, 313v: 12 *etc.*
mercuried *ppl.* 'turned into mercury,' 'joined with mercury'(?) 310v: 29*
mercurized *ppl.* 'joined with mercury,' 'treated with mercury' 311v: 4*, 5, 312v: 34 *etc.*
mettalin, mettaline *adj.* 'metallic,' 'pertaining to metals' 320v: 11, 12, 16 *etc.*
milk, milke *n.* 'mercury,' 'mercurial water' 320v: 11, 19 (x2) *etc.*; *in the phrase* "maidens milke": 'mercury [one of the components of all metals]' 303r: 13, 303v: 19

mineral, minerale, minerall, mynerall *adj.* 'pertaining to or having the nature of a mineral' 299r: 43, 304r: 20, 313v: 6 *etc.*; 'pertaining to the mineral stone' 304r: 24(?), 310v: 35, 312v: 12 *etc.*; *in phrases with* stone *or* elixir: 'one of the three kinds of the philosophers' stone, said to be able to turn base metals into silver and gold' 297r: 17, 304r: 2, 13 *etc.*

mineralls, minerales *n. pl.* 'minerals,' 'inorganic substances' 302r: 12, 303r: 9, 313r: 13*

miter *n.* '[poetic] meter' 317r: 17

moist, moiste, moyste *n.* 'one of the four principal qualities of the elements,' 'moistness' 295v: 28, 298v: 12, 302r: 40 *etc.*

moiste *adj. in phrases with* fire *or* heat: 'a water bath or balneum' 302r: 14, 317v: 11, 322r: 18

moisteth *v.* 'conveys moisture' 303r: 13

moistnes *n.* 'moisture,' 'a liquid' 307v: 20, 24, 25 *etc.*

moistur, moisture *n.* 'one of the four principal qualities of the elements,' 'moistness' 299r: 10, 304v: 7, 24 *etc.*; 'a liquid,' 'a fluid' 308r: 13, 319v: 17, 320v: 6 *etc.*

moone *n.* 'silver' 295r: 34, 323v: 29

morever *adv.* 'moreover' 324r: 10

mortall, mortalle *adj.* 'deadly' 304r: 21, 312v: 25, 313v: 5

mortification *n.* 'alteration of the form' 308v: 17

mortified, mortifyed *ppl.* 'altered in form,' 'dissolved' 300v: 9, 23, 303v: 7 *etc.*

moulte *v.* 'melt' 313r: 45

multipli, multiplied, multiply *v.* 'increase in quantity and/or potency' 311r: 32, 38, 39 *etc.*

multiplicable *adj.* 'multipliable,' 'capable of being increased in quantity and/or quality' 312v: 27, 315v: 24

multiplication *n.* 'the action or process of increasing the quantity and/or potency' 309v: 10, 311r: 37, 38

mutations *n. pl.* 'transmutations,' 'procedures to transform one substance into another' 303r: 19

naughty *adj.* 'evil' 299v: 14

nere *adj. (comp.)* 'nearer,' 'closer' 298v: 17

nether *conj.* 'neither' 297v: 26

nip *v.* 'close up [by pressing together the heated end of the neck or tube]' 311v: 21

nothing, nothinge *adv.* 'not at all,' 'in no way' 294r: 37, 319r: 1, 322r: 27

nyter *n.* 'saltpeter' 304r: 21

of *prep.* 'as regards,' 'concerning' 312v: 7, 313v: 5

off *prep.* 'of' 297v: 20

office *n.* 'function' 310v: 22

olefied *ppl.* 'turned into an oil' 305v: 22

on, ons *num./pron.* 'one' ded. l. 125, 294r: 36, 296v: 30 *etc.*
one *prep.* 'on' ded. l. 94, 307v: 18, 311r: 7 *etc.*
one *adj.* 'own' 297v: 21
open, opened, openinge, openithe *v.* 'reveal,' 'disclose,' 294v: 1, 306r: 28, 323r: 34; 'decipher' 301v: 6
openinge *v. n.* 'revelation,' 'disclosure' 301v: 19
operatione *n.* 'procedure of being created' 297v: 31*
operatiue *adj.* see "vertue"
ordayned, ordeyned *ppl.* 'selected,' 'appointed' ded. l. 117, 317r: 21*
orderly *adv.* 'in due order' 321v: 2
orpiment *n.* 'one of the substances known as spirits,' 'arsenic trisulfide'(?) 297r: 30
ouer *pron.* 'our' 298r: 31, 35
outward *adj.* 'from the outside,' 'external' 322r: 3
outwardly *adv.* 'on the outside,' 'externally' 304v: 24, 313v: 40, 320v: 34
overcom, overcome, overcommeth, overcommethe *v.* 'prevail over,' 'defeat,' 'surpass' 302v: 5, 30, 308r: 34 *etc.*; 'prevail,' 'be successful' 302v: 29; 'obtain,' 'get at'(?) 302v: 47

pace *v.* 'pass'(?) 324r: 10*
palle *adj.* 'undefined,' 'unclear' 301v: 7
paradiscicall *adj.* 'like paradise,' 'celestial' 322r: 37
partaking *ppl.* (*constr.* "of") 'sharing the nature of,' 'having some of the qualities or characteristics of' 309r: 12
participatinge *v. n.* (*constr.* "with") 'sharing the nature of'(?) 307r: 32*
pass, passe *n. in the phrase* "bring/brought to pass(e)": 'bring/brought about,' 'bring/brought to accomplishment,' 'accomplish(ed)' ded. l. 121, l. 191, 316v: 18 *etc.*
pass, passe, passeth *v.* 'go through,' 'proceed' 304r: 18, 312v: 22, 319v: 36 *etc.*; 'cause to go through' 312v: 36; 'surpass' 303r: 2
passion *n.* 'the fact or condition of being acted upon,' 'passivity' 302v: 24
paste *n.* 'the purified base material for the philosophers' stone' 309v: 30*
peace, peaces *n.* 'piece' 301r: 26, 307v: 23, 27 *etc.*
peaceably *adv.* 'in peace,' 'peacefully' 309v: 19
pear, peare *n. in the phrase* "pear(e) glasse": 'pear-shaped receptacle made of glass' 312r: 4, 5
penni, penny *n. in the phrase* "penni/penny weighte": 'one-twentieth of an ounce' 312r: 6, 12
perelles, perells, perrelles, perrells *n. pl.* 'pearls' 311r: 20, 316v: 23, 317v: 15 *etc.*
performe, performed *v.* 'turn,' 'change' 305r: 45, 305v: 14
permanent, permanente *adj.* see "aqua" and "water"
perseth *v.* 'pierces,' 'penetrates' 303r: 11
phillet *n.* 'a raised rim [on a table]' 310r: 21

philosofacall, philosoficall *adj.* 'related to alchemy,' 'alchemical,' 'as used in alchemical procedures' 305v: 25, 320v: 7
philosofer, philosofers, philosopheres, philosophers, philozofers, philozofors, philozophors *n.* 'alchemist' 294r: 13, 295r: 26, 295v: 12 *etc.*; for "stone of philosofers" and "philosofers stone," see "ston"
philosoficall *adj.* see "philosofacall"
philosofically *adv.* 'in an alchemical way,' 'in the way of an alchemist,' 'alchemically' 305v: 27, 318r: 17, 31 *etc.*
philosofie, philosophy, philozofie *n.* 'alchemy' ded. l. 104, 294r: 18, 296r: 25 *etc.*; *in the phrase* "naturall philosofie": 'the study of nature,' 'natural science,' 'alchemy' 294r: 11, 301v: 4
phisick, phisicke *n.* 'medicinal substance,' 'medical treatment' *ded.* l. 148, l. 161
phisicke *adj.* 'medicinal' *ded.* l. 150
plumbe *n. in the phrase* "allam plumbe": 'feather alum,' 'term used for a variety of minerals' *ded.* l. 96
porosity *n.* 'the quality or fact of being porous,' 'porousness' 299r: 7
posse *n.* 'pass'(?) 294v: 17*. See "pass" *n.*
potable *n. in the phrase* "potable gould": 'drinkable gold,' 'the elixir,' 'a powerful medicine' 322v: 23, 26. See "aurum"
potentiall *adj.* 'powerful,' 'potent' 321r: 17, 321v: 21
pouer *n.* 'power' 298r: 36. See "poware"
powar, poware, power *v.* 'pour' 299v: 31, 34, 311v: 12 *etc.*
poware *n.* 'power' 299r: 43, 299v: 8, 23 *etc.*; 'potentiality' 305r: 28; *in the phrase* "in poware": 'in potentiality' 302r: 31, 303v: 13, 304v: 17 *etc.*
powlders *n. pl.* 'powders' 295r: 22*
practike *adj.* 'practical' 307r: 2
preceptuall *adj.* 'pertaining to precepts' 295r: 12
precipitate *n.* 'solid substance [deposited from a solution]' 301r: 22, 27
precipitate, precipitated *v.* 'deposit in a solid form from solution in a liquid' 310v: 28, 312v: 37, 314r: 22 *etc.*; 'settle as a precipitate or solid substance,' 'become a solid substance' 301r: 24, 25
preparard *ppl.* 'prepared' 313v: 27*
prescribed *ppl.* 'ordered' 319r: 24
presently *adv.* 'immediately' 303v: 32
preuitie, preuitye, priuity *n.* 'secret knowledge,' 'secret procedure,' 'secret' 297v: 7, 9, 303v: 15
preuye, pryuie *adj.* 'hidden,' 'secret' 294v: 5, 304v: 45
primer *adj.* 'first' 321v: 29
prince *n.* 'queen,' 'female sovereign' *ded.* l. 32*, l. 35
priuity *n.* see "preuitie"
procreat, procreate *ppl.* 'procreated,' 'produced,' 'generated' 302v: 4, 304v: 6, 323v: 4
profitable *adj.* 'useful,' 'serviceable' 296r: 26, 298v: 22, 37 *etc.*
profiteth *v.* 'is of use' 300v: 12

profound *adj.* 'immense,' 'enormous' 301v: 7
profundity *n.* 'innermost part' 305r: 33
proiecte *ppl.* 'projected,' 'thrown' 300v: 44
proiection *n.* 'the process of applying (literally *throwing*) the philosophers' stone upon a metal with the result of transmutation into gold or silver' 295r: 5 (x2), 7 *etc.*
prolixiti *n.* 'long or wearisome duration' 303v: 31
prosed *v.* 'proceed' 310v: 26
prossequte *v.* 'prosecute,' 'carry out' *ded.* l. 36
protecte *ppl.* 'protected' *ded.* l. 81
pryuie *adj.* see "preuye"
puerly, purely *adv.* 'so as to make pure' 319v: 10, 320r: 2, 323r: 38
purgatoryall *adj.* 'of the nature of purgatory' 322r: 35
purpose *n.* 'result,' 'end' 310r: 25
putrffacon *n.* see "putrifaction"
putrifaction *n.* 'disintegration,' 'decomposition' 295r: 3, 300v: 39, 303r: 26 *etc.*
putrifi, putrifie, putrified, putrifiyinge, putrify *n.* 'subject to any decomposing or destructive process,' 'cause to decompose into its separate elements' 305r: 1, 11, 306r: 20 *etc.*; 'decompose chemically' 308v: 34, 314r: 6, 314v: 27 *etc.*
putrificacion *n.* see "putrifaction"

qallities *n. pl.* see "qualiti"
qualiti, qualitie, quality, quallity, qualities, quallities, quallitis *n.* 'one of the primary qualities (*coldness, dryness, heat, moistness*)' 297v: 26(?), 299v: 20, 302r: 27 *etc.*; 'character,' 'nature' 295v: 26, 305r: 34, 306r: 6 *etc.*; 'attribute,' 'property' 305v: 34 (x2); 'quality in the abstract,' 'the category of quality' 304v: 17, 305r: 36, 37; 'degree of goodness' 312r: 24, 314r: 7, 321r: 8
quantitie, quantity *n.* 'size [in three dimensions],' 'the category of quantity' 304v: 18, 305r: 36, 37
quick *adj.* 'living,' 'of life' 303v: 40
quickneth *v.* 'gives or restores life' 303r: 14
quicksiluer, quicksilver, quicsilver, quiksiluer *n.* '[the metal] mercury,' 'mercury [one of the components of all metals]' 295r: 26, 29, 296v: 11 *etc.*; *in the phrase* "quicksiluer sublimatum": 'a corrosive,' 'a dissolvent'(?) 295r: 17
quinta essentia, quintaescentia, quintaescentie, quintaessencia, quintaessentia, quintaessentiae, quintaessentie, quintescens, quintessence *n.* 'the fifth essence or element supposed to be latent in all things,' 'the most essential part of any substance' 295v: 4, 304r: 11, 304v: 45 *etc.*

radicaulle *adj. in the phrase* "radicaulle humydities": see "humiditie"
ravens, ravens *n. in the phrase* "ravens hed": 'raven's head' 308v: 30*, 320v: 22
rear, reare, reared *v.* 'raise,' 'lift up,' 'elevate' 308r: 18, 308v: 12, 309v: 4 *etc.*

receptakell *n.* 'vessel,' '[receiving] organ' 300r: 32
receptorie, receptory *n.* 'vessel' 310v: 13, 18, 321v: 37 *etc.*
rectification *n.* 'refinement,' 'purity' 323r: 14
rectifie, rectified, rectifiynge, rectifyed *v.* 'purify,' 'refine [by distillation]' 311r: 6, 313r: 22, 25 *etc.*
reduciable *adj.* 'reducible' 304v: 17
reedyfiid *v. pt.* 'rebuilt' 319r: 16
refeirid *ppl.* 'reserved'(?) 303r: 30*
refraineth *v.* 'restrains,' 'holds back' *ded.* l. 154
regimente *n.* 'management,' 'control' 303r: 40
rehers, rehearsed, rehersed *v.* 'enumerate,' 'repeat' 301r: 36, 321r: 11, 322v: 2
relation *n. in the phrase* "make relation . . . of": 'give an account of,' 'discuss' 295r: 14
relife *n.* 'relief' 301v: 34
remoue, remoued, remove, removed *v.* 'move,' 'transfer from one place to another' 301r: 23, 313r: 10, 318r: 35 *etc.*
reninge *n.* 'rennet,' 'a substance used to curdle milk' 309v: 30
rente *v. pt.* 'smashed,' 'broke' 301r: 26
repugn *v.* 'contend,' 'be contradictory' 323r: 26
repugnant *adj. (constr.* "against") 'opposed to,' 'standing against' 320v: 21
residence *n. in the phrase* "take residence": 'settle,' 'stay' 307v: 6
resolatiue, resolutiue, resolutyue *adj.* see "menstrue"
resoluble *adj.* see "menstrue"
resolue, resolued *v.* 'dissolve,' 'reduce into a liquid condition' 300r: 14, 37, 307v: 6 *etc.*; 'carry out the process of dissolution' 308r: 29
resolution *n.* 'dissolution,' 'the process of reducing into a liquid condition' 321r: 15
respire *v.* 'pass out,' 'transpire' 307r: 18
retorned *ppl.* 'turned away' *ded.* l. 66
reuegitated, reuegitatid *ppl.* 'revived' 300v: 23, 305v: 18*
reuell *v.* 'reveal' 306r: 30
reuolution *n.* 'the process of turning one element into another' 299v: 20, 303r: 33. See "rotation" and "whell"
roch, roche *n. in the phrase* "roch allam": 'rock alum,' 'potash alum' 307r: 38, 307v: 37
Romain *adj.* see "vitreall"
ron, ronninge, runn, runinge, runninge *v.* 'run,' 'flow' 295v: 12, 302r: 20, 23 *etc.*
rotation *n.* 'the process of turning one element into another' 305v: 23. See "reuolution" and "whell"
rote *v.* 'rot' 308v: 33
rote, rotes *n.* 'root' 302r: 12, 15, 302v: 45 *etc.*
rubified *ppl.* 'made red' 305v: 19
rude *adj.* 'basic,' 'inelegant' 294r: 29
ruell *n.* 'rule' 314v: 42
runn, runinge, runninge *v.* see "ron"

sal armonack, sal armoniak, sal armonake, sal armoniak *n*. 'sal ammoniac,' 'ammonium chloride,' 'one of the substances known as spirits' 296r: 22, 299v: 10, 300r: 9 *etc*.
sal combust *n*. 'salt treated with fire or heat' 324r: 12
sapoure *n*. 'fragrance,' 'odor' 311v: 34; 'value,' 'merit' 311r: 34
sault armoniack *n*. see "sal armonack"
sauoreth *v*. (*constr*. "of") 'shows traces of the presence or influence of,' 'has some of the characteristics of' 321r: 20
savoure, sauours *n*. 'odor' 309v: 32, 321v: 26
scarchly *adv*. 'scarcely' 305v: 3
scriptur, scripture, scriptures *n*. 'book,' 'writing' 294r: 3, 7, 10 *etc*.
sea *n*. see "water"
sealed, sealled *ppl*. 'closed securely,' 'enclosed,' 'shut in' 301v: 16, 314r: 6, 314v: 27 *etc*.
seam, seame, seameth *v*. 'seem' 298v: 21, 305v: 36, 312v: 8 *etc*.; 'appear,' 'be seen' 302r: 18, 309r: 9
season *n*. 'an indefinite period,' 'some time' 300v: 39, 317v: 10, 323r: 40
seath, seathing *v*. 'seethe,' 'boil' 299v: 32, 35, 300v: 8 *etc*.
seathinge *v*. *n*. 'seething,' 'boiling' 299v: 12, 311v: 27, 319v: 17
seller *n*. 'cellar' 310r: 24
sending *v*. (*constr*. "to") 'sending a message or messenger to' 294v: 15
sentur *n*. 'center' 297r: 1
separation *n*. 'division,' 'dissolution' 302r: 41, 303v: 23, 308v: 5
seruiage *n*. 'slavery'(?) 308r: 16*
setrine *adj*. 'citrine,' 'yellow' 296r: 10
seuerally *adv*. 'separately,' 'individually' 324r: 32, 325r: 22
shapt *adj*. 'sharp'(?) 307r: 39*
sharp, sharpe, sharpeste *adj*. 'biting [in taste]' 310v: 11, 313r: 11, 320r: 31 *etc*.
sharpnes *n*. 'acidity,' 'tartness' 313r: 27, 41, 314r: 37 *etc*.; 'intensity' 308r: 42
sheninge *ppl*. 'shining' 322r: 38
sheurly, showarly, showerly *adv*. see "surely"
showe, showinge *v*. 'become visible' 306r: 36, 317v: 30(?)
sight, sighte *n*. *in the phrase* "in sight(e)": 'in appearance,' 'visibly' 299v: 23, 316r: 32
signifie *v*. 'make known,' 'intimate' 306r: 11
signs *n*. *pl*. '[star] signs,' 'constellations' 319r: 33, 323r: 23
similitudes *n*. *pl*. 'likenesses,' 'images' 299v: 5; 'conterparts,' 'equals'(?) 320v: 15*
simple *adj*. 'uncompounded,' 'not composite in nature' 302v: 29, 309v: 12, 320v: 12
simplicity, simplissitye *n*. 'the state of being unmixed,' 'the condition of existing separately' 298r: 25, 313v: 42, 316r: 30 *etc*.
simplified *ppl*. 'reduced into a simple, uncompounded condition' 322r: 40
sirculate *ppl*. see "circulate"
slainte *ppl*. 'slain' 303r: 14*, 316r: 31, 318r: 30
slaunder *n*. 'source of shame or dishonor' *ded*. l. 84

soer *adv.* 'tightly,' 'firmly' 298v: 18

soferantly *adv.* 'sovereignly,' 'to a supreme degree' 309v: 32

Sol, Soll *n.* 'gold' 295r: 37, 295v: 3, 310v: 23 *etc.*

sole, soull, soulle, sowle, sowles *n.* 'soul,' 'a medium that reunites the body and the spirit' 302r: 23, 304r: 41, 304v: 2 *etc.*; 'a volatile substance' 295r: 37, 298r: 21. See "bodi" and "sprit"

soler *adj.* 'pertaining to gold,' 'golden' 305r: 40

solucion, solution *n.* 'dissolution,' 'the process of reducing into a liquid condition,' 'separation' 294v: 26, 298v: 41, 299r: 11 *etc.*

solued, soluid *ppl.* 'dissolved,' 'reduced into a liquid condition,' 'separated' 294v: 9, 295r: 8, 309v: 9

som *n.* 'sum,' 'the essential part of something,' 'essence' 294r: 7

sone *adv.* 'soon' 300r: 25, 26, 312v: 9 *etc.*

sonn, sonne *n.* 'gold' 295r: 34, 315r: 36, 323v: 29 *etc.* See "Sol"

sorte *n. in the phrase* "in this sorte": 'in this way' 305v: 35

soware *adj.* 'sour'; *in the phrase* "soware doughe": 'yeast,' 'ferment.' See "ferment"

specie *n.* 'a category composed of individuals having some common qualities or characteristics,' 'a subcategory of a larger class or genus' 296v: 26, 27, 30 *etc.*

spending *n.* 'means of support,' 'money' 294v: 23

sperites *n. pl.* see "sprit"

spirit *n.* see "sprit"

spirituall, spiritualle *adj.* 'volatile,' 'vaporous,' 'consisting of a pure essence or spirit' 306r: 6, 308r: 9, 309v: 25

springe, springeth, springinge *v.* 'grow,' 'arise or develop by growth' 302r: 14, 302v: 33, 305v: 15 *etc.*

sprit, sprite, sprites *n.* 'a volatile substance' (often referring to one of the five substances *mercury, sulfur, arsenic, orpiment,* and *sal ammoniac*) 295r: 22, 35, 295v: 1 *etc.*; 'a vapor or volatile substance released when the base material is dissolved' 302r: 22, 24, 25 *etc.*; 'the highly-refined substances or fluids believed to permeate the human body' 323r: 23(?). See "bodi" and "sole"

stand *ppl.* 'stood' 314v: 19, 316r: 41, 320r: 14

stainge *adj.* 'staying'(?) 308r: 15*

stannes *n. pl.* 'stones'(?) 307r: 37*

starue *v.* 'kill,' 'destroy' ded. 1. 125

stedfast, stedfaste *adj.* 'unchanging,' 'stable in form' 296v: 24, 302v: 27, 40 *etc.*

still, stille, styll *n.* 'an apparatus or vessel used for distillation' 310v: 5, 9, 311r: 8 *etc.*; *in the phrase* "still bodie": 'a vessel used for distillation' 310r: 27, 321v: 33; *in the phrase* "still hed": 'the cover or top of a distillation vessel' 311v: 18, 320r: 28

still *v.* 'distil,' 'purify' 325r: 2, 4; 'trickle down or fall in minute drops,' 'distil' 325r: 9

stillatorie, stillatory *n.* 'a still,' 'a vessel used for distillation' 314r: 31, 321v: 9, 16; *in the phrase* "stillatory vessel": 'a still,' 'a vessel used for distillation' 314v: 16

stinckinge *adj.* 'base,' 'corrupt' 298v: 21

ston, stone, stones *n.* 'the philosophers' stone,' 'the substance that transforms base metals into silver or gold, heals bodily diseases, and prolongs life' 294r: 13, 294v: 8, 295r: 16 *etc.*; *often in phrases such as* "stone of philosofers" *and* "philosofers stone." See "animal," "mineral," *and* "vegital"

stop, stope, stoped, stopped *v.* 'plug,' 'close up' 301r: 5, 7, 307v: 27 *etc.*

stopple *n.* 'an appliance for closing the orifice of a vessel,' 'a stopper' 317v: 8

straightway, straighteway *adv.* 'immediately,' 'at once' 313v: 31, 317r: 36, 324r: 32

strength *v.* 'strengthen' 316v: 21, 317v: 15, 318v: 17 *etc.*

striue, stryuinge *v.* 'contend,' 'struggle' 320v: 48, 321r: 9

strongly *adv.* 'firmly,' 'securely' 307r: 19, 314v: 41; 'emphatically,' 'efficiently' 320r: 5, 323v: 19

styll *n.* see "still"

studie, studieth *v.* 'endeavor' 302v: 46; (*constr.* "in") 'study' 303r: 45

subiect *n.* 'the substance of which a thing consists or from which it is made' 304r: 33, 36, 304v: 5 *etc.*

sublimacion, sublimacon, sublimaon, sublimation, sublimatione, sublimations *n.* 'the process of refining a substance [by heating it to the point of vaporization],' 'refining' 294v: 7, 26, 295v: 2 *etc.*

sublimatores, sublimatories, sublimatoryes *n.* 'vessels used for sublimation' 295r: 22, 307v: 21, 32 *etc.*

sublimatum *n.* 'a product of sublimation' 310r: 18. See "quicksiluer"

sublime, sublimed, sublimeth, subliminge, sublyme, sublymed *v.* 'refine [through the process of sublimation]' 295r: 6, 296r: 11, 299v: 10 *etc.*; 'carry out the process of sublimation' 294v: 12, 296r: 23; 'undergo sublimation' 307r: 34, 307v: 39

subliminge, sublyminge *v. n.* 'sublimation,' 'refining' 300r: 40, 320v: 37

substract, substracte, substracted *v.* 'subtract,' 'draw off,' 'remove' 313r: 20, 314v: 20, 316r: 36 *etc.*

subtill *adj.* 'thin,' 'pure,' 'light' 295v: 13, 33, 296v: 21 *etc.*; 'fine,' 'finely powdered' 318v: 5

subtilly *adv.* 'acutely,' 'shrewdly' 300v: 7

suerly *adv.* see "surely"

suete *n.* 'suit,' 'plea' *ded.* l. 78

suffer, suffred, sufferinge *v.* 'endure,' 'stand' 303v: 31, 308r: 42, 310v: 15 *etc.*; 'allow' 295v: 17, 307r: 29, 310v: 17 *etc.*; 'be passive' 302v: 25(?)*

sufficient *adv.* 'sufficiently' 312v: 20, 313r: 36

sulfur, sulfure, sulpher, sulphur, sulphure, sulfurs *n.* 'sulfur,' 'one of the substances known as spirits,' 'sulfur [one of the components of all metals]' 294v: 32, 34, 295v: 14 *etc.*; *in the phrase* "sulfur viue": 'sulfur [one of the components of all metals]' 316v: 28

sulfurous, sulfurus *adj.* 'of or pertaining to sulfur' 302v: 20, 303v: 24, 310v: 31 *etc.*

surely *adv.* 'securely,' 'tightly' 310v: 6, 13, 19 *etc.*
sweete *n.* 'sweat,' 'drops of moisture' 307v: 26, 320v: 49
sweete *v.* 'sweat,' 'labor' 303v: 25

tain, taine, tayne, taineth, tayneth, tayned, tayninge, teyned *v.* 'dye,' 'give color,' 'transmute' 297r: 31, 298r: 39, 303r: 20 *etc.*
tarrie, tarry *v.* 'stay,' 'abide' 295r: 6, 297r: 33, 303v: 43
tartur *n.* 'tartar,' 'a reddish acid compound deposited from wines' *ded.* l. 103
tella *n.* 'terra,' 'earth' 319v: 24*
temperat, temperate, temperatt *adj.* 'moderate' 296r: 3, 311r: 14, 309v: 22 *etc.*; 'mixed' 298v: 13
teyned *ppl.* see "tain"
the, y^e *pron.* 'they' *ded.* l. 87, 299v: 4
thear, theare *pron.* 'their' *ded.* l. 83, l. 91, l. 105 *etc.*
thend *n.* 'the end' 301r: 12, 317r: 17
thentent *n.* 'the intent' 298r: 6, 305v: 2
theorica, theorik, theorika, theorike *n.* 'theory,' 'theoretical argument or part' 301v: 2, 39, 304r: 33 *etc.*
ther, there *pron.* 'their' 295v: 2, 32, 296v: 32
thicknes *n.* 'a thick substance' 302v: 4, 323r: 40
thorowe, throwe *prep.* 'through' 312v: 36, 315r: 18, 319r: 29 *etc.*
thriall *n.* 'thrall,' 'slave' 294v: 21
throughly *adv.* 'fully,' 'completely' 307v: 17
tincbir *n.* 'tincture'(?) 298r: 39*
tincture, tincturs, tyncture, tyncturs *n.* 'a spiritual substance or principle,' 'the quintessence,' 'the elixir,' 'a dyeing agent,' 'color' 298v: 6, 10, 302r: 21 *etc.*
tinturouse *adj.* 'full of tincture' 298r: 21
too *num.* 'two' 294v: 24, 295r: 14, 299r: 17 *etc.*
totcheth *v.* 'concerns,' 'pertains to' 295r: 16
totchinge *n.* 'treatment,' 'discussion' 302r: 36
totchinge, towching *prep. in the phrase* "as totchinge/towching": 'concerning,' 'as regards' 297r: 7, 321v: 25
transmueth *v.* 'transmutes,' 'changes [into another substance]' 315r: 8
transmuptation *n.* 'transmutation,' 'the process of changing one substance into another, usually a base metal into silver or gold' 324v: 27*
transmutable *adj.* 'possible to transmute,' 'possible to change from one substance into another' 323v: 28
transmutation *n.* 'the process of changing one substance into another, usually a base metal into silver or gold' 304r: 5, 317r: 30, 318v: 3 *etc.*
transmute *v.* 'change [into another substance]' 323v: 21
troblinge *v. n.* 'stirring up'(?) 311r: 18*
tryall *n.* 'test,' 'testing' 313v: 8

tuthie *n.* 'tutty,' 'zinc oxide' 297r: 30
two *prep.* 'to' *ded.* l. 146

vain, vains *n.* 'vein' 304r: 20, 304v: 22
vaine *adj.* 'useless,' 'worthless' 303v: 39; 'empty' 304v: 12
vapored, vapoured *ppl.* 'removed through vaporization' 311v: 5, 312v: 40; 'turned into vapor' 312v: 41; (*constr.* "away") 'evaporated' 307v: 24
vegetable, vegitable, vigitable *adj.* 'characterized by, exhibiting or producing, the phenomena of life and growth' 295v: 3, 296v: 10, 299r: 39 *etc.*; *in the phrase* "vegitable stone": see "vegital"
vegital, vegitall, vegitalle *n. in the phrase* "vegital stone": 'one of the three kinds of the philosophers' stone,' 'a stone said to be able to transmute base metals into silver and gold and to cure diseases' 297r: 16, 24, 304r: 3 *etc.* See "vegetable"
venum *n.* 'venom,' 'sulfur [one of the components of all metals]' 302v: 10
Venus *n.* 'copper' 295r: 33, 300r: 18, 317r: 43 *etc.*
verdegreece *n.* 'verdigris,' 'rust on copper and brass' 296r: 25
verie, very *adj.* 'true,' 'proper' 298r: 35, 36, 39 *etc.*
vermelon, vermilion *n.* 'vermilion,' 'cinnabar,' 'mercuric sulphide' 297r: 19, 313r: 5, 7 *etc.*
vertue *n.* 'power,' 'property,' 'quality' 295r: 10, 299v: 23, 308r: 27 *etc.*; 'efficacious quality,' 'medicinal potency' 298v: 36, 309v: 8; 'value,' 'worth,' 'good quality' 303r: 26, 311r: 34, 311v: 35 *etc.*; *in the phrase* "attractiue vertue": 'the power to attract or absorb' 314v: 2, 321v: 24; *in the phrase* "working(e) vertue" or "vertue operatiue": 'the active power,' 'the active property' 307r: 25, 28, 32 *etc.*; *in the phrase* "naturall vertues": 'physical powers [governing the body]' 323r: 24; *in the phrase* "by vertue": 'by the help of,' 'through' 323r: 17, 324v: 5
vesture *n.* 'garment,' 'piece of clothing' 316r: 20
vicinall *n.* 'vitriol'(?) 304r: 20*
ville *adj.* 'vile,' 'of little worth' 305v: 7
violence *n.* 'great force,' 'intensity' 309v: 23, 313r: 2, 318r: 9
viscus *adj.* 'viscous,' 'thick' 304v: 22
vitreall, vitrial, vitriall, vitryall *n.* 'vitriol,' 'a sulfate' 295r: 18, 297r: 11, 19 *etc.*; *in the phrase* "Romain vitriall": 'Roman vitriol,' 'iron or copper sulfate' 314v: 30
vnclarly *adv.* 'unclearly' *ded.* l. 8
vnctuositie, vnctuosity *n.* 'viscosity,' 'oiliness,' 'oily matter or substance' 305r: 34, 313v: 11, 322v: 9 *etc.*
vnctuous, vnctuouse *adj.* 'oily' 304r: 27, 320v: 5, 16 *etc.*
vnderstand *ppl.* 'understood' 302r: 40
vndon *v.* 'ruined,' 'destroyed' 294v: 19
vnkindlie *adv.* 'unnaturally,' 'to an unnatural extent' 297r: 27, 297v: 3
vnperfect, vnperfecte *adj.* 'base,' 'impure' 295r: 31, 295v: 22, 26 *etc.*
vnressononable *adj.* 'unreasonable' 297v: 15*

volatiue *adj.* 'volatile,' 'characterized by a natural tendency to dispersion in fumes or vapor' 319v: 49

volatylle *adj.* 'characterized by a natural tendency to dispersion in fumes or vapor' 295r: 37

vtter *adj.* 'outer' 302v: 34

vulgar, vulgare, vulger *adj.* 'common,' 'ordinary' 301r: 31, 305r: 29, 30 *etc.*

wanting *n.* 'lack,' 'shortage' 298v: 2

warely *adv.* 'cautiously,' 'carefully' 307v: 31, 314v: 4, 9 *etc.*

wasteth *v.* 'destroys' 294v: 11

water, watter, waters *n.* 'water,' 'one of the four elements' 300r: 15, 302r: 40, 41 *etc.*; '[watery] liquid,' 'watery substance or element,' 'water' 295v: 12, 296v: 20, 299r: 31 *etc.*; 'acid,' 'corrosive water' 300v: 32, 35, 310r: 23 *etc.*; *in the phrase* "ardent water": 'ardent spirits,' 'alcohol' 297r: 18, 313r: 6, 14 *etc.*; *in the phrase* "strong water": 'corrosive water,' 'acid' 296r: 8, 300v: 31, 34; *in the phrases* "our water," "water of life," "water permanente," "water of the sea": 'mercury [one of the components of all metals]' 303r: 12, 13, 315r: 17, 320v: 11 *etc.*

waternes *n.* 'watery constituent,' 'watery element' 295v: 16

watri, watry *adj.* 'possessing the essential nature or characteristics of water as an element,' 'humid' 303v: 10, 304v: 7, 308r: 26 *etc.*

watrines *n.* 'watery constituent,' 'watery element' 318r: 27

wax, waxeth *v.* 'grow,' 'turn' 307r: 23, 35, 315r: 31 *etc.*

wayned *ppl.* 'weaned' 317v: 32

well *n.* 'well,' 'spring' 314r: 35

wen *v.* 'think,' 'expect' *ded.* l. 104

wet *n.* 'one of the four principal qualities of the elements,' 'moistness' 303r: 37. See "moist"

whell, whelle *n.* 'wheel'; *in the phrase* "whell of philosofie": 'the process of turning elements into each other' 304r: 17*, 305v: 33, 322r: 40 *etc.*

wher *adv.* 'whether' 301v: 22

whotnes *n.* 'hotness' 324r: 11*

wild, wilde *adj.* 'uncontrolled,' 'unrestrained' 295v: 12*, 303r: 25

willith *v.* 'requires' 297r: 25

winde, wynd *n.* 'mercurial fumes or vapors' 303v: 6*, 310v: 39

wine *v.* 'profit' (?) *ded.* l. 109*

without *adv.* 'on the outside' 312r: 27, 320v: 44

wond *n.* 'wound' 307v: 41

wonderes *adv.* 'wondrously,' 'wonderfully' *ded.* l. 133

worckinge *v. n.* see "working"

wordle *n.* 'world' 294r: 14, 298r: 9, 298v: 9 *etc.*; *in the phrase* "leess wordle": 'the microcosm,' 'the philosophers' stone' 294r: 14*, 319r: 31, 323r: 22

wordly *adj.* 'worldly' 319r: 10
work, worke, worketh, workinge *v.* 'carry out,' 'perform' 297v: 19, 20, 301v: 33 *etc.*
worke, workes *n. in the phrases* "worke of alkamie," "worke of philosofie," and "worke of the stone": 'the procedure designed to produce the philosophers' stone' 307r: 16, 40, 310v: 35 *etc.*; *in the phrase*s "great work" and "greatest worke": 'an alchemical procedure involving silver and gold' 295r: 2, 320v: 20 (x2); *in the phrase* "leesse worke": 'an alchemical procedure involving turning copper into silver or gold' 295r: 33, 323v: 26, 324r: 2 *etc.*
workeman, workman, workemen *n.* 'practitioner,' 'alchemist' 294r: 12, 18, 37 *etc.*
worker *n.* 'practitioner,' 'alchemist' 294v: 36
working, workinge *n.* 'operation,' 'performance,' 'action' 296r: 6, 297v: 27, 299v: 23 *etc.*; 'operation,' 'work,' 'procedure' 294r: 12, 297v: 23, 298r: 6 *etc.*; 'production,' 'preparation' 295r: 17, 312v: 26, 315v: 18 *etc.*; *in the phrase* "working vertue": see "vertue"; *in the phrase* "workinge of the stone": see "worke"
wouchsaf *v.* 'permit,' 'allow' 296v: 4
wright *v.* 'write' *ded.* l. 114
wynd *n.* see "winde"

yerth *n.* 'earth,' 'solid body' *ded.* l. 139, 297r: 5, 297v: 3. See "earth"
ymbibed *v.* see "embibe"
ymbibition *n.* see "imbibition"
yssue *n.* 'exit,' 'passage out' 310r: 23

Proper Names

Alkamus Alkamus, mythical inventor of alchemy. 324v: 25*
Antazagurras Antazagurras, unidentified alchemist. 321r: 31. See p. 51.
Aristotle Pseudo-Aristotle. 323r: 24*
Avicen, Avycen (Pseudo?-)Avicenna (980–1036), Arabic scholar and physician. 306r: 16, 321v: 3, 325r: 13
Guido Guido Montaynor or Guido de Monte(?). 305r: 17, 38, 306r: 12, 325r: 15. See p. 43.
Hermes Hermes Trismegistos, legendary founder of alchemy. 303v: 14, 24, 309v: 16, 324r: 10
Machabeus Judas Maccabeus. 319r: 16. See EN 319r: 14–17.
Morien, Moryen Morienus Romanus, seventh-century alchemist(?). 316r: 8, 11, 316v: 13, 43, 319r: 9, 43
Moyses Moses. 319r: 15, 23, 24. See EN 319r: 14–17. For "the syster of Moyses," see EN 319r: 22–23.
Noye Noah. 319r: 14, 19. See EN 319r: 14–17.

Raymon, Raymond, Raymone (Pseudo-)Raymundus Lullus or Raymond Lull (*c.* 1232/3–1315/6), one of the most famous alchemical authorities in the Middle Ages and early modern period. 305r: 17, 27, 305v: 6, 306r: 5, 8, 317v: 18, 20, 23, 320v: 7, 321r: 12, 30, 324r: 9, 37, 40. See Pereira, *The Alchemical Corpus*.

Reply, George Reply George Ripley (1415?–1490?), English alchemist. 306r: 17, 317r: 17*. See p. 45.

Rhasis (Pseudo)al-Razi (*c.* 825–924/5), Arabic physician and philosopher. 305r: 29

Sallomon Solomon. 319r: 15. EN 319r: 14–17.

References

Abraham, Lyndy. *Harriot's Gift to Arthur Dee: Literary Images from an Alchemical Manuscript*. The Durham Thomas Harriot Seminar, Occasional Paper 10. Durham: Durham University Library, 1993.

———, ed. *Arthur Dee:* Fasciculus Chemicus, *Translated by Elias Ashmole*. New York: Garland, 1997.

———. *A Dictionary of Alchemical Imagery*. Cambridge: Cambridge University Press, 1998.

The Alchemy Website. www.levity.com/alchemy/home.html. (As accessed in 2002–2006.)

Appleby, John H. "Arthur Dee and Johannes Banfi Hunyades: Further Information on their Alchemical and Professional Activities." *Ambix* 24 (1977): 96–109.

Arnold of Villanova. *Hec sunt opera Arnaldi de Villa Nova nuperrime recognita ac emendata diligentique opere impressa que in hoc volumine continentur*. Basel, 1509.

Artis Auriferae, quam alchemiam vocant, volumen secundum. Basel, 1572.

Ash, Eric H. *Power, Knowledge, and Expertise in Elizabethan England*. Baltimore: Johns Hopkins University Press, 2004.

Ashmole, Elias. *Theatrum Chemicum Britannicum, Containing Severall Poeticall Pieces of our Famous English Philosophers*. London, 1652; repr. Kila, MT: Kessinger Publishing, 1997.

Bailey, Richard W. "The Need for Good Texts: The Case of Henry Machyn's Day Book, 1550–1563." In *Studies in the History of the English Language II: Unfolding Conversations*, ed. Anne Curzan and Kimberly Emmons, 217–28. Berlin: Mouton de Gruyter, 2004.

Batho, G. R. "The Library of the 'Wizard' Earl: Henry Percy Ninth Earl of Northumberland (1564–1632)." *The Library*, 5th ser. 15 (1960): 246–61.

Bayer, Penny. "Lady Margaret Clifford's Alchemical Receipt Book and the John Dee Circle." *Ambix* 52 (2005): 271–84.

Beal, Peter. *In Praise of Scribes: Manuscripts and their Makers in Seventeenth-Century England*. Oxford: Clarendon Press, 1998.

Black, William H. *A Descriptive, Analytical, and Critical Catalogue of the Manuscripts Bequeathed unto the University of Oxford by Elias Ashmole*. Oxford: Oxford University Press, 1845.

Bobrick, Benson. *Fearful Majesty: The Life and Reign of Ivan the Terrible*. New York: G. P. Putnam's Sons, 1987.
Borgnet, Auguste, and Émile Borgnet, eds. *B. Alberti Magni, Ratisbonensis Episcopi, Ordinis Praedicatorum, Opera omnia*. 38 vols. Paris: L. Vives, 1890–1899.
Braekman, Willy L., ed. *Studies on Alchemy, Diet, Medecine [sic] and Prognostication in Middle English*. Scripta 22. Brussels: Omirel, 1986.
Briquet, Charles-Moïse. *Les filigranes*. The New Briquet, Jubilee Ed., 4 vols. Amsterdam: Paper Publications Society, 1968.
Buck, W. S. B. *Examples of Handwriting 1550–1650*. London: Society of Genealogists, 1996.
Carlson, David. "The Writings and Manuscript Collections of the Elizabethan Alchemist, Antiquary, and Herald Francis Thynne." *Huntington Library Quarterly* 52 (1989): 203–72.
Carroll, Ruth. "The Middle English Recipe as a Text-Type." *Neuphilologische Mitteilungen* 100 (1999): 27–42.
———. "Middle English Recipes: Vernacularisation of a Text-Type." In *Medical and Scientific Writing in Late Medieval English*, ed. Irma Taavitsainen and Päivi Pahta, 174–91. Cambridge: Cambridge University Press, 2004.
Churchill, William A. *Watermarks in Paper in Holland, England, France etc. in the XVII and XVIII Centuries and their Interconnection*. Amsterdam: Menno Hertzberger, 1935.
Clay, John W., ed. *Familiae Minorum Gentium*. 4 vols. Publications of the Harleian Society 37–40. London: The Harleian Society, 1894–1896.
Colby, Frederic T., ed. *The Visitation of the County of Devon in the Year 1620*. Publications of the Harleian Society 6. London: The Harleian Society, 1872.
Colvin, H. M., John Summerson, Martin Biddle, J. R. Hale, and Marcus Merriman. *The History of the King's Works*, vol. 4, *1485–1660 (Part II)*. London: Her Majesty's Stationery Office, 1982.
Cook, Judith. *Dr Simon Forman: A Most Notorious Physician*. London: Chatto & Windus, 2001.
Craven, James B. *Doctor Heinrich Khunrath: A Study of Mystical Alchemy*, limited ed. Glasgow: Printed and Bound by Adam McLean, 1997.
Crisciani, Chiara. "Alchemy and Medicine in the Middle Ages." *Bulletin de philosophie médiévale* 38 (1996): 9–21.
Crosby, Allan J., ed. *Calendar of State Papers, Foreign Series, of the Reign of Elizabeth, 1566–8*. London: Longman, 1871.
———, ed. *Calendar of State Papers, Foreign Series, of the Reign of Elizabeth, 1572–74*. London: Longman, 1876.
Crosland, Maurice P. *Historical Studies in the Language of Chemistry*. Cambridge, MA: Harvard University Press, 1962.

Debus, Allen G. *The Chemical Philosophy: Paracelsian Science and Medicine in the Sixteenth and Seventeenth Centuries*. 2 vols. New York: Science History Publications, 1977.
Dick, Oliver L., ed. *Aubrey's Brief Lives*. London: Secker and Warburg, 1949.
DML = *Dictionary of Medieval Latin from British Sources*, ed. R. E Latham. Oxford: Oxford University Press, 1975–.
Dobbs, Betty Jo T. *The Foundations of Newton's Alchemy or "The Hunting of the Greene Lyon."* Cambridge: Cambridge University Press, 1975.
Dobson, Eric J. *English Pronunciation 1500–1700*. 2 vols. Oxford: Clarendon Press, 1968.
Edmond, Mary. "Simon Forman's Vade-Mecum." *The Book Collector* 26 (1977): 44–60.
Edwards, A. S. G., and Douglas Moffat. "Annotation." In *A Guide to Editing Middle English*, ed. Vincent P. McCarren and Douglas Moffat, 217–36. Ann Arbor, MI: University of Michigan Press, 1998.
Faulkner, Kevin. "Scintillae Marginila [sic]: Sparkling Margins — Alchemical and Hermetic Thought in the Literary Works of Sir Thomas Browne." Available at http://levity.com/alchemy/sir_thomas_browne.html. (As accessed in 2006.)
Feingold, Mordechai. "The Occult Tradition in the English Universities of the Renaissance: A Reassessment." In *Occult and Scientific Mentalities in the Renaissance*, ed. Brian Vickers, 73–94. Cambridge: Cambridge University Press, 1984.
Figurovski, N. A. "The Alchemist and Physician Arthur Dee (Artemii Ivanovich Dii)." *Ambix* 13 (1965): 35–51.
Gentleman's Magazine 62 (1792): 798–801.
Geoghegan, D. "A Licence of Henry VI to Practise Alchemy." *Ambix* 6 (1957): 10–17.
Gerson, Armand J. "The Organization and Early History of the Muscovy Company." In *Studies in the History of English Commerce in the Tudor Period*, 1–122. New York: D. Appleton, 1912.
Gettings, Fred. *Dictionary of Occult, Hermetic and Alchemical Sigils*. London: Routledge and Kegan Paul, 1981.
Gordon, Ian A. *The Movement of English Prose*. London: Longmans, 1966.
Görlach, Manfred. "Text-Types and Language History: The Cookery Recipe." In *History of Englishes: New Methods and Interpretations in Historical Linguistics*, ed. Matti Rissanen, Ossi Ihalainen, Terttu Nevalainen, and Irma Taavitsainen, 736–61. Berlin: Mouton de Gruyter, 1992.
Gotti, Maurizio. *Robert Boyle and the Language of Science*. Milan: Guerini scientifica, 1996.
Grosses universal Lexicon. Halle and Leipzig, 1732–1754.

Grund, Peter. Review of *Scientific and Medical Writings in Old and Middle English: An Electronic Reference*, by Linda E. Voigts and Patricia D. Kurtz. *ICAME Journal* 26 (2002): 160–64.

———. "In Search of Gold: Towards a Text Edition of an Alchemical Treatise." In *Middle English from Tongue to Text: Selected Papers from The Third International Conference on Middle English: Language and Text*, ed. Peter J. Lucas and Angela M. Lucas, 265–79. Frankfurt am Main: Peter Lang, 2002.

———. "The Golden Formulas: Genre Conventions of Alchemical Recipes in the Middle English Period." *Neuphilologische Mitteilungen* 104 (2003): 455–75.

———. Review of *The Alchemy Reader: From Hermes Trismegistus to Isaac Newton*, ed. Stanton J. Linden. *Studia Neophilologica* 75 (2003): 211–12.

———. "'Misticall Wordes and Names Infinite': An Edition of Humfrey Lock's Treatise on Alchemy, with an Introduction, Explanatory Notes and Glossary." Ph. D. diss., Uppsala University, 2004.

———. "Albertus Magnus and the Queen of the Elves: A 15th-Century English Verse Dialogue on Alchemy." *Anglia* 122 (2004): 640–62.

———. "'ffor to make Azure as Albert biddes': Medieval English Alchemical Writings in the Pseudo-Albertan Tradition." *Ambix* 53 (2006): 21–42.

———. "A Previously Unrecorded Fragment of the Middle English *Short Metrical Chronicle* in Bibliotheca Philosophica Hermetica M199." *English Studies* 87 (2006): 277–93.

———. "Manuscripts as Sources for Linguistic Research: A Methodological Case Study Based on the *Mirror of Lights*." *Journal of English Linguistics* 34 (2006): 105–25.

Haage, Bernhard D. *Alchemie im Mittelalter: Ideen und Bilder, von Zosimos bis Paracelsus*. Düsseldorf: Artemis and Winkler, 1996.

Halleux, Robert. *Les textes alchimiques*. Typologie des sources du Moyen Âge occidental 32. Turnhout: Brepols, 1979.

Halliwell, James O., ed. *The Private Diary of Dr. John Dee and the Catalogue of his Library of Manuscripts*. London: John Bowyer Nichols and Son, 1842.

Halversen, Marguerite A., ed. "*The Consideration of Quintessence*: An Edition of a Middle English Translation of John of Rupescissa's *Liber de Consideratione de Quintae Essentiae* [sic] *Omnium Rerum* with Introduction, Notes, and Commentary." Ph. D. diss., Michigan State University, 1998.

Hamel, Joseph. *England and Russia; Comprising the Voyages of John Tradescant the Elder, Sir Hugh Willoughby, Richard Chancellor, Nelson, and Others, to the White Sea*. London: Richard Bentley, 1854.

Hannaway, Owen. *The Chemists and the Word: The Didactic Origins of Chemistry*. Baltimore: Johns Hopkins University Press, 1975.

Harkness, Deborah E. "'Strange' Ideas and 'English' Knowledge: Natural Science Exchange in Elizabethan London." In *Merchants and Marvels: Com-*

merce, Science, and Art in Early Modern Europe, ed. Pamela H. Smith and Paula Findlen, 137–60. New York: Routledge, 2002.

Harmoniae inperscrutabilis chymico-philosophiae sive philosophorum antiquorum consentientivm. Decas I. Frankfurt, 1625.

Harvey, E. Ruth. *The Inward Wits: Psychological Theory in the Middle Ages and the Renaissance*. Warburg Institute Surveys 6. London: The Warburg Institute, 1975.

Heawood, Edward. *Watermarks, Mainly of the 17th and 18th Centuries*. Monumenta Chartae Papyraceae Historiam Illustrantia 1. Hilversum: The Paper Publications Society, 1950.

Hirsch, Rudolf. "The Invention of Printing and the Diffusion of Alchemical and Chemical Knowledge." In *The Printed Word: Its Impact and Diffusion*, ed. idem, 115–41. London: Variorum Reprints, 1978.

Histoire littéraire de la France. 42 vols. Paris: Imprimerie Nationale, 1773–.

Holmyard, Eric J. *Alchemy*. Harmondsworth: Penguin, 1957; repr. New York: Dover, 1990.

Hughes, Edward. "The English Monopoly of Salt in the Years 1563–71." *English Historical Review* 40 (1925): 334–50.

———. *Studies in Administration and Finance 1558–1825, with Special Reference to the History of Salt Taxation in England*. Manchester: Manchester University Press, 1934.

Hughes, Jonathan. *Arthurian Myths and Alchemy: The Kingship of Edward IV*. Phoenix Mill, UK: Sutton, 2002.

Hunt, Tony, ed. *Three Receptaria from Medieval England: The Languages of Medicine in the Fourteenth Century*. Medium Aevum Monographs n. s. 21. Oxford: Society for the Study of Medieval Languages and Literature, 2001.

Johnston, Stephen. "Mathematical Practitioners and Instruments in Elizabethan England." *Annals of Science* 48 (1991): 319–44.

Josten, C. H., ed. *Elias Ashmole (1617–1692): His Autobiographical and Historical Notes, his Correspondence, and Other Contemporary Sources Relating to his Life and Work*. 5 vols. Oxford: Clarendon Press, 1966.

Kassell, Lauren. "How to Read Simon Forman's Casebooks: Medicine, Astrology, and Gender in Elizabethan London." *Social History of Medicine* 12 (1999): 3–18.

———. "'The Food of Angels': Simon Forman's Alchemical Medicine." In *Secrets of Nature: Astrology and Alchemy in Early Modern Europe*, ed. William R. Newman and Anthony Grafton, 345–84. Cambridge, MA: MIT Press, 2001.

———. *Medicine and Magic in Elizabethan London: Simon Forman: Astrologer, Alchemist, and Physician*. Oxford: Clarendon Press, 2005.

Keiser, George R. *A Manual of the Writings in Middle English 1050–1500*, vol. 10, *Works of Science and Information*. New Haven: Connecticut Academy of Arts and Sciences, 1998.

———. "Editing Scientific and Practical Writings." In *A Guide to Editing Middle English*, ed. Vincent P. McCarren and Douglas Moffat, 109–22. Ann Arbor, MI: University of Michigan Press, 1998.

———. "Preserving the Heritage: Middle English Verse Treatises in Early Modern Manuscripts." In *Mystical Metal of Gold: Essays on Alchemy and Renaissance Culture*, ed. Stanton J. Linden, 189–214. AMS Studies in the Renaissance 42. New York: AMS Press, 2006.

Kibre, Pearl. "Albertus Magnus on Alchemy." In *Albertus Magnus and the Sciences: Commemorative Essays 1980*, ed. James A. Weisheipl, 187–202. Toronto: Pontifical Institute of Mediaeval Studies, 1980.

———. *Studies in Medieval Science: Alchemy, Astrology, Mathematics and Medicine*. London: Hambledon Press, 1984.

Kytö, Merja, Terry Walker, and Peter Grund. "English Witness Depositions 1560–1760: An Electronic Text Edition." *ICAME Journal* 31 (2007): 65–85.

Lass, Roger. "*Ut Custodiant Litteras*: Editions, Corpora and Witnesshood." In *Methods and Data in English Historical Dialectology*, ed. Marina Dossena and Roger Lass, 21–48. Bern: Peter Lang, 2004.

Linden, Stanton J., ed. *George Ripley's* Compound of Alchymy *(1591)*. Aldershot: Ashgate, 2001.

———, ed. *The Alchemy Reader: From Hermes Trismegistus to Isaac Newton*. Cambridge: Cambridge University Press, 2003.

Locke, John G. *Book of the Lockes: A Genealogical and Historical Record of the Descendants of William Locke of Woburn with an Appendix Containing a History of the Lockes in England*. Boston: James Munroe, 1853.

Love, Harold. *Scribal Publication in Seventeenth-Century England*. Oxford: Clarendon Press, 1993.

MacDonald, Michael. *Mystical Bedlam: Madness, Anxiety, and Healing in Seventeenth-Century England*. Cambridge: Cambridge University Press, 1981.

Mäkinen, Martti. "On Interaction in Herbals from Middle English to Early Modern English." *Journal of Historical Pragmatics* 3 (2002): 229–51.

Manget, Jean-Jacques, ed. *Bibliotheca chemica curiosa*. 2 vols. Geneva, 1702.

Manzalaoui, M. A., ed. Secretum Secretorum: *Nine English Versions*. EETS OS 276. Oxford: Oxford University Press, 1977.

Marotti, Arthur F. *Manuscript, Print, and the English Renaissance Lyric*. Ithaca: Cornell University Press, 1995.

Matheson, Lister M., ed. *Popular and Practical Science of Medieval England*. East Lansing: Colleagues Press, 1994.

McConchie, Roderick W. *Lexicography and Physicke: The Record of Sixteenth-Century English Medical Terminology*. Oxford: Clarendon Press, 1997.

McKitterick, David. *Print, Manuscript, and the Search for Order, 1450–1830*. Cambridge: Cambridge University Press, 2003.

McVaugh, Michael "The 'Humidum Radicale' in Thirteenth-Century Medicine." *Traditio* 30 (1974): 259–83.
MED = *The Middle English Dictionary*. Available at http://ets.umdl.umich.edu. (As accessed in 2006.)
Mooney, Linne R. *The Index of Middle English Prose. Handlist XI: The Library of Trinity College, Cambridge*. Cambridge: D. S. Brewer, 1995.
Moorat, S. A. J. *Catalogue of Western Manuscripts on Medicine and Science in the Wellcome Historical Medical Library*, vol. 1, *Mss. Written before 1650 A.D*. London: Wellcome Historical Medical Library, Wellcome Institute of the History of Medicine, 1962.
Moran, Bruce T. *Distilling Knowledge: Alchemy, Chemistry, and the Scientific Revolution*. Cambridge, MA: Harvard University Press, 2005.
Multhauf, Robert P. *The Origins of Chemistry*. London: Oldbourne, 1966.
Mustanoja, Tauno F. *A Middle English Syntax, Part I: Parts of Speech*. Mémoires de la Société Néophilologique de Helsinki 23. Helsinki: Société Néophilologique, 1960.
Newman, William R. "The Genesis of *Summa Perfectionis*." *Archives internationales d'histoire des sciences* 35 (1985): 240–300.
———, ed. *The* Summa Perfectionis *of Pseudo-Geber: A Critical Edition, Translation and Study*. Leiden: Brill, 1991.
———. "*Decknamen* or Pseudochemical Language? Eirenaeus Philalethes and Carl Jung." *Revue d'histoire des sciences* 49 (1996): 159–88.
———, and Lawrence M. Principe. "Alchemy vs. Chemistry: The Etymological Origins of a Historiographic Mistake." *Early Science and Medicine* 3 (1998): 32–65.
Norri, Juhani. "Notes on the Origin and Meaning of Chemical Terms in Middle English Medical Manuscripts." *Neuphilologische Mitteilungen* 92 (1991): 215–36.
———. *Names of Sicknesses in English, 1400–1550: An Exploration of the Lexical Field*. Annales Academiae Scientiarum Fennicae: Dissertationes Humanarum Litterarum 63. Helsinki: Academia Scientiarum Fennica, 1992.
———. *Names of Body Parts in English, 1400–1550*. Annales Academiae Scientiarum Fennicae 291. Helsinki: Academia Scientiarum Fennica, 1998.
———. "Entrances and Exits in English Medical Vocabulary, 1400–1550." In *Medical and Scientific Writing in Late Medieval English*, ed. Taavitsainen and Pahta, 100–43.
Obrist, Barbara, ed. *Constantine of Pisa: The Book of the Secrets of Alchemy*. Leiden: Brill, 1990.
ODNB = *Oxford Dictionary of National Biography*. Available at www.oxforddnb.com. (As accessed in 2006.)
OED = *The Oxford English Dictionary*. Available at www.dictionary.oed.com. (As accessed in 2006.)

O'Farrell-Tate, Una, ed. *The Abridged English Metrical Brut: Edited from London British Library MS Royal 12 C. XII.* Heidelberg: C. Winter, 2002.
Ogilvie-Thomson, S. J. *The Index of Middle English Prose. Handlist VIII: Oxford College Libraries.* Cambridge: D. S. Brewer, 1991.
OLD = *The Oxford Latin Dictionary.* Oxford: Clarendon Press, 1968–1982.
Oldireva Gustafsson, Larisa. *Preterite and Past Participle Forms in English 1680–1790: Standardisation Processes in Public and Private Writing.* Studia Anglistica Upsaliensia 120. Uppsala: Acta Universitatis Upsaliensis, 2002.
Pahta, Päivi, ed. *Medieval Embryology in the Vernacular: The Case of* De Spermate. Mémoires de la Société Néophilologique de Helsinki 53. Helsinki: Société Néophilologique, 1998.
———. "Code-Switching in Medieval Medical Writing." In *Medical and Scientific Writing in Late Medieval English*, ed. Taavitsainen and eadem, 73–99.
———, and Irma Taavitsainen. "Noun Phrase Structures in Early Modern English Medical Writing." Paper presented at the Twelfth International Conference on English Historical Linguistics, Glasgow, 21–26 August 2002.
Patai, Raphael. *The Jewish Alchemists: A History and Source Book.* Princeton: Princeton University Press, 1994.
Payen, J. "*Flos Florum* et *Semita Semite*: Deux traités d'alchimie attribués à Arnaud de Villeneuve." *Revue d'histoire des sciences* 12 (1959): 289–300.
Pereira, Michela. *The Alchemical Corpus Attributed to Raymond Lull.* Warburg Institute Surveys and Texts 18. London: The Warburg Institute, 1989.
———. "*Mater Medicinarum*: English Physicians and the Alchemical Elixir in the Fifteenth Century." In *Medicine from the Black Death to the French Disease*, ed. Roger French, Jon Arrizabalaga, Andrew Cunningham, and Luis García-Ballester, 26–52. Aldershot: Ashgate, 1998.
———. "Alchemy and the Use of Vernacular Languages in the Late Middle Ages." *Speculum* 74 (1999): 336–56.
———, and Barbara Spaggiari, eds. *Il Testamentum alchemico attribuito a Raimondo Lullo: Edizione del testo latino e catalano dal manoscritto Oxford, Corpus Christi College, 244.* Florence: Sismel, Edizione del Galluzzo, 1999.
Petti, Anthony G. *English Literary Hands from Chaucer to Dryden.* London: Edward Arnold, 1977.
Principe, Lawrence M. *The Aspiring Adept: Robert Boyle and His Alchemical Quest.* Princeton: Princeton University Press, 1998.
———. "Decknamen." In *Alchemie: Lexikon einer hermetischen Wissenschaft*, ed. Claus Priesner and Karin Figala, 104–6. Munich: C. H. Beck, 1998.
———, and William R. Newman. "Some Problems with the Historiography of Alchemy." In *Secrets of Nature: Astrology and Alchemy in Early Modern Europe*, ed. William R. Newman and Anthony Grafton, 385–431. Cambridge, MA: MIT Press, 2001.
Pritchard, Allan. "Thomas Charnock's Book Dedicated to Queen Elizabeth." *Ambix* 26 (1979): 56–73.

Raymond, Stuart A. *Devon: A Genealogical Bibliography*. 2nd ed. Exeter: Raymond Genealogical Bibliographers, 1994.
Read, Conyers. *Lord Burghley and Queen Elizabeth*. London: Jonathan Cape, 1960.
Read, John. *Prelude to Chemistry: An Outline of Alchemy, its Literature and Relationships*. London: G. Bell, 1939.
Reidy, John, ed. *Thomas Norton's* Ordinal of Alchemy. EETS OS 272. Oxford: Oxford University Press, 1975.
Reiter, Raimond. "Die 'Dunkelheit' der Sprache der Alchemisten." *Muttersprache* 97 (1987): 323–26.
Ripley, George. *Georgii Riplaei Canonici Angli opera omnia chemica*. Kassel, 1649.
Rissanen, Matti. "Syntax." In *The Cambridge History of the English Language*, vol. 3, *1476–1776*, ed. Roger Lass, 187–331. Cambridge: Cambridge University Press, 1999.
Ritson, Joseph. *Bibliographia Poetica: A Catalogue of Engleish* [sic] *Poets of the Twelfth, Thirteenth, Fourteenth, Fifteenth, and Sixteenth, Centurys* [sic]. London, 1802.
Roberts, Gareth. *The Mirror of Alchemy: Alchemical Ideas and Images in Manuscripts and Books from Antiquity to the Seventeenth Century*. London: The British Library, 1994.
Roberts, Julian, and Andrew G. Watson. *John Dee's Library Catalogue*. London: Bibliographical Society, 1990.
Rulandus, Martinus. *Lexicon alchemiae sive dictionarium alchemisticum*. Frankfurt, 1612.
Ruska, Julius. *Tabula Smaragdina: Ein Beitrag zur Geschichte der Hermetischen Literatur*. Heidelberg: Carl Winter, 1926.
———., ed. *Turba Philosophorum: Ein Beitrag zur Geschichte der Alchemie*. Berlin: Julius Springer, 1931.
Ryan, William F. "Alchemy and the Virtues of Stones in Muscovy." In *Alchemy and Chemistry in the 16th and 17th Centuries*, ed. Piyo Rattansi and Antonio Clericuzio, 149–59. Dordrecht: Kluwer, 1994.
———. "Magic and Divination: Old Russian Sources." In *The Occult in Russian and Soviet Culture*, ed. Bernice Glatzer Rosenthal, 35–58. Ithaca: Cornell University Press, 1997.
———. *The Bathhouse at Midnight: An Historical Survey of Magic and Divination in Russia*. Phoenix Mill, UK: Sutton, 1999.
Salmon, William, ed. *Medicina Practica, or Practical Physick*. London, 1692.
Saunders, A. D. "The Building of Upnor Castle, 1559–1601." In *Ancient Monuments and their Interpretation*, ed. M. R. Apted, R. Gilyard-Beer, and idem, 263–83. London: Phillimore, 1977.
Schuler, Robert M. *English Magical and Scientific Poems to 1700: An Annotated Bibliography*. New York: Garland, 1979.

———, ed. *Alchemical Poetry 1575–1700, from Previously Unpublished Manuscripts*. New York: Garland, 1995.

Schütt, Hans-Werner. "Sprache der Alchemie." In *Alchemie: Lexikon einer hermetischen Wissenschaft*, ed. Claus Priesner and Karin Figala, 340–42. Munich: C.H. Beck, 1998.

Scott, Edward John L. *Index to the Sloane Manuscripts in the British Museum*. London: The British Museum, 1904.

Sherman, William H. *John Dee: The Politics of Reading and Writing in the English Renaissance*. Amherst: University of Massachusetts Press, 1995.

Shirley, John W. "The Scientific Experiments of Sir Walter Ralegh, the Wizard Earl, and the Three Magi in the Tower 1603–1617." *Ambix* 4 (1949): 52–66.

Singer, Dorothea W. *Catalogue of Latin and Vernacular Alchemical Manuscripts in Great Britain and Ireland, Dating from before the 16th Century*. 3 vols. Brussels: Maurice Lamertin, 1928, 1930, 1931.

Steele, Robert, ed. *Three Prose Versions of the* Secreta Secretorum. EETS ES 74. London: Kegan Paul, 1898.

———. "Alchemy." In *Shakespeare's England*. 1: 462–74. 2 vols. Oxford: Clarendon Press, 1917.

Taavitsainen, Irma. *The Index of Middle English Prose. Handlist X: Scandinavian Collections*. Cambridge: D. S. Brewer, 1994.

———. "Dialogues in Late Medieval and Early Modern English Medical Writing." In *Historical Dialogue Analysis*, ed. Andreas H. Jucker, Gerd Fritz, and Franz Lebsanft, 243–68. Amsterdam: John Benjamins, 1999.

———. "Scientific Language and Spelling Standardisation 1375–1550." In *The Development of Standard English 1300–1800: Theories, Descriptions, Conflicts*, ed. Laura Wright, 131–54. Cambridge: Cambridge University Press, 2000.

———. "Metadiscursive Practices and the Evolution of Early English Medical Writing 1375–1550." In *Corpora Galore: Analyses and Techniques in Describing English*, ed. John M. Kirk, 191–207. Amsterdam: Rodopi, 2000.

———. "Middle English Recipes: Genre Characteristics, Text Type Features and Underlying Traditions of Writing." *Journal of Historical Pragmatics* 2 (2001): 85–113.

———. "Transferring Classical Discourse Conventions into the Vernacular." In *Medical and Scientific Writing in Late Medieval English*, ed. eadem and Pahta, 37–72.

———, and Päivi Pahta. "Vernacularisation of Medical Writing in English: A Corpus-Based Study of Scholasticism." *Early Science and Medicine* 3 (1998): 157–85.

———, and Päivi Pahta, eds. *Medical and Scientific Writing in Late Medieval English*. Cambridge: Cambridge University Press, 2004.

———, Päivi Pahta, Noora Leskinen, Maura Ratia, and Carla Suhr. "Analysing Scientific Thought-Styles: What Can Linguistic Research Reveal about the History of Science?" In *Variation Past and Present:* VARIENG *Studies on English for Terttu Nevalainen*, ed. Helena Raumolin-Brunberg, Minna Nevala, Arja Nurmi, and Matti Rissanen, 251–70. Mémoires de la Société Néophilologique de Helsinki 61. Helsinki: Société Néophilologique, 2002.

———, Päivi Pahta, and Martti Mäkinen. *Middle English Medical Texts*. CD-ROM. Amsterdam: John Benjamins, 2005.

Tannenbaum, Samuel A. *The Handwriting of the Renaissance*. New York: Columbia University Press, 1930.

Tanselle, G. Thomas. "Classical, Biblical, and Medieval Textual Criticism and Modern Editing." *Studies in Bibliography* 36 (1983): 21–68.

Taylor, Frank S. "Thomas Charnock." *Ambix* 2 (1946): 148–76.

———. *The Alchemists: Founders of Modern Chemistry*. New York: Henry Schuman, 1949.

Thorndike, Lynn. *A History of Magic and Experimental Science*. 8 vols. New York: Columbia University Press, 1923–1958.

———, and Pearl Kibre. *A Catalogue of Incipits of Mediaeval Scientific Writings in Latin: Revised and Augmented Edition*. London: Mediaeval Academy of America, 1963.

Tilton, Hereward. *The Quest for the Phoenix: Spiritual Alchemy and Rosicrucianism in the Work of Count Michael Maier (1569–1622)*. Berlin: Walter de Gruyter, 2003.

Tolstoy, George, ed. *The First Forty Years of Intercourse between England and Russia. 1553–1593*. St. Petersburg, 1875.

Traister, Barbara H. *The Notorious Astrological Physician of London: Works and Days of Simon Forman*. Chicago: University of Chicago Press, 2001.

Venn, J. A. *Alumni Cantabrigienses*. 4 vols. Cambridge: Cambridge University Press, 1922–1927.

Visser, Fredericus Th. *An Historical Syntax of the English Language*. 4 vols. Leiden: Brill, 1963–1973.

Voigts, Linda E. "Anglo-Saxon Plant Remedies and the Anglo-Saxons." *Isis* 70 (1979): 250–68.

———. "Scientific and Medical Books." In *Book Production and Publishing in Britain 1375–1475*, ed. Jeremy Griffiths and Derek Pearsall, 345–402. Cambridge: Cambridge University Press, 1989.

———. "The Character of the *Carecter*: Ambiguous Sigils in Scientific and Medical Texts." In *Latin and Vernacular: Studies in Late-Medieval Texts and Manuscripts*, ed. Alastair J. Minnis, 91–109. Cambridge: D. S. Brewer, 1989.

———. "The 'Sloane Group': Related Scientific and Medical Manuscripts from the Fifteenth Century in the Sloane Collection." *British Library Journal* 16 (1990): 26–57.

———. "What's the Word? Bilingualism in Late-Medieval England." *Speculum* 71 (1996): 813–26.

———. "The Master of the King's Stillatories." In *The Lancastrian Court: Proceedings of the 2001 Harlaxton Symposium*, ed. Jenny Stratford, 233–52. Harlaxton Medieval Studies 13. Donington: Shaun Tyas, 2003.

———, and Patricia D. Kurtz. *Scientific and Medical Writings in Old and Middle English: An Electronic Reference*. CD-ROM. Ann Arbor, MI: University of Michigan Press, 2000.

Webster, Charles. *The Great Instauration: Science, Medicine and Reform 1626–1660*. New York: Holmes and Meier, 1976.

———. "Alchemical and Paracelsian Medicine." In *Health, Medicine and Mortality in the Sixteenth Century*, ed. idem, 301–34. Cambridge: Cambridge University Press, 1979.

Willan, T. S. *The Early History of the Russia Company 1553–1603*. Manchester: Manchester University Press, 1956.

Wilson, W. J. "An Alchemical Manuscript by Arnaldus de Bruxella." *Osiris* 2 (1936): 220–405.

Woudhuysen, H. R. *Sir Philip Sidney and the Circulation of Manuscripts 1558–1640*. Oxford: Clarendon Press, 1996.

Wretts-Smith, Mildred. "The English in Russia during the Second Half of the Sixteenth Century." *Transactions of the Royal Historical Society*, 4th ser. 3 (1920): 72–102.

Wyckoff, Dorothy. *Albertus Magnus: Book of Minerals*. Oxford: Clarendon Press, 1967.

Wyld, Henry C. *A History of Modern Colloquial English*. 3rd ed. Oxford: Basil Blackwell, 1936.

Zetzner, Lazarus, ed. *Theatrum chemicum, praecipuos selectorum auctorum tractatus de chemiae et lapidis philosophici antiquitate, veritate, iure, praestantia & operationibus, continens*. 4 vols. Strasbourg, 1659.